T0272019

Understanding Surface and Thin Film Science

This book is a conceptual overview of surface and thin film science, providing a basic and straightforward understanding of the most common ideas and methods used in these fields. Fundamental scientific ideas, deposition methods, and characterization methods are all examined.

Relying on simple, conceptual models and figures, fundamental scientific ideas are introduced and then applied to surfaces and thin films in the first half of the book. Topics include vacuum and plasma environments, crystal structure, atomic motion, thermodynamics, electrical and magnetic properties, optical and thermal properties, and adsorbed atoms on surfaces. Common methods of gas-phase thin film deposition are then introduced, starting with an overview of the film growth process and then a discussion of both physical and chemical vapor deposition methods. This is followed by an overview of a wide range of characterization techniques including imaging, structural, chemical, electrical, magnetic, optical, thermal, and mechanical techniques.

Thin film science is a natural extension of surface science, especially as applications involve thinner and thinner films; distinct from other literature in the field, this book combines the two topics in a single volume. Simple, conceptual models and figures are used, supported by some mathematical expressions, to convey key ideas to students as well as practicing engineers, scientists, and technicians.

Understanding Surface and Thin Film Science

Thomas M. Christensen

CRC Press
Taylor & Francis Group
Boca Raton London New York

CRC Press is an imprint of the
Taylor & Francis Group, an **informa** business

First edition published 2023
by CRC Press
6000 Broken Sound Parkway NW, Suite 300, Boca Raton, FL 33487-2742

and by CRC Press
4 Park Square, Milton Park, Abingdon, Oxon, OX14 4RN

CRC Press is an imprint of Taylor & Francis Group, LLC

© 2023 Taylor & Francis Group, LLC

ISBN: 9781482233032 (hbk)
ISBN: 9781032392158 (pbk)
ISBN: 9780429194542 (ebk)

DOI: 10.1201/9780429194542

Typeset in Palatino
by codeMantra

To Ann

whose support and patience

made this possible.

Contents

Preface..xiii
Author..xv

1. Surfaces and Thin Films...1
   1.1   Introduction...1
   1.2   Atomic Scale Models..4
   1.3   Overview..6
   References..8

2. Vacuum and Plasma Environments..9
   2.1   Kinetic Theory of Gases..9
        2.1.1   Ideal Gas Law...9
        2.1.2   Pressure and Vacuum...10
        2.1.3   Interactions in the Gas...10
        2.1.4   Interactions with Surfaces...15
        2.1.5   Vapor Pressure...16
        2.1.6   Gases vs. Solids..16
   2.2   Plasmas..17
        2.2.1   Creating a Plasma..17
        2.2.2   Characterizing a Plasma...20
        2.2.3   Motion of Charges...21
        2.2.4   Plasma Sources...24
        2.2.5   Chemistry in Plasmas...25
        2.2.6   Applications of Plasmas..26
   2.3   Cleaning Surfaces...27
   References..29
   Problems...30

Part I   Surfaces

3. Crystal Structure..35
   3.1   Structure of Crystals..35
   3.2   Reciprocal Lattice...43
   3.3   Bonds in Crystals...44
   3.4   Defects in Crystals...46
   3.5   Ideal Surfaces..49
   3.6   Surface Reconstructions..50
   3.7   Minimizing Energy...52
   3.8   Surface Roughness...53
   3.9   Non-Crystalline (Amorphous) Solids...54
   3.10 Characterization of Structure...54
   References..55
   Problems...55

4. Atomic Motion: Vibrations, Waves, and Diffusion in Solids 59
  4.1   Thermal Vibrations 59
  4.2   Elastic Waves and Phonons 61
       4.2.1   Elastic Waves 61
       4.2.2   Phonons 65
       4.2.3   Surface Waves and Phonons 66
  4.3   Diffusion 68
       4.3.1   Bulk Diffusion 68
       4.3.2   Surface and Interface Diffusion 73
       4.3.3   Surface Roughness 74
  4.4   Characterization of Atomic Motion 76
  References 76
  Problems 76

5. Thermodynamics 81
  5.1   Thermodynamics in Solids 82
  5.2   Phase Diagrams 85
       5.2.1   One-Component Systems 85
       5.2.2   Two-Component Systems 87
       5.2.3   Binary Solid Solutions 87
       5.2.4   Binary Eutectics 88
       5.2.5   Other Considerations 89
  5.3   Surface Thermodynamics 89
       5.3.1   Surface Energy, Tension, and Stress 89
       5.3.2   Minimizing Energy 91
       5.3.3   Segregation in Two-Component Systems 93
  5.4   Characterization of Thermodynamics 94
  References 95
  Problems 95

6. Electrical, Magnetic, Optical, and Thermal Properties 99
  6.1   Electrical Properties 99
       6.1.1   Free Electron Gas Model 99
       6.1.2   Jellium and Nearly Free Electron Model 101
       6.1.3   Other Electronic Models 103
       6.1.4   Semiconductors 105
       6.1.5   Resistivity and Conductivity 106
       6.1.6   Surfaces 107
       6.1.7   Surface States 110
       6.1.8   Excitons and Plasmons 112
  6.2   Magnetic Properties 114
       6.2.1   Magnetic Materials 114
       6.2.2   Ferromagnetic Energies 117
       6.2.3   Surface Magnetism 118
  6.3   Optical Properties 119
       6.3.1   Electromagnetic Waves in Materials 119
       6.3.2   Optical Properties at Surfaces 122
       6.3.3   Adsorbates on Surfaces 123
       6.3.4   Films 124

6.4   Thermal Properties 127
    6.4.1   Specific Heat 128
    6.4.2   Thermal Conductivity 128
    6.4.3   Thermal Expansion 128
6.5   Characterization of Surface and Film Properties 129
References 129
Problems 130

7. Adsorbed Atoms on Surfaces 135
7.1   Thermodynamics of Adsorbed Atoms 135
7.2   Ordered Structures 139
7.3   Molecular Adsorption 144
7.4   Adsorbate Motions 145
7.5   Characterization of Adsorbed Atoms 147
References 147
Problems 147

Part II   Thin Films

8. Overview of Thin Film Growth 151
8.1   Introduction 151
8.2   Homogeneous Nucleation and Growth 152
8.3   Steps in Film Formation 158
    8.3.1   Thermal Accommodation 158
    8.3.2   Binding and Desorption 159
    8.3.3   Surface Diffusion 160
    8.3.4   Heterogeneous Nucleation 161
    8.3.5   Island Growth 167
    8.3.6   Island Coalescence 169
    8.3.7   Thicker Films – Zone Models 170
8.4   Deviations from Non-Ideal Structure 173
8.5   Advanced Modelling 174
8.6   Summary and Characterization of Thin Films 175
References 175
Problems 176

9. Physical Vapor Deposition 179
9.1   Evaporation 179
    9.1.1   Source 179
    9.1.2   Transport 180
    9.1.3   Deposition 183
    9.1.4   Evaporation Parameters and Processes 186
9.2   Sputter Deposition 188
    9.2.1   Source 188
    9.2.2   Transport 190
    9.2.3   Deposition 191
    9.2.4   DC (Diode) Sputter Deposition 192

        9.2.5   RF Sputter Deposition 194
        9.2.6   Magnetron Sputter Deposition 195
    9.3   Modifications to Physical Vapor Deposition 196
        9.3.1   Ion-Assisted Deposition 196
        9.3.2   Reactive Deposition 196
        9.3.3   Comparison of Evaporation and Sputtering 197
    9.4   Molecular Beam Epitaxy and Epitaxial Films 197
    9.5   Arc Vaporization – Cathodic Arc Deposition 200
        9.5.1   Source 200
        9.5.2   Transport 200
        9.5.3   Deposition 201
    9.6   Pulsed Laser Deposition – Laser Ablation 201
        9.6.1   Source 201
        9.6.2   Transport 202
        9.6.3   Deposition 202
    References 203
    Problems 203

10. **Chemical Vapor Deposition** 209
    10.1   Overview and Chemical Reactions 209
    10.2   Source 213
    10.3   Transport 213
    10.4   Deposition 215
    10.5   Modifications of CVD 218
        10.5.1   Low-Pressure CVD 218
        10.5.2   Plasma Enhanced CVD 219
        10.5.3   Laser Enhanced CVD 220
        10.5.4   Hot Wire CVD 220
        10.5.5   Metalorganic CVD 221
    10.6   Atomic Layer Deposition 222
    References 223
    Problems 224

Part III   Characterization of Surfaces and Thin Films

11. **Characterization: Overview and Imaging Techniques** 227
    11.1   Overview of Characterization 227
    11.2   Imaging Techniques 229
        11.2.1   Optical Microscopes 229
        11.2.2   Scanning Electron Microscope 232
        11.2.3   Transmission Electron Microscope 237
        11.2.4   Low Energy Electron Microscope 239
        11.2.5   Scanning Probe Microscopes 240
    References 246
    Problems 247

12. Characterization: Structural Techniques..249
    12.1 X-Ray Diffraction 249
    12.2 Low Energy Electron Diffraction 256
    12.3 Reflection High Energy Electron Diffraction.......................262
    12.4 X-Ray Reflectivity..263
    12.5 Scattering Techniques 265
    12.6 Stylus Profilometer...266
    12.7 Quartz Crystal Microbalance..268
    12.8 Film Density Measurements 270
    References 270
    Problems..271

13. Characterization: Chemical and Elemental Techniques 273
    13.1 Auger Electron Spectroscopy 273
    13.2 Energy and Wavelength Dispersive X-Ray Analysis.........280
    13.3 X-Ray Photoelectron Spectroscopy.....................................281
    13.4 Ultraviolet Photoelectron Spectroscopy 286
    13.5 Near-Edge X-Ray Absorption Fine Structure 287
    13.6 Secondary Ion Mass Spectrometry.....................................288
    13.7 Scattering Techniques 290
    13.8 Fourier Transform Infrared Spectroscopy 292
    13.9 Raman Spectroscopy 295
    13.10 Electron Energy Loss Spectroscopy 296
    References 299
    Problems..300

14. Characterization: Electrical, Magnetic, and Optical Techniques...........303
    14.1 Electrical Characterization 303
        14.1.1 Hall Effect 303
        14.1.2 Resistivity: Four-Point Probe..............................304
    14.2 Magnetic Characterization 307
        14.2.1 Magneto-Optical Kerr Effect...............................307
        14.2.2 Spin-Polarized Electron Techniques 308
        14.2.3 Magnetic Force Microscopy 309
        14.2.4 Brillouin Light Scattering 310
        14.2.5 Magnetometers...311
        14.2.6 Ferromagnetic Resonance....................................312
        14.2.7 X-Ray Magnetic Circular Dichroism...................313
    14.3 Optical Characterization...314
        14.3.1 Reflectance 315
        14.3.2 Ellipsometry 317
        14.3.3 Light Scattering 321
        14.3.4 Interferometry 322
    References 324
    Problems..324

15. Characterization: Thermodynamic, Thermal, and Mechanical Techniques 327
    15.1   Thermodynamic Characterization 327
        15.1.1 Surface Tension 327
        15.1.2 Surface Adsorption and Desorption 328
        15.1.3 Differential Scanning Calorimetry and Thermogravimetric
                Analysis 331
        15.1.4 Diffusion 332
    15.2   Thermal Characterization 333
        15.2.1 Micro-Thermal Microscopy 333
        15.2.2 Photothermal Analysis 334
        15.2.3 Cross-Plane Thermal Analysis 334
    15.3   Mechanical Characterization 335
        15.3.1 Brillouin Light Scattering 335
        15.3.2 Stress 335
        15.3.3 Friction 337
        15.3.4 Microindentation – Nano-indentation 338
        15.3.5 Adhesion 339
    References 343
    Problems 343

Appendix 1: Physical Constants and Unit Conversions 345

Appendix 2: Acronyms and Abbreviations 347

Appendix 3: Basic Vacuum Technology 349

Index 353

Preface

This book is a conceptual overview of surface and thin film science aimed at providing students as well as practicing engineers, scientists, and technicians with a basic understanding of the most common ideas and methods used in these fields. Thin film science is a natural extension of surface science especially as applications involve thinner and thinner films. Relying on simple, conceptual models and figures, fundamental scientific ideas are introduced and then applied to surfaces and thin films in the first half of the book. Common methods of gas-phase thin film deposition are then introduced followed by an overview of a wide range of characterization techniques.

The book arose from a series of technical short courses offered through the American Vacuum Society and the Society of Vacuum Coaters as well as upper-division undergraduate and introductory graduate courses offered at the University of Colorado Colorado Springs. The influence of other books on these subjects will be apparent throughout the book. The thin film book by Ohring and the surface science book by Prutton were particularly influential. I attempted to include the many sources which I consulted in the references at the end of each chapter, but have undoubtedly missed some where I got an idea but did not document it well enough in my lecture notes over the past 33 years of teaching.

I have used this book in upper-division undergraduate and introductory graduate classes in surface science and in thin film science. For the surface science course, I typically cover chapters 1–7 and selected material on characterization from chapters 11–15. For the thin films class, I select material from chapters 1–6 before covering chapters 8–10 and selected material from chapters 11–15. I have taught both integrating the characterization material into the class as I go through the earlier chapters and also separating it out and covering it as a unit at the end of the class. I have separated the material out in this book, but have tried to provide enough internal cross-referencing to allow either course organization.

I thank my colleagues at the University of Colorado Colorado Springs for the many conversations over the years from which I learned so much. My thesis adviser, Jack Blakely, and fellow graduate students at Cornell University shaped my early thinking about surface science. While on sabbatical at the University of Maine, team teaching a thin films class with Bob Lad and David Frankel further developed these ideas. I greatly appreciate the hospitality of the faculty at Macalester College while I was on sabbatical writing an early draft of this book. The patience and guidance of Luna Han and others from Taylor and Francis publishers helped keep this project on track. Finally, I thank my family. My parents set me on the path of learning. My wife, Ann, and daughters, Emily and Mary, were understanding and supportive throughout the many stages of my academic career.

Author

Thomas M. Christensen joined the faculty of the University of Colorado – Colorado Springs (UCCS) Department of Physics and Energy Science in 1989. He served the campus as a faculty member, department chair, associate dean, dean, and Provost. He co-directed the UCCSTeach program for preparing future secondary science and math teachers. Dr. Christensen received both the College (1993) and campus (1996) Outstanding Teaching Awards, the Chancellor's Award (2003) to recognize his service and teaching, and the University of Colorado Excellence in Leadership Award (2015). He received his B.S. degrees in Physics and Astrophysics from the University of Minnesota and his Ph.D. in Applied Physics from Cornell University. He was a Member of the Technical Staff at Sandia National Labs prior to joining the UCCS faculty. Dr. Christensen's research in experimental surface physics has led to many published papers in international science journals and presentations at scientific meetings. He has been the principal or co-principal investigator on over $2 million in research grants and contracts for work in surface physics and in science education. He taught 25 different classes at UCCS at all levels from introductory classes for non-majors to graduate-level classes. He has served his primary professional society (AVS) on national Education and Diversity committees. In his spare time, Dr. Christensen plays string bass with the Pikes Peak Philharmonic orchestra and bass guitar with the Physics Classic Rock and Roll Orchestra.

1

Surfaces and Thin Films

1.1 Introduction

Surface science and thin film science are industrially and fundamentally important areas of materials science. The broad term "science" is appropriate here in that these areas are difficult to assign to just one discipline within the sciences and engineering. Practitioners include people with training in physics, chemistry, biology, materials science, and many areas of engineering. Practical knowledge of thin films has existed for thousands of years and has been chronicled in a series of papers by Greene (2014, 2015). The study of surfaces and thin films draws insights from many different scientific and engineering approaches. The two topics, while often treated as separate, are natural complements of one another. Thin films are composed of thin (typically under 10^{-6} m thick) materials that are deposited onto the surface of a substrate material. The initial stage of film formation, with atoms forming a sub-monolayer arrangement on the surface, has long been considered an important aspect of surface science.

In this chapter and throughout the book, we integrate these two aspects of materials science and explore them on an atomic level to try to gain a greater conceptual understanding of the processes that we work with in experimental research and production environments. Surface properties (as opposed to bulk properties) become more important as we reduce the size of components in manufacturing and electronics. Catalysis of chemical reactions is a surface phenomenon with broad industrial applications. The properties of materials depend fundamentally on atoms and their positions. As a result, this book takes an atomistic approach and introduces basic concepts of atomic structure and dynamics, thermodynamics, electricity, and magnetism applied to thin films and surfaces. After establishing these concepts, we explore the most common processes involved in the physical and chemical deposition of thin films in vacuum/gas environments. To understand these processes and the products of thin film deposition, we introduce many of the most common techniques for characterizing surfaces and thin films.

This chapter establishes some examples of surface behavior and introduces several simple models, which will be used throughout the book. Before getting into the models, it is valuable to reflect briefly on why we expect surfaces and thin films to behave differently from bulk materials and what observed properties we seek to understand in studying these topics.

To begin, note that solid-state physicists who study crystalline materials typically assume that these materials have atoms arranged in well-characterized positions that display both short-range and long-range order and extend infinitely in all directions. The existence of a surface violates this assumption of infinite extent and also breaks the symmetry of the system by introducing a place where the structure changes from a well-characterized

DOI: 10.1201/9780429194542-1

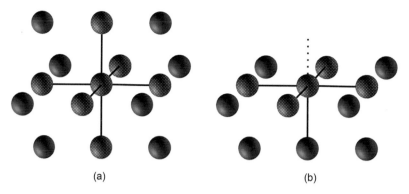

FIGURE 1.1
Simple cubic structures. (a) Atoms in the bulk have six nearest-neighbor bonds. (b) Atoms on a surface have five nearest-neighbor bonds.

arrangement of atoms to no atoms at all. This simple observation already suggests that surfaces are different from the bulk.

Extending this idea, we imagine atoms in a crystalline solid arranged in a cubic lattice as in Figure 1.1a. In the bulk, each atom is surrounded by six nearest-neighbor atoms. An atom on a surface as in Figure 1.1b, however, is missing the atom above it. It only has bonds with five nearest-neighbor atoms. This changes the bonding environment of surface atoms. In response to this changed environment, they might relax slightly outward from the surface or be pulled below the surface. These changes alter the surface atomic spacing. The atomic spacing is a fundamental parameter in understanding the mechanical properties of materials. The surface would be expected, then, to have different mechanical properties from the bulk. This change in the equilibrium position of the surface atoms would also change the vibrational properties of that atom. The presence of a bond that is not satisfied could lead to greater chemical reactivity of the surface as well.

Chemically, the surface is where the material meets the outside environment. Interesting chemistry is expected where a surface interacts with the gas environment above it or where a film might be deposited onto it. Chemical reactions at a surface could dramatically change the mechanical, electrical, and optical properties, for example, of the surface. We will also find that alloy materials may experience the segregation of one element to a surface resulting in a very different chemical composition at the surface when compared to the bulk composition. These chemical composition effects are demonstrated in Figure 1.2 and will be discussed in Chapter 5.

Electrical properties, such as electrical resistance, can be modeled as arising from a process where electrons are scattered off of ionic cores in the solid inhibiting their motion and giving rise to electrical resistance. The electrons are also able to scatter from surfaces and interfaces with thin films. The result is that electrons near a surface or interface are expected to scatter more frequently than electrons in the bulk leading to changes in the electrical properties of a material near the surface or near an interface. The average distance that an electron travels between collisions (the mean free path) in the bulk is typically in the 1–10 nm range in many materials. If an electron is closer to the surface than this distance, it may experience scattering sooner as represented in Figure 1.3.

The result of this brief discussion of qualitative concepts is that the "surface" is viewed as not just a plane but rather a region in which the properties of a material differ from the bulk. This surface region is often the top 5–10 atomic layers of a material. With miniaturization

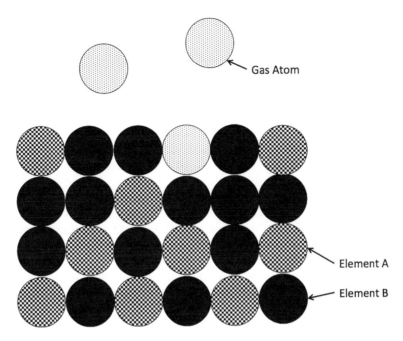

FIGURE 1.2
A material with a composition of 50% element A and 50% element B in the bulk shows segregation of element B to the surface and surface contamination from the gas.

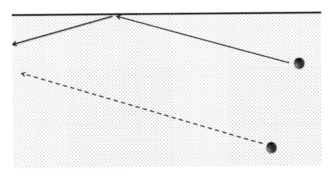

FIGURE 1.3
Electrons closer to the surface may travel a shorter distance (solid line) before undergoing a collision compared with electrons farther from the surface (dashed line).

now reaching the nanometer scale typical of atomic dimensions, an understanding of these differences becomes increasingly important.

In addition to these surface and interface-related effects, the processes involved in creating thin films can give them different properties from a bulk sample of the same material. Thin film deposition can result in a film whose crystalline structure, porosity, grain structure, and defects will differ from the bulk. The presence of a substrate can produce chemical and stress-related effects in the film. Thin films might not form a continuous layer on a surface which can give them dramatically different properties from bulk. We will explore these concepts in more detail in later chapters.

1.2 Atomic Scale Models

Since we are adopting an atomic-level examination of surfaces and thin films, we consider here simple models of how to treat atoms. The simplest models treat atoms as very small, hard, spheres interacting only by collisions. Other models might attach springs between the hard spheres to model bonding. The main differences in models will arise from how the atoms interact with one another. Simple models will be adequate for a conceptual understanding in many cases. We now explore three simple models: (1) hard sphere collision models, (2) interacting hard sphere (mass and spring) models, and (3) statistical energy barrier models.

When treating atoms in a gas, the kinetic theory of gases and the equation of state for an ideal gas (ideal gas law) assume that atoms are hard spheres and interact only through elastic collisions. All other forces are neglected. The size of the atoms is assumed to be much smaller than the distance between atoms, and so the number density of gas atoms is much smaller than the number density of atoms in a solid. In this model, no internal structure is attributed to molecules vs. atoms, and so the two words can be used interchangeably and molecules are also pictured as hard spheres. The consequences of this model will be developed in Chapter 2.

When treating atoms in a solid, we need to consider a longer range interaction between the atoms in order to explain phenomena such as bonding and crystalline structure of solids. The simplest model to consider is to treat atoms as being connected by springs. The spring constant of the spring describes the relative strength of the bonding interactions. This simple picture will help us gain understanding of vibrations, bonding, energy transfer, crystalline structure and many other concepts.

Masses on springs are a standard topic in introductory physics courses. If a mass on a spring is displaced a distance Δx from the equilibrium position (where all forces are balanced and there is no net force), then a spring will exert a force

$$F = -C\,\Delta x \tag{1.1}$$

where C is the force constant of the spring. The negative sign indicates that the force will always attempt to move the mass back to the equilibrium position. The potential energy, U, associated with this spring and mass system is

$$U = \tfrac{1}{2}C(\Delta x)^2 \tag{1.2}$$

which is just a parabola with a minimum at the equilibrium position and is the harmonic oscillator potential energy. Any deformation away from the equilibrium position results in an increase in energy of a system.

This model might seem somewhat artificial, but it is important to recognize that it is well justified mathematically. If we consider any arbitrary potential energy distribution arising from a more sophisticated modeling of atoms in a solid, but accept that there is an equilibrium position with minimum energy, then we can always perform a Taylor's series expansion of that arbitrary potential $U(x)$ about that equilibrium point, $x=x_0$.

$$U(x) = U(x_0) + (\Delta x)\frac{dU}{dx}\Big|_{x_0} + \frac{1}{2}(\Delta x)^2\left(\frac{d^2U}{dx^2}\right)\Big|_{x_0} + \text{higher order terms} \tag{1.3}$$

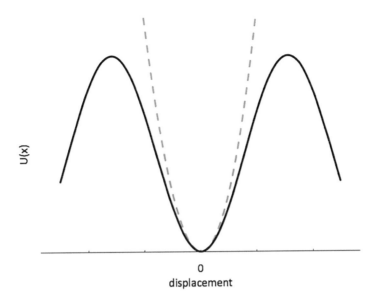

FIGURE 1.4
A sine or cosine potential energy (solid line) is well modeled by a second order harmonic oscillator potential (dashed line) when near the minimum.

Since we can always add an arbitrary constant to any potential energy, we choose $U(x_0)$ to be zero which eliminates the first term on the right side of the equation. The second term is the derivative of the function evaluated at the minimum point. From calculus, we know that this is also zero, and so the second term is eliminated. The third term has exactly the form of the energy in our mass and spring system. So we see that the mass and spring system is the first-order approximation to any arbitrary potential energy that has a minimum at the equilibrium position. If we always stay near the equilibrium position, this model will be a good approximation to the actual energy function. This explains why this model is both conceptually and mathematically useful. We will employ this model in various situations throughout the book. An example comparing the potential energy function $U(x)=\sin(x)$ with the harmonic oscillator function near the equilibrium point (minimum) is provided in Figure 1.4.

If this potential energy is related to a thermal energy ($k_B T$ where k_B is Boltzmann's constant), then the thermodynamic equipartition theorem predicts that the average displacement of an atom vibrating in a solid is proportional to $\sqrt{T}$. Not surprisingly, the higher the temperature, the farther the atoms in a solid will vibrate from their equilibrium positions.

In some cases of atoms in solids or on solid surfaces, we are interested in longer-range motions well beyond the equilibrium position. For these motions, the mass and spring model is not useful. In these cases, it is simplest to consider some sort of energy barrier to the motion and put all of the various interactions into an energy barrier term.

Consider an example that combines the two models. A surface atom is held into the solid by bonds to neighboring atoms that can be represented as springs. This atom will be vibrating, and we find that the amplitude of the vibration depends on the temperature of the atom. One could think about the possibility of the atom vibrating so violently that it breaks the springs and escapes from the surface. This could be modeled using classical mechanics to calculate forces and assuming that the energy function does not extend to infinity but rather is truncated at some energy above which we declare the atom to have escaped from the surface. This analysis from classical mechanics, however, ignores the statistical nature of the situation. If the surface has an average temperature, T, we find that atoms will be vibrating with a range of amplitudes corresponding to a well-defined statistical distribution of temperatures. Instead of trying to solve the classical mechanics problem for each atom, we assume that there is some energy barrier, ΔG, that an atom must overcome to escape from the solid. We use the symbol G for the energy here since it is a Gibbs free energy.

The problem is now a question of statistical probability. What is the probability that an atom will escape from a surface whose average temperature is T? This will depend on three factors. The first is how frequently that atom tries to escape. This would depend on the frequency of the vibration (which could be expressed in terms of the mass and spring model). Each time the atom vibrates outward away from the surface, it is trying to escape. The second factor is the size of the energy barrier, ΔG. If the barrier is large, the probability of escape should be small. The third factor is how much energy the atom has. If the energy of the atom is much smaller than the barrier, the probability of escape would be small. Typically, this energy is a thermal energy, represented by $k_B T$.

From statistical mechanics, we find that the correct way to combine the two energy terms is in an exponential form, called a Boltzmann factor, that looks like $e^{-\Delta G/k_B T}$. This term has the correct behavior in that the probability gets smaller if $\Delta G \gg k_B T$. This is equivalent to considering the probability that an atom will possess some amount of energy ΔG compared to the average available energy $k_B T$. The probability that a particular atom will get significantly more than its share of energy is very low.

We put these terms together and express the probability of escape per unit time as

$$p = \nu e^{-\Delta G/k_B T} \tag{1.4}$$

where ν is the natural frequency of vibration (how often the atom attempts to move) and the exponential term describes how likely it is to be successful in overcoming the energy barrier for a given temperature, T.

The combination of these simple models, the kinetic theory of gases, the harmonic mass and spring system in solids, and energy barriers to inhibit motion, will allow us to gain a conceptual and often an approximate mathematical understanding of surfaces and thin films on an atomic scale.

1.3 Overview

With these general concepts and simple models in place, we next (in Chapter 2) expand our understanding of two important environments in surface and thin film science. The vacuum and plasma environments are very common in many applications. In the following chapters, we examine the properties of materials and how they differ at surfaces. In Chapter 7, we introduce the presence of small numbers of atoms on top of the surface that begins the transition from surface science into thin film science. After exploring thin films, the final section of the book introduces techniques that can be used to characterize surfaces and thin films.

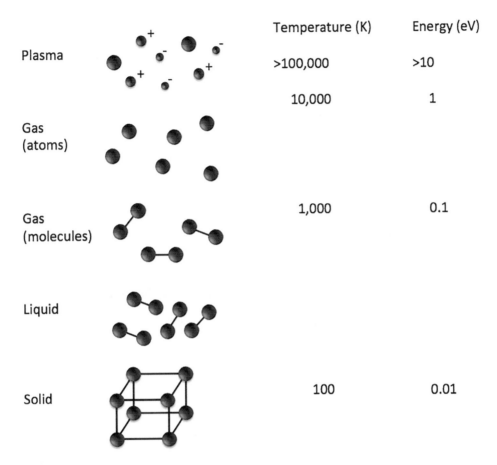

	Temperature (K)	Energy (eV)
Plasma	>100,000	>10
	10,000	1
Gas (atoms)		
Gas (molecules)	1,000	0.1
Liquid		
Solid	100	0.01

FIGURE 1.5
Phases of matter with order of magnitude temperatures and corresponding energies.

Throughout the book, we will be talking about different states of matter. It is useful to have a sense of the temperatures and energies associated with these different states as presented in Figure 1.5. The energies associated with bonds in solids are typically only a few tenths of an eV meaning that at temperatures in the hundreds of Kelvins, the solids can be broken apart into liquids. At higher temperatures, the liquids can evaporate into gases. At still higher temperatures (tens of thousands of Kelvins), the atoms can be ionized to form plasmas.

From a practical viewpoint, we seek to understand how the properties of surfaces and thin films connect to their structure and composition. The origins of the structure and composition, however, are best understood by looking deeper at how atomic level kinematics and dynamics, electronic structure, and thermodynamics determine structure and composition. These same fundamental concepts help us understand the properties that would be anticipated from specific structures and compositions.

The references (Blakely 1973, Chopra 1969, Eckertova 1977, Frey and Khan 2015, Hudson 1998, Ibach 2006, Luth 2015, Ohring 2002, Prutton 1994, Somorjai 2010, Venables 2000, Zangwill 1998) listed at the end of this chapter are just a sampling of the many excellent books available that cover aspects of surface and thin film science. In many cases, they will discuss certain topics in greater detail than in this book.

References

Blakely, J. M. 1973. *Introduction to the Properties of Crystal Surfaces*. Oxford: Pergamon Press.
Chopra, K. L. 1969. *Thin Film Phenomena*. New York: McGraw-Hill Book Company.
Eckertova, L. 1977. *Physics of Thin Films*. New York: Plenum Press.
Frey, H. and H. R. Khan, eds. 2015. *Handbook of Thin-Film Technology*. Berlin: Springer Verlag.
Greene, J. E. 2014. Tracing the 5000-year recorded history of inorganic thin films from 3000 BC to the early 1900s AD. *Appl. Phys. Rev.* 1, 041302: 1–36.
Greene, J. E. 2015. Tracing the 4000 year history of organic thin films: From monolayers on liquids to multilayers on solids. *Appl. Phys. Rev.* 2, 011101: 1–11.
Hudson, J. 1998. *Surface Science: An Introduction*. New York: John Wiley & Sons, Inc.
Ibach, H. 2006. *Physics of Surfaces and Interfaces*. Berlin: Springer.
Luth, H. 2015. *Solid Surfaces, Interfaces and Thin Films*. 6th ed. Berlin: Springer.
Ohring, M. 2002. *Materials Science of Thin Films*. 2nd ed. San Diego, CA: Academic Press.
Prutton, M. 1994. *Introduction to Surface Physics*. Oxford: Clarendon Press.
Somorjai, G. A. 2010. *Introduction to Surface Chemistry and Catalysis*. 2nd ed. New York: John Wiley & Sons, Inc.
Venables, J. A. 2000. *Introduction to Surface and Thin Film Processes*. Cambridge: Cambridge University Press.
Zangwill, A. 1988. *Physics at Surfaces*. Cambridge: Cambridge University Press.

2

Vacuum and Plasma Environments

Surfaces are where a material meets the environment, which is often a gas or plasma. We examine these environments here and refer back to this chapter as needed throughout the book. Although there are thin-film deposition techniques done in liquid- or even solid-phase environments, we focus our attention in this book on gas-phase deposition techniques. The closely related subjects of liquid-solid interfaces and electrochemical thin-film growth have been examined by other authors (Erbil 2006, Paunovic 2006). With the atomic scale model from Chapter 1 in mind, we explore first a simple, but useful, model of atoms and molecules in vacuum environments. We define "vacuum" as the realm where the molecular density is less than the atmosphere, and we examine several characteristics of vacuum environments with a focus on the kinetic theory of gases. Following this, we examine the behavior of charged particles in plasma environments and conclude with an examination of how to apply these ideas to produce clean surfaces. Additional details are available in the references (Delchar 1993, Frey and Khan 2015, Kauzmann 1966, O'Hanlon 2003, Tompkins 1997).

2.1 Kinetic Theory of Gases

2.1.1 Ideal Gas Law

Many elements of vacuum technology can be understood by examining the equation of state of an ideal gas – more commonly known as the ideal gas law. When considering an atomic model, the ideal gas law is expressed as

$$PV = Nk_BT \tag{2.1}$$

where P is absolute pressure, V is volume, N is number of gas molecules, k_B is Boltzmann's constant, and T is temperature (in K). We often define $n = N/V$ as the number density of gas molecules and rewrite the ideal gas law

$$P = nk_BT \tag{2.2}$$

The ideal gas law can also be written as $PV = NRT$ where N is the number of moles of gas and R is the universal gas constant. $R = N_Ak_B$ where N_A is Avagadro's number.

The existence of an equation of state, which relates P, V, T, and mass, m, or N, tells us that any other property of the gas must be able to be described by knowing three of these four parameters. This provides a powerful theoretical base on which to build a model of gas behavior.

In this chapter, we will typically assume that the gas is composed of a single component. If more than one type of molecule exists in the same volume at thermodynamic

DOI: 10.1201/9780429194542-2

9

equilibrium so that they have the same temperature, then we can treat each type of gas molecule independently and they each will obey the ideal gas law $P_iV=N_ik_BT$. The partial pressure, P_i, is defined as the pressure of a particular gas component. The sum of the partial pressures of each gas component will yield the total pressure

$$P = \sum_i P_i$$

$$\text{and } N = \sum_i N_i.$$

The ideal gas law is an approximation that is good at low gas densities (low pressures). For pressures below 1 atm, we can write a more accurate equation of state as:

$$\frac{PV}{Nk_BT} = 1 + B'(T)P \tag{2.3}$$

The correction term, $B'(T)$, for non-ideal gases gets smaller as the pressure decreases. For common gases at temperatures typical of laboratory environments, the error in using the idea gas law rather than this corrected equation of state is less than 0.1% at pressures around 100 Pa (1 torr) and gets proportionally smaller as we continue down in pressure.

2.1.2 Pressure and Vacuum

We defined vacuum as being an environment where the number density of gas is less than that found in the atmosphere at sea level. We see from the ideal gas law, Equation 2.2, that the pressure of the gas is proportional to the gas number density and so we typically report the pressure rather than the gas density. Although the SI unit of gas is the Pascal (N/m^2), many other units are in common use and Appendix 1 provides conversion factors between units.

We use different terms for the vacuum in different pressure ranges. Ultra-high vacuum (UHV) refers to pressures below about 10^{-6} Pa. High vacuum describes pressures from 10^{-6} up to about 10^{-2} Pa. Medium vacuum ranges from 10^{-2} up to 1000 Pa and low vacuum covers the range from 1000 Pa up to atmospheric pressure (10,100 Pa). A basic introduction to vacuum technology, pumps, and gauges is provided in Appendix 3.

It is important to recognize that the process of creating a vacuum can also change the chemical composition of the gas in a chamber. The pumps used to achieve pressures below the atmosphere will remove some elements more efficiently than other elements. As a result, the composition of the gas will change as a chamber filled with air is pumped down from the atmosphere. Table 2.1 shows the common components of dry air at atmospheric pressure and at UHV in an ion-pumped chamber. Even though the chamber started with mainly N_2 and O_2 and a trace of H_2, at UHV, the composition is primarily H_2 with only a trace of N_2 and O_2.

2.1.3 Interactions in the Gas

A simple model of molecules in a gas (introduced in Chapter 1) is to treat them as very small, hard, spheres that only interact when they collide. We assume that there are a large number of them so that we can apply statistical methods to analyze their behavior. By

TABLE 2.1

Composition of Dry Air and Typical Ion-Pumped Vacuum

Component	Volume % in Dry Air	Volume % in Ion-Pumped Chamber at 10^{-9} torr
N_2	78%	Trace
O_2	21%	Trace
Ar	0.93%	Trace
CO_2	0.03%	4%
CH_4	Trace	3%
H_2O	Trace	5%
CO	Trace	6%
H_2	Trace	80%

"very small" we mean that their diameter is much less than the typical distance between molecules. Since they only interact by collisions, they will move in straight lines between collisions. This physical model, known as the kinetic theory of gases, is extremely useful for picturing the behavior of gases in vacuum and provides a quantitative model that reasonably explains much of the observed behavior. We assume here that there is only one type of molecule in the gas so that the mass and size of the molecules in the gas are all the same.

While we do not derive the formulas discussed below, the derivations typically arise from noting that pressure is force/area and that the force is the change in momentum when hard sphere molecules collide with each other or with walls. Combining this with the cylindrical volume swept out by a molecule of diameter, d, moving with some average speed during a given time, allows us to derive most of the formulas in this chapter.

In Figure 2.1, we freeze all of the molecules except one and examine the motion of that one molecule. From this picture, we can see that the molecule interacts both with other gas molecules and with the wall of the chamber containing the gas. The more gas molecules present (the higher the pressure), the more collisions with other gas molecules and the less important will be the collisions with the wall. If there are very few other gas molecules (lower pressure) then collisions with the wall will constitute a higher percentage of the collisions and the walls become an important part of the system. This suggests that we will see some differences in gas behavior in a vacuum at different pressures. We will explore these differences after considering a few other basic properties.

Since we are applying statistical methods to examine the behavior of the gas particles, we recognize that the particles will not all have the same velocity. The distribution of velocities in this model is well characterized and is described by the Maxwell velocity distribution. Figure 2.2a shows a Maxwell velocity distribution for two inert gases at the same temperature. Figure 2.2b shows a Maxwell velocity distribution for one gas at three different temperatures. Notice that the curve shifts to the right (higher speed) and spreads out for higher temperatures or lower masses.

Since the distribution is not symmetric, the average velocity will not be at the peak of the curve. It is common to define a root mean square velocity as being representative of the speed of the molecules.

$$v_{rms} = \sqrt{\frac{3k_B T}{m}} \tag{2.4}$$

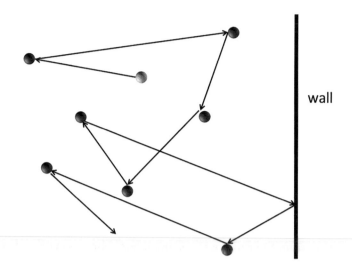

FIGURE 2.1
Motion of one atom showing hard sphere collisions with other atoms and the wall in the kinetic theory of gases.

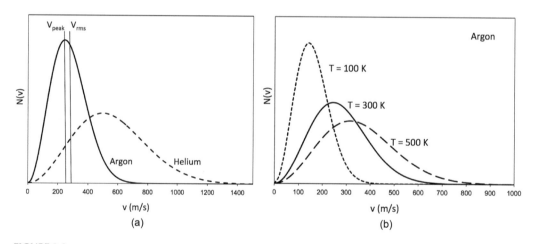

FIGURE 2.2
(a) Maxwell velocity distribution at $T=300$ K for Ar and He. (b) Maxwell velocity distribution for Ar at $T=100$, 300, and 500 K.

where m is the mass of a molecule. This can be compared to the value of the velocity at the peak of the distribution.

$$v_{peak} = \sqrt{\frac{2k_B T}{m}} \tag{2.5}$$

Alternately, one can define a standard average value of the velocity yielding a value of

$$v_{ave} = \sqrt{\frac{(8/\pi)k_B T}{m}} \tag{2.6}$$

TABLE 2.2

Root Mean Square Velocities for Several Gases at Room Temperature in Several Systems of Units

Molecule	v_{rms} (m/s)	v_{rms} (miles/hour)	v_{rms} (km/hour)
H_2	1700	3790	6120
N_2	450	1000	1620
Ar	380	850	1368

These formulas behave as we might expect in that higher temperatures and lower masses both result in faster motion. The values for v_{peak} and v_{rms} are indicated in Figure 2.2a. The value for v_{ave} lies between v_{peak} and v_{rms}.

To get a sense for these speeds, Table 2.2 provides v_{rms} values for three gases at room temperature. Since many of us do not really have a feel for speeds in m/s, the values are also converted to miles/hour and km/hour. The molecules in the air around you (mainly N_2) are traveling at about 1000 miles/hour since the velocity is independent of the gas pressure. What changes with pressure is the distance they travel before experiencing a collision. Near atmospheric pressure, the molecules travel a much shorter distance before experiencing a collision than at UHV.

This distance that a molecule travels between collisions can be statistically described by the mean free path, λ_{mfp}, which is the average distance that a particle travels between collisions. The kinetic theory of gases can be used to show that this depends on the diameter of the molecule, d, and the number density of molecules, n.

$$\lambda_{mfp} = \frac{1}{\sqrt{2}\pi d^2 n} \tag{2.7}$$

This result gives us the reasonable model that the distance will be smaller if the molecules are larger or if they have a higher number density. The distance between molecules and number density, however, are not variables that we typically measure in the laboratory. Using the ideal gas law, we can express the mean free path (in meters) for air at room temperature as

$$\lambda_{mfp} = \frac{5 \times 10^{-5}}{P} \tag{2.8}$$

where the pressure is in torr.

Although only numerically correct for molecules with molecular weight near 28 and temperatures near 300 K, formulas for air at room temperature will frequently be presented here to allow greater insight into gas molecule behavior and to allow quick estimates to be calculated.

Table 2.3 provides some typical values for air at room temperature. These values confirm the picture we had in Figure 2.1. At atmospheric pressures, the mean free path is very small, and molecules collide with one another much more frequently than with the walls of the chamber. By the time we reach high vacuum and UHV, however, the mean free paths are many meters or even kilometers in length. If the chamber is only a meter across, the molecules will collide with the chamber walls much more frequently than with each other.

This picture of gas interactions suggests defining three regimes. (1) Viscous flow where the mean free path is much less than some typical size (D) of the chamber containing the gas. (2) Intermediate (transitional) flow where the mean free path is comparable to the size

TABLE 2.3

Particle Mean Free Paths for Room Temperature Air at Several Pressures

Pressure	Mean Free Path
1 atm	6.7×10^{-8} m
1 torr	5×10^{-5} m
1 mtorr	0.05 m
10^{-6} torr	50 m
10^{-9} torr	50,000 m

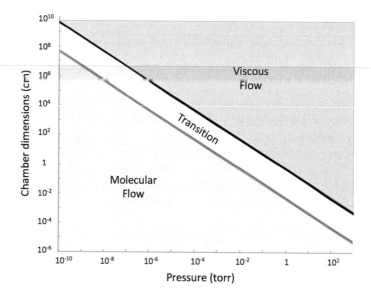

FIGURE 2.3

Gas flow regimes as a function of chamber dimension and pressure.

of the chamber. (3) Molecular flow where the mean free path is much greater than the size of the chamber.

Intermediate flow is the most complicated of these, but fortunately is not typical of operating conditions in deposition or characterization vacuum systems. Viscous flow will be dominated by gas molecules colliding with other gas molecules. As a result, molecules will tend to "drag" one another along in the flow forming a collective motion. The condition for having viscous flow can be defined as

$$D(\text{cm})P(\text{torr}) > 0.5 \tag{2.9}$$

for air at room temperature.

Molecular flow will be dominated by gas molecule–wall interactions and the gas molecules will move independently of one another with each gas molecule acting as though it was the only molecule in the chamber. The condition for having molecular flow can be defined as

$$D(\text{cm})P(\text{torr}) < 0.005 \tag{2.10}$$

for air at room temperature. The boundaries of the flow regimes are indicated in Figure 2.3.

2.1.4 Interactions with Surfaces

We finish our discussion of the kinetic theory of gases by exploring some relationships that will be useful when considering atoms on surfaces and film deposition. Consider how many gas molecules are striking a surface each second. The kinetic theory of gases yields the following result for the flux, Φ, of gas molecules incident on a surface:

$$F = 0.25nv_{rms} \tag{2.11}$$

This expression can be rewritten in terms of directly measurable properties as

$$\Phi = \frac{P}{\sqrt{2\pi mk_BT}} = 3.513 \times 10^{22} \frac{P}{\sqrt{MT}} \tag{2.12}$$

where the constant has been chosen so that Φ will be in molecules/cm^{2}s when P is measured in torr and T in K. M is the molecular weight of the gas molecule in atomic mass units.

For example, consider N_2 with a molecular weight of 28 at room temperature (293 K) and at a pressure of 1×10^{-7} torr. The flux of molecules striking the surface is 3.88×10^{13} molecules/cm^{2}s.

With this flux of molecules striking the surface, assuming that all of them stick, we determine the time it would take to form a complete layer of gas on the surface. This time to form a monolayer of gas is denoted as t_m.

$$t_m = \frac{4}{nv_{rms}d^2} \tag{2.13}$$

For air at room temperature, this can be expressed as

$$t_m = 1.86 \times 10^{-6} / P \tag{2.14}$$

where P is the pressure in torr.

Table 2.4 provides the times to form a monolayer of air at room temperature for several pressures. Note that these calculations assume that the molecules are moving in random directions. Some deposition techniques have a directed flow toward the substrate, which may result in a higher flux of atoms on the surface. These calculations apply to both intentional deposition and background gas contamination. If you are depositing a film at 10^{-6} torr with a technique that has a flux of film molecules producing a monolayer of film every 10 seconds, the background gas contamination striking the surface is producing a layer of contaminant every 2 seconds. While not quantitatively correct due to the number of approximations made, this does indicate the need for good vacuum conditions in order to produce films with low contaminant levels.

TABLE 2.4

Approximate Time for Room Temperature Air to Form a Monolayer at Several Pressures

Pressure	t_m
1 atm	2×10^{-9}s
10^{-6} torr	2 s
10^{-9} torr	30 minutes

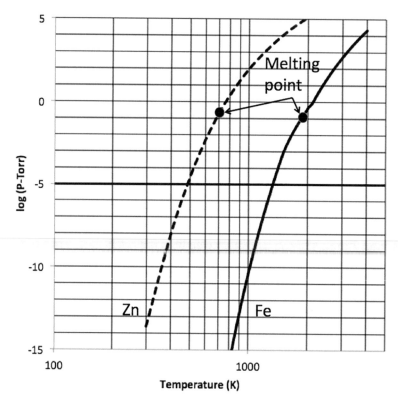

FIGURE 2.4
Approximate vapor pressure curves for Zn and Fe.

2.1.5 Vapor Pressure

Any solid material in equilibrium in a vacuum will have atoms of that material in the gas phase above the solid. Statistical mechanics does not allow a sharp transition from a large number density of atoms of an element in the solid to no atoms in the vapor adjacent to it. We describe the statistically expected density of atoms above the solid by a vapor pressure that is dependent on temperature. Figure 2.4 shows typical vapor pressure curves for Zn and Fe. Note that Zn reaches a vapor pressure of 10^{-5} torr at a temperature around 485 K. Fe does not reach this same vapor pressure until reaching a temperature of approximately 1320 K. Vapor pressure considerations enter strongly into designing fixtures that will be used in vacuum systems and in understanding evaporation processes. The higher vapor pressure of Zn at low temperatures could lead to significant Zn contamination and so Zn is typically not used in vacuum systems. For instance, if you heat Zn to only 200°C (476 K) the vapor pressure of Zn is approximately 5×10^{-6} torr.

2.1.6 Gases vs. Solids

We have been discussing here the behavior of gases. In much of this book, we will be examining solids. In thin film deposition, we are often interested in the interaction of gases with solids. At the atomic level, it is important to recognize that we treat these two environments with very different models in most cases. The kinetic theory of gases assumes no interactions between atoms unless they collide. We discussed in Chapter 1 treating atoms in a solid as connected by springs. This provides both an attractive interaction (if

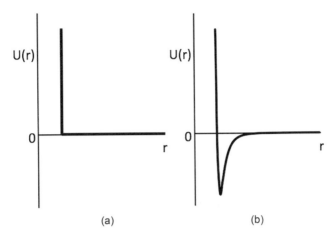

FIGURE 2.5
Two interatomic potentials: (a) Hard sphere potential used in the kinetic theory of gases and (b) Lennard-Jones potential used in solids.

two atoms are far apart) and a repulsive interaction (if the atoms are close together). We will discuss slightly more advanced models for the interactions between atoms in a solid in later chapters but they will always combine attractive and repulsive terms. In Figure 2.5, we show the potential energy for two atoms interacting under the kinetic theory of gases (hard sphere potential) and under a typical solid model (Lennard-Jones potential). Note that differences in the shapes of these potential energy graphs will lead to significantly different behaviors of atoms in solids and in the gas phase.

2.2 Plasmas

Plasmas are used in a variety of applications in surface and thin film science. We cover the basic concepts here and reference this section throughout the book. Additional details are available in references (Heald and Marion 1995, Lorrain, Corson and Lorrain 1988, Reitz, Milford and Christy 2008). A plasma is a dilute, partially ionized gas. Plasmas are sometimes described as the fourth state of matter and the highest energy state as pictured in Figure 1.5. At low energies the bonds between atoms are intact and solids form. As more energy (typically thermal) is applied to the solid it transitions into a liquid where the ordered crystalline structure is lost. Even more energy separates the atoms a greater distance from one another into the gas phase. Finally, applying even higher energies can cause atoms to ionize creating a plasma. While electrically neutral, a plasma contains free negatively charged electrons and positively charged ions as well as neutral particles. The electrons have a much lower mass than the ions, which makes them far more mobile in the plasma environment. We exploit this fact in a large number of applications in surface and thin film science.

2.2.1 Creating a Plasma

If we start with a gas of neutral atoms, we create a plasma by removing electrons from atoms, leaving a positively charged ion core. Removing one electron from a neutral atom requires an energy typically in the 1–20 eV range. This energy is the first ionization energy

TABLE 2.5

First and Second Ionization Energies for Several Elements

Element	First Ionization Energy	Second Ionization Energy
Argon	15.76 eV	27.76 eV
Oxygen	13.62	34.93
Sodium	5.14	47.0
Cesium	3.89	25.16

for the atom. Removing a second electron from the ion core requires higher energies given by the second ionization energy. Typical ionization energies are shown in Table 2.5.

We can relate these energies to temperatures by considering the thermal energy, $k_B T$, that would be equivalent to the ionization energy. 1 eV corresponds to about 11,600 K. Given the range of ionization energies, it is typically not practical to achieve ionization by thermal processes. Instead, we rely on electron collisions with atoms. These electrons, however, must be accelerated to have sufficient energy to ionize atoms. This is typically done by applying a high voltage across the gas region between an anode (more positive electric potential) and a cathode (more negative electric potential).

Consider a single electron in a gas. If it has sufficient energy to ionize a neutral atom (X), we can get the reaction, $e^- + X \rightarrow 2e^- + X^+$. We now have two free electrons and a positive ion. If the released electrons can be accelerated to sufficient energy, they can ionize two other neutral atoms releasing two additional electrons from each atom, $2e^- + 2X \rightarrow 4e^- + 2X^+$. This cascade process can potentially double the number of electrons in each step rapidly leading to significant ionization of a gas and the formation of a plasma. The effect will depend on the energy of the electrons and on which elements are present in the gas. Additional secondary electrons may also be released by ions colliding with the high-voltage cathode.

The result of this cascade process is an exponential increase in the number of electrons and thus in the discharge current, I_D, between the high-voltage electrodes. This discharge current will clearly depend on both the ionization process in the gas and the secondary electron emission process at the cathode. The Townsend equation describes the resultant current.

$$I_D = I_0 \frac{e^{\alpha_T d}}{\left[1 - \alpha_{T2}\left(e^{\alpha_T d} - 1\right)\right]} \tag{2.15}$$

where I_0 is the initial current before the voltage is turned on (arising from various sources including photoelectric effect and thermionic emission), α_T is the Townsend ionization coefficient, which represents the number of ion pairs per unit length created by a negative ion as it moves across the gap between the electrodes, α_{T2} is the Townsend secondary electron emission coefficient, which represents the number of secondary electrons released from a surface per incident positive ion, and d is the spacing between the electrodes.

We can explore some of the underlying science. The ionization coefficient, α_T, will depend on the probabilities of a collision occurring per unit length between an electron and an ion and on the probability of an ionization occurring for a given collision. We defined the mean free path, λ_{mfp}, in Equation 2.8 as the mean distance a particle will travel between collisions. So $1/\lambda_{mfp}$ is just the probability of a collision per unit length. The probability of ionization can be described by a Boltzmann factor (as introduced in Chapter 1) where the energy barrier is V_1, the first ionization energy of an atom, and the available energy

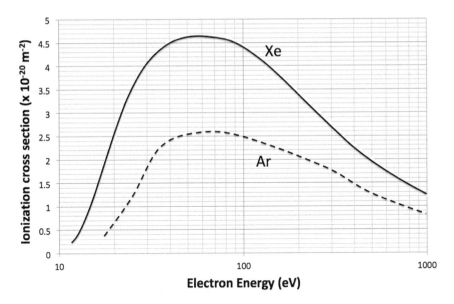

FIGURE 2.6
Electron impact ionization cross sections of Ar and Xe have maxima below 100 eV.

TABLE 2.6

Electron Impact Ionization Efficiency Varies with Element

Element	Number of Ions Formed per cm of Travel in 10^{-2} torr Gas for Electrons with about 100 eV
He	0.015
Ne	0.025
H_2	0.040
N_2	0.100
Ar	0.110
Hg	0.210

depends on the applied electric field, E, the charge of the electron, q_e, and the path length through which the electron travels without a collision, λ_{mfp}. Putting this all together,

$$\alpha_T = \frac{1}{\lambda_{mfp}} e^{V_1/q_e E \lambda_{mfp}} \tag{2.16}$$

Notice that this coefficient, and thus the Townsend equation, depends on the element present in the gas through V_1 and on the gas density (or pressure) through λ_{mfp}. It can be shown, due to the very low mass of the electron, that the electron mean free path is $4\sqrt{2}$ times the mean free path of the atoms in the gas according to the kinetic theory of gases.

Experimentally we find that electrons ionize atoms by collision most effectively for energies around 100 eV as shown in Figure 2.6 for Ar and Xe gases. Ar has a first ionization energy of 15.76 eV from Table 2.5. The ionization energy represents a threshold energy that must be exceeded, but the ionization process becomes more efficient as energies increase up to around 100 eV before decreasing again at higher energies.

The efficiency of ionization also depends on what elements are present in the gas. Table 2.6 compares the number of ions formed per cm of travel of 100 eV electrons traveling in a 10^{-2} torr gas. The ability to ionize the gas may vary by more than an order of magnitude between different gases.

In addition to ionization, electron collisions can also produce excited (but not ionized) atoms. Ionization can also occur when an excited atom collides with an atom in the ground state. If the excitation energy of the excited atom exceeds the ionization energy of the neutral atom, this may result in the ionization of the ground state atom through a process known as Penning ionization.

The requirements of ionization discussed here provide some practical guidelines for designing plasma systems. The electrons must be accelerated through a sufficient distance without collisions to gain enough energy to ionize an atom when they collide. This requires that the high-voltage electrodes be separated by a sufficient distance (typically a few cm). The gas density must also be low enough that the electrons can travel the required distance without collisions. The gas density, however, must not be too low since we want frequent collisions to generate a cascade effect of new electrons to keep the gas ionized.

2.2.2 Characterizing a Plasma

Plasmas are typically characterized by temperature (or energy), electron density, n_e, and particle density, n, or neutral atom density, n_0. In surface and thin film processing, we typically operate with cold plasmas having particle energies of 1–100 eV. The low mass electrons are far more mobile than the ions resulting in the electron temperature being generally greater than the ion temperature, especially in a dilute plasma. Since the electrons and ions are not in thermal equilibrium with each other, the electrons and ions have independent velocity distributions. Note that defining temperature is only sensible if a thermodynamic equilibrium occurs. We need enough collisions of electrons with other electrons (and ions with other ions) that equilibrium is achieved over short time scales. For the electron and ion temperatures to be different, we assume that there are not enough collisions between electrons and ions to create thermodynamic equilibrium between the species during the typical lifetime of an ion (before it is neutralized by recombination with an electron.)

Typical total particle densities (electrons, ion cores, and neutral atoms) in a plasma at about 5×10^{-3} torr pressure are on the order of 10^{19} particles/m³. Since a plasma is electrically neutral, the number density of electrons and of ions should be the same. If the plasma is only weakly ionized (fewer atoms have been ionized), the ion and electron densities will be around 10^{14} particles/m³. If the plasma is strongly ionized, the ion and electron densities will be higher with values in the 10^{18}–10^{26} particles/m³ range for various gas pressures.

Letting q_e, T_e, n_e, and m_e be the electron charge, temperature, number density and mass, and ϵ_0 be the permittivity of free space, these parameters are often grouped as follows:

$$\text{Debye length in meters: } \lambda_D = \sqrt{\frac{\varepsilon_0 k_B}{q_e^2}} \sqrt{\frac{T_e}{n_e}} \tag{2.17}$$

$$\text{degree of ionization: } \alpha_i = n_e/(n_e + n_0) \tag{2.18}$$

$$\text{plasma frequency: } \omega_P = \sqrt{\frac{q_e^2}{\varepsilon_0 m_e}} \sqrt{n_e} \tag{2.19}$$

where n_0 is the number density of neutral atoms and $n_e = n_{\text{ion}}$. We now briefly explore each of these parameters.

The Debye length can be regarded as the length over which a charged particle will influence the behavior of the plasma. Equivalently, this is the distance over which significant charge separation can occur in the electrically neutral plasma. In a plasma with an electron density of 10^{16} electrons/m^3 and an electron thermal energy of 50 eV (about 580,000 K), the Debye length is about 5×10^{-4} m.

If the degree of ionization exceeds a critical value, $\alpha_C = 1.73 \times 10^{16} \, \sigma_{eA} \, T_e^2$ where σ_{eA} is the electron-atom collision cross section in m^2 (typically $10^{-20} - 10^{-19}$), then the plasma will behave as though it is fully ionized. For many thin film processes, the degree of ionization is typically on the order of 10^{-4}. A plasma formed from a room temperature gas at 10^{-2} torr, would have a neutral particle density around 10^{20} m^{-3} (from Equation 2.2) and so we would expect that $n_e = n_{ion} = 10^{16}$ m^{-3}.

The plasma frequency is a natural frequency of the plasma system. When electromagnetic waves with frequencies less than the plasma frequency are incident on the plasma (a typical situation), the plasma is opaque to the radiation. If the frequency of an incident electromagnetic wave is greater than the plasma frequency, the plasma is transparent. For an electron density of 10^{16} electrons/m^3, the plasma frequency is about 10^9 Hz which is in the radio wave part of the electromagnetic spectrum.

2.2.3 Motion of Charges

Plasmas are typically used in environments involving applied electric and magnetic fields. We explore first the behavior of charged particles in static fields (electrostatics) before turning our attention to electrodynamics. While we just assumed that collisions between electrons and between ions are frequent enough to create thermodynamic equilibrium, we now neglect collisions in order to get a sense of how charged particles move in a dilute plasma. We consider several special geometries of fields.

Start with a constant electric field, $\vec{E}$, with no magnetic field ($B=0$). In a plasma, the positive and negative charges will be separated by this applied electric field resulting in the creation of an additional electric field which will leave the interior of the plasma electric field free and approximately neutral. A sheath region is formed around the plasma in which the charged particles are separated as represented in Figure 2.7. For conditions typical of DC sputter deposition (described in Chapter 9) with $n_e = 10^{16} - 10^{18}$ electrons/m^3, the thickness of this sheath region will be about 10^{-5} to 10^{-3} m. The applied electrical potential is approximately constant in the plasma region with most of the change in potential from the applied electric field being in the sheath region. The size of the sheath region depends on the potential applied and on the surface area of the electrode. Higher potentials and smaller electrodes will have thicker sheath regions.

FIGURE 2.7
The plasma and sheath in a simple diode system.

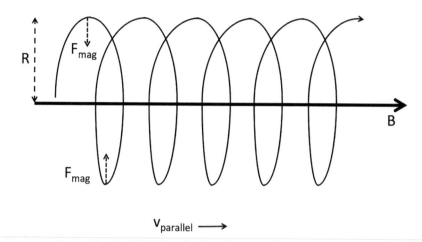

FIGURE 2.8
FIGURE 2.8
Path of a charged particle in a uniform magnetic field, B.

Now consider a constant magnetic field, $\vec{B}$, with no electric field ($E=0$). In this case, charged particles will follow a spiral path around the magnetic field lines. This is very important in many applications. Instead of a relatively short straight line path from one side of a region to the other, a charged particle will follow a much longer spiral path. This longer path length will increase the probability of collisions in this region.

Explore this motion by breaking up the velocity of a charged particle into two components parallel and perpendicular to a magnetic field line.

$$\vec{v} = \vec{v}_{\text{parallel}} + \vec{v}_{\text{perp}} \tag{2.20}$$

The magnetic field exerts a force on the charged particle, $\vec{F}_{\text{mag}} = q\vec{v} \times \vec{B}$. The cross product will pick out the perpendicular component, $F_{\text{mag}} = qv_{\text{perp}}B$. This force will be in the radial direction, as indicated in Figure 2.8, and so is a centripetal force.

$$F_{\text{centripetal}} = \frac{mv_{\text{perp}}^2}{R} = qv_{\text{perp}}B \tag{2.21}$$

Solving this for R, yields the Larmor radius

$$R_L = \frac{mv_{\text{perp}}}{qB} \tag{2.22}$$

The angular frequency of the circular motion is known as the cyclotron frequency.

$$\omega_{\text{cyclotron}} = \frac{qB}{m} \tag{2.23}$$

Note that electrons have much lower mass than ions and so the electrons will have smaller radii (for the same perpendicular velocity) and higher frequencies of circular gyration around the magnetic field lines.

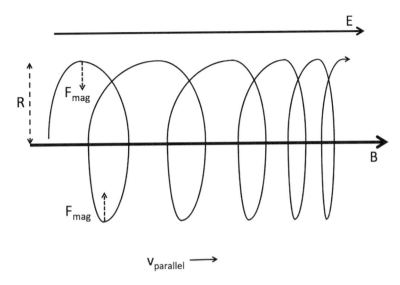

FIGURE 2.9
Path of an electron in parallel uniform electric and magnetic fields.

Next consider uniform, constant $\vec{E}$ and $\vec{B}$ that are parallel to one another. The magnetic field still creates circular motion (if the charged particle has a velocity component perpendicular to the magnetic field) and the electric field causes a change in the parallel velocity component of the charged particle. This will stretch out or compress the helical motion with position depending on the sign of the charge of the particle. Figure 2.9 demonstrates the effect for an electron traveling in parallel electric and magnetic fields with a velocity component, $v_{parallel}$, in the direction of the fields.

Finally, consider the combination of a uniform, constant $\vec{E}$ and $\vec{B}$ but require that $\vec{E}$ is perpendicular to $\vec{B}$. This produces a very complicated motion if the velocity of the charged particle is in any arbitrary direction. An interesting case arises if we place a charge at rest in crossed fields. Consider the electric field as being in the y direction and the magnetic field in the z-direction. Initially, the charged particle will experience no force from the magnetic field since its velocity is zero. The electric field will begin to accelerate it in the y direction (or $-y$ depending on the charge of the particle). Now the charge has velocity and so the magnetic force will accelerate the particle out of the plane of the crossed fields and it will have a velocity component in the x direction. The result of this motion is a constant drift velocity component, $v_{drift}=E/B$, in the x direction perpendicular to E and B and a cycloid motion in the xy plane caused by the particle attempting to have a circular motion about the magnetic field lines. This motion is shown in Figure 2.10. Notice that if we added an initial velocity component in the direction parallel to B, there is no force in that direction and so this would just add a constant velocity component $v_{parallel}$ in the direction of B to the cycloidal motion. Note v_{drift} does not depend on mass or charge, so all charged particles will drift together.

Moving from electrostatics to electrodynamics, consider the motion of charged particles when the applied fields are time-varying. If an electromagnetic wave is incident on the plasma, the time-varying electric field exerts a much stronger force than the time-varying magnetic field and so we focus our attention on time-varying electric fields. Particle motion can be described with a damped oscillator model for the charged particles similar to a driven mass on a spring with frictional damping.

Consider the forces acting on charged particles:

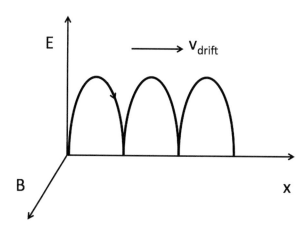

FIGURE 2.10

Cycloid motion of a charged particle in crossed electric and magnetic fields.

- driving force from varying E field
- no restoring force since the charged particles are not bound (spring constant=0)
  - this is not true if charge separation in the plasma leads to electrostatic restoring forces
- damping force from collisions (this could be large)

We can then use Newton's 2nd Law, $F = ma$, in the differential form to write down an equation of motion for the charged particles. Using typical values for parameters in the resulting equation of motion indicates that at low frequencies (<50 kHz) both ions and electrons oscillate in the time-varying field. At higher frequencies (>50 kHz), however, the heavy ions do not change directions quickly enough to follow the time-varying electric fields as they switch direction. The result is that electrons oscillate at higher frequencies while ions are relatively stationary. We make use of this in applications of plasmas in thin-film deposition.

2.2.4 Plasma Sources

We briefly outline various sources used to generate plasmas. The most common are simple plate electrodes with a high voltage applied across them. These produce relatively low plasma densities (10^{15}–10^{16} charged particles/m^3) and require relatively high gas densities. They are commonly used in sputter deposition and will be discussed further in Chapter 9. Other plasma sources typically make use of ways to couple energy more efficiently into the gas system in order to create higher electron densities at lower gas densities (pressures). This allows lower kinetic energy of the ions that can minimize damage to samples. Several plasma sources allow separate control of the power going into the plasma and the voltage applied across the plasma.

Inductively coupled plasma (ICP) sources can be used to generate high plasma densities (10^{17}–10^{18} charged particles/m^3) while operating well at relatively lower gas pressures ($<5\times10^{-2}$ torr). The technique can be used, however, even up to atmospheric pressures and beyond. The key concept is the coupling of radio frequency (RF) energy inductively into the plasma that behaves like a lossy electrical conductor producing more efficient ionization of the gas. Figure 2.11a shows a wire coil electrode wound around the plasma. A time-varying

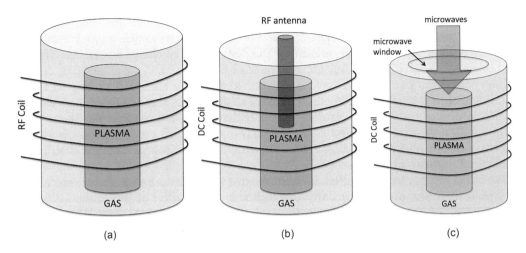

FIGURE 2.11
Plasma sources (a) ICP, (b) helicon, (c) ECR.

RF current (0.5–27 MHz) in the coil creates a time-varying magnetic field which induces azimuthal electric currents in the gas that contribute to the ionization of the gas.

A helicon wave (magnetron ion etching) design, represented in Figure 2.11b, also can be used to generate high plasma densities (10^{17}–10^{19} charged particles/m^3) while operating at lower gas pressures ($<5 \times 10^{-2}$ torr). While similar to ICP, in the helicon design RF energy is radiated into the plasma in the presence of a constant magnetic field (100–300 G) directed along the antenna axis, so that the RF frequency is tuned for resonant absorption again producing more efficient ionization.

Electron cyclotron resonance (ECR) produces high plasma densities (10^{18}–10^{19} charged particles/m^3) while operating at lower gas densities (down to 10^{-4} torr). A static magnetic field (875 G) is applied to the plasma region. In this case, microwave energy is coupled into electrons by matching the microwave frequency to the electron cyclotron frequency (Equation 2.21) to again produce more efficient ionization as indicated in Figure 2.11c. The plasma density can be controlled by adjusting the microwave power and gas pressure. We can also control which ion species will be created such as O_2^+ vs. O^+.

2.2.5 Chemistry in Plasmas

Some surface and thin film applications that involve plasmas make use of the complicated chemical environment that can exist in a plasma rather than just the presence of charged species. The environment is not just one of a "hot" gas where thermally driven equilibrium processes dominate. Plasmas are typically not in thermodynamic equilibrium with electron temperatures being significantly higher than ion or neutral gas temperatures. While we typically characterize the energy of a plasma by average electron and ion temperatures, the electron and ion velocities follow a Maxwell distribution, as in Figure 2.2 and particles with energies considerably higher than the average will be present in the plasma. Collisions between these high-energy electrons and atoms, molecules and ionized atoms and molecules can lead to a rich variety of chemical reactions driven at rates higher than would be expected in a thermally activated process. The presence of a plasma may, therefore, allow chemical reactions to proceed at a higher rate while at a lower temperature. This is often desirable in certain material processing steps.

Electron collision reactions can be broadly grouped into three overlapping categories. The first category is excitation where collision with an electron excites a particle but does not change the chemical species present or the distribution of charged particles. The second category, which has been our focus in most of this chapter, is the creation or neutralization of charged particles. We have focused on ionization where electron impact creates a positive ionized particle and releases an extra electron. It is also possible for an electron to attach to a neutral particle resulting in a negatively ionized particle. An electron can collide with a positively charged particle which may result in recombination leading to the two charged particles forming a neutral particle. All of these change the distribution of charged particles in the plasma.

The third set of electron collision reactions are those that change the chemical species present in the plasma typically through the dissociation of molecules. When an electron strikes a molecule, the molecule may dissociate in several ways. The molecule may break up into its component atoms with each of the atoms (which may be in excited states) being electrically neutral. One of the component atoms from the molecule may be ionized in the dissociation process leading to a change in both the chemistry and the charged particle distribution. It is also possible for the charge in the molecule to redistribute on dissociation so that both a positive and a negative ion are formed upon dissociation.

Collisions among the atoms, molecules, and ionized species are also possible. We mentioned earlier the example of Penning ionization where an excited atom collides with a neutral atom resulting in the ionization of the neutral atom. Similarly, the collision of two excited atoms might cause the ionization of one of the atoms. The collision of an ionized atom with a neutral atom can cause charge to be transferred from one atom to the other. Even though the overall charge distribution does not change in this reaction, if the two atoms are different elements, this will change the distribution of charged atoms of these particular elements.

Gas-phase chemistry is normally dependent on thermodynamic rate constants that assume that temperature is the driving force causing chemical reactions. As we can see from the discussion in this section, plasma chemistry is considerably more complicated with a large number of non-thermal processes driving the chemistry.

2.2.6 Applications of Plasmas

We will encounter plasma environments frequently throughout this book. Some examples are in the cleaning and etching of surfaces, as a source for sputter deposition, in the bombardment of growing thin films during deposition to modify film properties and in the activation of reactive gases (for instance in Chemical Vapor Deposition). In any of these applications, we tend to be making use of one or both of two important properties of plasmas. One of these properties is simply the presence of charged particles that can be controlled by the application of electric and magnetic fields. This will allow us to have some control over which collisions are encouraged or discouraged and the energies of those collisions. The second property is the unique chemical environment found in the plasma that can allow us to drive certain chemical reactions to occur more readily at temperatures below where they would be expected to be significant in thermal processes.

All of the processes of interest in this book also involve the presence of a solid surface that itself can further influence the chemistry both on the surface and in the environment immediately above it.

2.3 Cleaning Surfaces

One consequence of the kinetic theory of gases, which leads to an application that may involve plasmas, is the need to clean surfaces inside a vacuum system. From Equation 2.10 we see that samples that have been exposed to atmospheric conditions before being loaded into a vacuum system will have been exposed to many monolayers of gas impinging on their surfaces. Although these layers will not all stick, the surface of these samples will certainly be covered with contaminants. If chemical reactions occur, this may be a well-defined layer. Si, for example, will form a thin (1–2 nm) layer of SiO_2 when exposed to air at room temperature. This oxide layer stops growing at that thickness leading to a passivated Si surface. The oxide layer, which is also exposed to the atmosphere, will have additional contamination on its surface. If the goal is to have a clean Si surface, either for direct study or as a substrate for thin film growth, we need to be able to clean off both the surface contamination and the oxide layer that has formed.

Many recipes for cleaning particular surfaces have been published and can provide guidance for specific applications. Laboratories frequently have standard wet chemical cleaning processes using acids, alcohols, water, and other chemicals for sample preparation prior to introduction into a vacuum system. We briefly review some of the most common methods for achieving clean surfaces after the sample is in a vacuum environment. The focus is on uniform samples and not on patterning or lithography.

In some cases, if a surface can be formed inside the vacuum chamber, no additional cleaning may be needed. An example of this is a crystal that can be cleaved in vacuum to expose the desired surface. This may require cooling the crystal to liquid nitrogen temperatures in order for it to be brittle enough to achieve a smooth surface. Another option might be to simply cover up the contamination by depositing a film layer on a contaminated surface. This might be a layer of the same material as the substrate or might be a completely different material that is the desired surface material.

If the contamination on the surface is not strongly bonded, the process of achieving high vacuum or UHV may be sufficient to remove surface contamination at room temperature. Stronger bonds will require adding heat in order to provide a source of energy for breaking those bonds and allowing the contaminants to desorb. Some of the contamination that desorbs will re-adsorb from the vapor and so the surface is never perfectly clean, but can be cleaned to a level below the detection limit of our characterization techniques. Heating will also encourage the diffusion of surface contamination into the bulk of the sample. This may be acceptable since the contamination concentration would be rather low when dissolved into the bulk and it does reduce the contamination at the surface. Heating, however, may also encourage the diffusion of bulk contamination to the surface, which is not desired.

Chemical reactions can also be used to clean surfaces, often accompanied by heating the sample to encourage chemistry. This can be facilitated by introducing a background gas into the vacuum system. For instance, O_2 gas can be used to clean C contamination from surfaces of oxide materials. H_2 gas can be used to remove an oxide layer by reduction of the oxide. Plasmas can also be used to encourage chemical reactions in processes such as reactive ion etching. Molecules in a plasma can be broken apart into more reactive species that can then react with elements on the sample surface to form volatile chemical compounds which desorb and are pumped away. For example, CF_4 in a plasma can be broken up to yield free F ions which are highly reactive. They can react with a Si surface to form SiF_4 which is a volatile species that can leave the surface and enter the gas phase. Plasma

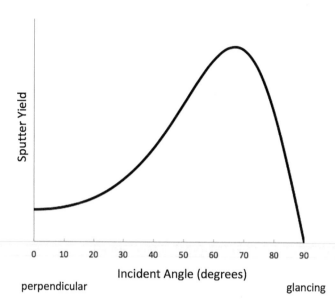

FIGURE 2.12
Sputter yield variation with incident angle.

etching is a common technique for selectively removing certain elements from a sample while allowing others to remain.

A more destructive technique for sample cleaning makes use of properties discussed in section 2.2. Inert gas ions can be used to remove contamination from the surface through collisions and momentum transfer. These atomic "cannon balls", usually inert gas ions such as Ar^+ or Xe^+, can be directed at a surface either through a large area ion etching process in which the sample is biased with a high voltage to attract the ions or by using an ion gun that directs a beam of ions at the surface. The disadvantage of this technique is that the surface structure is typically damaged by the impact of the ions. The surface structure may often be healed by annealing the sample. The annealing process, however, may also encourage diffusion from the bulk. Contaminants dissolved in the bulk material may come to the surface causing a need for additional sputtering.

Imagining this process as hard sphere collisions, one can quickly see that having the ions directed into the surface from a direction perpendicular to the surface will mainly drive the surface contamination into the sample without removing it. Having the ions impact the surface at glancing incidence will not efficiently transfer much momentum and so will not knock off as much contaminant. Experimental and theoretical work suggests that an optimum angle of incidence for many materials is about 20–30° relative to the plane of the surface. This is indicated in Figure 2.12, which examines the sputter yield, S, as a function of the incident angle. Sputter yield is defined as

$$S = \frac{\text{number of atoms ejected}}{\text{number of ions incident}} \tag{2.24}$$

The sputter yield will depend on the mass of the atoms in the sample that are being sputtered as well as the mass of the incident ion. The sputter yield typically increases with the mass of the incident ion. Xe ions are more efficient in removing contamination, but the extra cost of Xe gas compared to Ar gas may not make this an economically viable substitution. The energy of the incident ion is also an important variable that we can

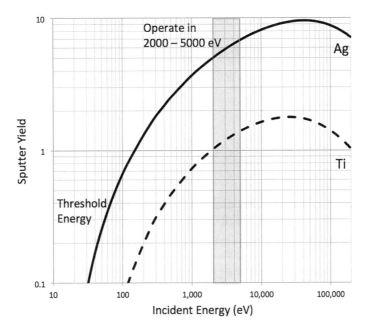

FIGURE 2.13
Sputter yield variation with incident energy.

control. The incident ions must reach some threshold energy before atoms will typically be removed from the surface. The sputter yield then increases up to an energy around 10 keV before decreasing slightly above that level as shown in Figure 2.13. Notice that the sputter yield at 2–5 keV incident energy is already quite high. Given the extra cost of a 10 kV power supply and the additional damage possible to the surface from these higher energy ions, most sputter cleaning is done in the 2–5 keV incident energy range.

In reactive ion etching, we combine the concepts of plasma etching with ion beam sputtering. Instead of using only inert ions for the sputtering, we now include highly reactive chemical species. The resultant rates of material removal (etch rates) are substantially higher than either sputtering or plasma etching alone.

References

Delchar, T. A. 1993. *Vacuum Physics and Techniques.* London: Chapman and Hall.

Erbil, H. Y. 2006. *Surface Chemistry of Solid and Liquid Interfaces.* Oxford: Blackwell Publishing Ltd.

Frey, H. and H. R. Khan, eds. 2015. *Handbook of Thin-Film Technology.* Berlin: Springer Verlag.

Heald, M. A. and J. B Marion. 1995. *Classical Electromagnetic Radiation.* 3rd ed. Fort Worth, TX: Saunders College Publishing.

Kauzmann, W. 1966. *Thermal Properties of Matter, Volume 1: Kinetic Theory of Gases.* New York, W. A. Benjamin, Inc.

Lorrain, P., D. R. Corson, and R. Lorrain. 1988. *Electromagnetic Fields and Waves.* 3rd ed. New York: W.H. Freeman & Co.

O'Hanlon, J. F. 2003. *A User's Guide to Vacuum Technology.* 3rd ed. Hoboken, NJ: John Wiley & Sons Inc.

Paunovic, M. and M. Schlesinger. 2006. *Fundamentals of Electrochemical Deposition*. 2nd ed. Hoboken, NJ: John Wiley and Sons, Inc.

Reitz, J. R., F. J. Milford and R. W. Christy. 2008. *Foundations of Electromagnetic Theory*. 4th ed. Reading, MA: Addison-Wesley.

Tompkins, H. G. 1997. *The Fundamentals of Vacuum Technology*. 3rd ed. New York: American Vacuum Society.

Problems

1. A 2.5 L chamber at room temperature is pumped to a pressure of 10^{-6} torr. How many gas atoms are in the chamber?

2. A room temperature, 1 L, cubical chamber has been completely pumped of all the gas present in the volume. The internal surfaces of the chamber, however, are covered with one atomic layer of molecular nitrogen corresponding to 1018 molecules/m². Assume that this monolayer of gas instantly and completely desorbs into the empty vacuum. Calculate the resulting pressure in the chamber.

3. Consider a cylinder of compressed N_2 gas with an interior height of 122 cm and an interior diameter of 19 cm. If the cylinder is pressurized to 2600 pounds/in² at room temperature, what is the mass of the N_2 gas in the cylinder? Assume that the N_2 can be treated as an ideal gas.

4. Calculate v_{peak}, v_{rms}, and v_{ave} for an oxygen molecule (O_2) at 300 K.

5. A turbomolecular pump operates with rapidly rotating, angled blades that strike the gas molecules and give them momentum. The process of momentum transfer is more efficient if the average linear velocity of the molecule is less than the linear velocity of the blade tip. Assuming that a particular pump has a linear blade velocity of 400 m/s, calculate the ratio of the average room temperature molecular velocity to the blade tip velocity for the following gases:

    a. Hydrogen (H_2)

    b. Nitrogen (N_2)

    c. Xenon (Xe)

    d. Is the turbomolecular pump better for pumping heavy elements or light elements?

6. For air at room temperature, how long would it take to form a monolayer of air on a surface, assuming everything sticks, at (a) atmospheric pressure, (b) 10^{-2} torr, (c) 10^{-5} torr, (d) 10^{-10} torr?

7. What is the mean free path for air at room temperature at a pressure of (a) 10^{-2} torr, (b) 10^{-5} torr, and (c) 10^{-10} torr?

8.

    a. Consider a steel tube with an inside diameter of 5 cm. Below what pressure would we expect room temperature air inside the tube to experience molecular flow?

    b. What would that limiting pressure be for a steel chamber with inside diameter of 50 cm?

9. A clean metal surface is exposed to Ar gas for 5 seconds at room temperature and a pressure of 10^{-6} Pa. Assume that every Ar atom that strikes the surface sticks. What fraction of the metal surface will be covered with Ar atoms assuming each Ar atom sticks to one metal atom and that the surface density of metal atoms is 10^{15} atoms/cm^2. How much time would it take to form a complete monolayer of Ar?

10. If we have two different types of atoms in a gas (with different size and mass), we can determine how often they will collide. Suppose that atom A is the lower mass atom and is described by a diameter, d_A, and speed, v_A. Similarly atoms of type B have d_B and v_B. If we assume that atom A is much lower mass than atom B, then $v_A \gg v_B$ and we can neglect the motion of the B atoms.

   a. Show, using a quick sketch or two, that the two atoms may collide during a time t if they both occupy a cylindrical volume $\pi d^2 v_A t$ where $d = \frac{1}{2}(d_A + d_B)$.

   b. If there exist n_B atoms of type B per unit volume, derive a formula for the number of collisions per second, z, which depends on d, v_A and n_B.

   c. Show that the mean free path for an atom of type A colliding with an atom of type B is given by $\lambda_A = 1/(\pi d^2 n_B)$.

   d. Now assume that atom A is actually an electron so that $d_A \ll d_B$. Show that the mean free path of an electron in a gas of atoms is $4\sqrt{2}$ times the mean free path of the gas atoms.

11. Collisional damping of the oscillating motion of charged particles in a plasma driven by a time-varying electric field $(E(t) = E_0 e^{-i\omega t})$ can be examined by determining the equation of motion from Newton's 2nd Law. Collisions of frequency, γ, cause a change in momentum (Δp) of the particles that results in a force $\gamma \Delta p$ that opposes the driving motion.

   a. Show that Newton's 2nd Law can be written $\dfrac{d^2 y}{dt^2} + \gamma \dfrac{dy}{dt} = \dfrac{q}{m} E_0 e^{-i\omega t}$.

   b. Determine the solution to this differential equation for $y(t)$.

12. Ionization in a plasma occurs mainly by electron collisions with ions. If we consider head-on, elastic, binary collisions between two particles with masses, M_A and M_B and assume that M_B is at rest before the collision, it can be shown that $\dfrac{E_{B-final}}{E_{A-initial}} = \dfrac{4 M_A M_B}{(M_A + M_B)^2}$.

   a. If the two particles have the same mass, what fraction of energy is transferred to particle B?

   b. If particle A is an electron and particle B is an Argon atom, what fraction of energy is transferred?

13. Consider an Argon plasma with a total particle density of 10^{16} particles/m^3. The electron temperature is 8 eV and $n_e = 5 \times 10^{12}$ electrons/m^3. Calculate values for:

   a. the mean velocity of the electrons

   b. the Debye length of the plasma

   c. the degree of ionization of the plasma

14. An Ar ion gun is often used to clean surfaces or remove surface layers during the characterization of thin films. This sputter process behaves exactly the same as the sputtering process we discussed for sputter deposition. An Ar ion beam of 1 keV is used to sputter clean crystalline Cu during AES depth profiling. The ion current is 10^{-8} A, the sputter yield is 2.85, and the area sputtered is 0.5 cm×0.5 cm.

  a. Calculate the sputter rate of Cu in units of monolayers/minute. (A monolayer is one atomic layer of the Cu). Assume that the (100) surface is exposed and that Cu is fcc with a lattice parameter of 3.61 Å.

  b. The beam energy is raised to 10 keV where the sputter yield is 6.25. If the ion current is now 10^{-6} A, calculate the sputter rate (in monolayers/minute) of Cu under these conditions.

Part I

Surfaces

3

Crystal Structure

This chapter examines the structure of crystals and clean surfaces. We focus our attention on crystalline materials, which have atoms arranged with both short-range and long-range order, as opposed to amorphous (non-crystalline) materials, which only have atoms demonstrating short-range order. After exploring the properties of bulk crystalline materials, which are often relevant in films, we will examine what happens when we terminate a bulk crystal to form a surface. Many properties of thin films and surfaces depend on the positions that atoms occupy and so this is an appropriate place to begin our investigation.

3.1 Structure of Crystals

Crystalline structure is discussed in detail in solid state physics books (Blakemore 1985, Kittel 1996, Moffatt, Pearsall and Wulff 1964). Application of these concepts to crystalline surfaces is explored in various surface physics books (Hudson 1998, Ibach 2006, Luth 2015, Prutton 1994, Somorjai 2010, Zangwill 1988). Three-dimensional crystalline materials can be described and classified by considering the long-range order of the atoms in the crystal. Crystal structure is built up from two aspects: (1) the *lattice*, which is a regular periodic array of points in space (not necessarily points that have an atom at them), and (2) the *basis*, which is a group of atoms located at specific positions referenced to each point in the lattice.

A small portion of an infinite, two-dimensional, rectangular lattice is presented in Figure 3.1. The points on the lattice are separated by primitive translation vectors, $\overline{1}$, that completely characterize the lattice. If you consider the lattice structure around some point $\vec{r}'$, it will be identical to the structure around a point $\vec{r}$ if $\vec{r}' = \vec{r} + u_1\vec{a}_1 + u_2\vec{a}_2 + u_3\vec{a}_3$ where the u_i are integers.

The simplest possible basis would be a single atom. In this case, the lattice would have an atom at each lattice point. The basis, however, is often a group of atoms with a well-defined spatial arrangement. Figure 3.2 shows a lattice and a two-atom basis that are combined to form a crystal structure.

With this introductory understanding of the relationship between a lattice, basis, and crystal, we now explore lattices in more detail. The primitive translation vectors can be used to define primitive lattice cells that can be formed around each lattice point.

These primitive lattice cells must obey several requirements:

1. be defined by primitive translation vectors
2. fill all space when repeated
3. have a minimum cell volume
4. contain exactly one lattice point

DOI: 10.1201/9780429194542-4

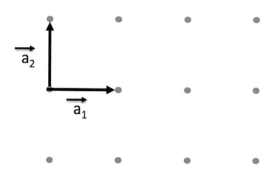

FIGURE 3.1
Rectangular two-dimensional lattice with primitive translation vectors.

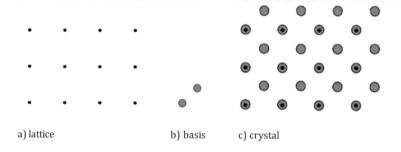

a) lattice   b) basis   c) crystal

FIGURE 3.2
Building a crystal: (a) Square lattice. (b) Two-atom basis. (c) Crystal formed by placing the two-atom basis on the square lattice.

An example of a rectangular primitive lattice cell in three dimensions is provided in Figure 3.3. Each side of the cell is defined by primitive translation vectors. The rectangular cell that results will fill all space if repeated. It is the smallest volume cell that can be created with this lattice. Initially, it might appear that the cell contains eight lattice points, but note that the lattice points are not completely enclosed by the cell. Each point is actually shared with 8 adjacent cells and so only 1/8 of each lattice point is contained within a single primitive cell. The number of lattice points in this cell is then $8 \times (1/8) = 1$ lattice point. The cell in Figure 3.3 satisfies all of the criteria for being a primitive lattice cell.

Primitive cells, however, are not unique for a particular lattice type. Figure 3.4 shows a two-dimensional example of three possible primitive cells that satisfy all of the criteria. The volumes (areas in two dimensions) of these cells are all the same and are the minimum possible volume that can be formed with this lattice.

Two-dimensional example

One way of forming well-defined primitive cells is the Wigner-Seitz primitive cell. To create a Wigner-Seitz primitive cell:

1. Choose a lattice point.
2. Draw lines from this point to neighboring lattice points.

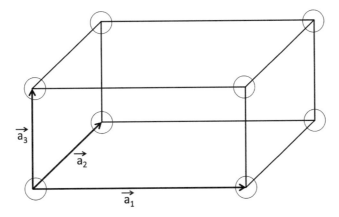

FIGURE 3.3
Rectangular lattice primitive cell. Note that only 1/8 of each lattice point is contained inside the primitive cell.

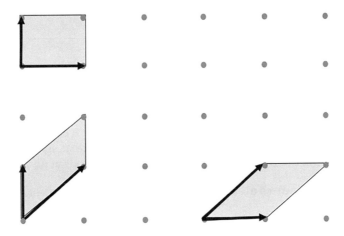

FIGURE 3.4
Three possible primitive cells for the same lattice.

3. At the midpoints of these lines, draw perpendicular lines (in two dimensions) or planes (in three dimensions).

4. Form the smallest region enclosed by those lines (in two dimensions) or planes (in three dimensions).

Figure 3.5 shows an example of constructing a two-dimensional Wigner-Seitz primitive cell.

When describing crystalline structures we do not typically use the primitive lattices since they are not unique. Bravais lattices provide a unique way of characterizing crystal lattices. Each point in an infinite Bravais lattice has identical surroundings. In three dimensions, there are 14 Bravais lattices (which are typically not primitive) as shown in Figure 3.6. These can be grouped into seven types depending on the lengths of the lattice vectors (a_1, a_2, a_3) and the angles between them (α, β, γ).

1. cubic $a_1=a_2=a_3$, $\alpha=\beta=\gamma=90°$
2. tetragonal $a_1=a_2\neq a_3$, $\alpha=\beta=\gamma=90°$
3. hexagonal $a_1=a_2\neq a_3$, $\alpha=\beta=90°$, $\gamma=120°$
4. orthorhombic $a_1\neq a_2\neq a_3$, $\alpha=\beta=\gamma=90°$
5. rhombohedral $a_1=a_2=a_3$, $\alpha=\beta=\gamma<120°$
6. monoclinic $a_1\neq a_2\neq a_3$, $\alpha=\beta=90°\neq\gamma$
7. triclinic $a_1\neq a_2\neq a_3$, $\alpha\neq\beta\neq\gamma$

Two of the cubic lattices, body-centered cubic, and face-centered cubic are so common that we abbreviate them as bcc and fcc respectively.

In two dimensions, important for surfaces, there exist only five Bravais lattices. They are:

1. square $a_1=a_2$ $\theta=90°$
2. hexagonal $a_1=a_2$ $\theta=120°$
3. rectangular $a_1\neq a_2$ $\theta=90°$
4. centered rectangular $a_1\neq a_2$ $\theta\neq90°$
5. oblique $a_1\neq a_2$ $\theta\neq90°$

These are pictured in Figure 3.7.

Cutting through a lattice creates many different planes that are defined by different sets of intersected lattice points. These lattice planes are very important in describing surfaces and are typically characterized using Miller indices. The Miller indices of a plane are defined by examining where the plane intersects the axes of a coordinate system formed by the primitive vectors. In Figure 3.8 we consider a cubic or rectangular lattice and we see an arbitrary plane that intersects the a_1 axis at $(x_0, 0, 0)$, the a_2 axis at $(0, y_0, 0)$, and the a_3 axis at $(0, 0, z_0)$. The Miller indices are $\left(\dfrac{j}{x_0}\ \dfrac{j}{y_0}\ \dfrac{j}{z_0}\right)$ where j is the smallest integer that makes

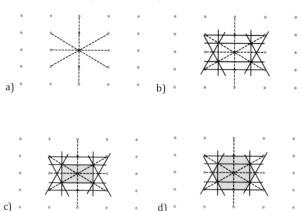

a)   b)

c)   d)

FIGURE 3.5
Wigner-Seitz construction of a primitive cell and next largest cell for a rectangular lattice. (a) Lines connecting lattice point to neighboring lattice points. (b) Perpendicular lines at midpoints. (c) Primitive Wigner-Seitz cell. (d) Next largest cell.

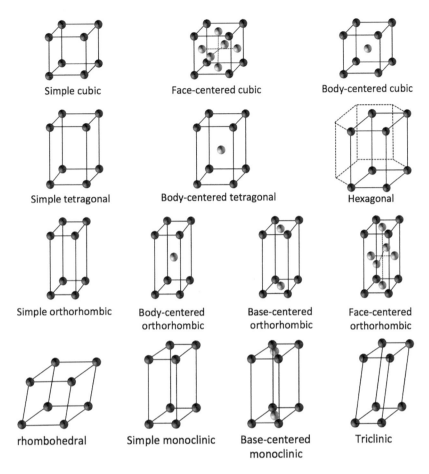

FIGURE 3.6
Three-dimensional Bravais lattices (corner atoms are shown darker than centered atoms for clarity).

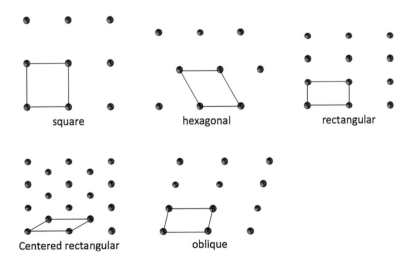

FIGURE 3.7
Two-dimensional Bravais lattices.

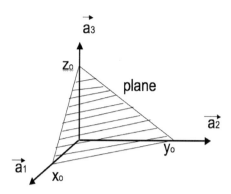

FIGURE 3.8
Arbitrary plane in a lattice.

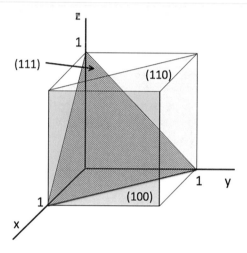

FIGURE 3.9
Miller indices of common planes in a cubic lattice.

each of the terms an integer. For example, if the axes' intersections are at $x_0=2$, $y_0=3$, and $z_0=1$, then we have $\left(\dfrac{j}{2} \quad \dfrac{j}{3} \quad \dfrac{j}{1}\right)$. $j=6$ is the smallest integer that will convert each of these into an integer and so the Miller indices for this plane are (3 2 6).

Figure 3.9 demonstrates the Miller indices of the three most commonly referenced planes of a cubic lattice. The (1 1 1) plane was determined from the intersection points at $x_0=1$, $y_0=1$, $z_0=1$. It cuts the cube along the diagonal from the upper left corner in Figure 3.9 to the bottom center. The (1 1 0) plane was determined from the intersection points at $x_0=1$, $y_0=1$, $z_0=\infty$ since the plane never intersects the z-axis. This plane divides the cube into two triangular prisms. The (1 0 0) plane intersects the axes at $x_0=1$, $y_0=\infty$, $z_0=\infty$. It is the front surface of the cube in Figure 3.9. The back surface of the cube, which is the face that would be opposite the (1 0 0) face, presents a problem. It intersects the x-axis at $x_0=0$. Since $1/0=\infty$, no integer j will make this ∞ into an integer. Consider, therefore, the next plane back that will intersect the axes at $x_0=-1$, $y_0=\infty$, $z_0=\infty$. The resulting Miller indices are (–1 0 0). Instead of using negative numbers, we typically notate a negative value using a bar over the number and so the Miller indices become $\left(\bar{1}\ 0\ 0\right)$. Note that if the choice of plane results in a set

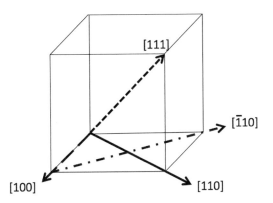

FIGURE 3.10
Direction vectors in a cubic lattice.

of integers that can be reduced to smaller integers having the same ratio, you should use the set containing the smallest integers. For instance, a (6 2 4) plane is equivalent to a (3 1 2) plane.

Miller indices are also used to determine directions in crystals. Directions are notated [h k l]. This is particularly easy in cubic structures where the direction [h k l] is perpendicular to the (h k l) plane. Figure 3.10 demonstrates how the [1 0 0], [1 1 1] and [$\bar{1}$ 0 0] directions are related to the corresponding planes.

The angle (α) between any two direction vectors [h_1 k_1 l_1] and [h_2 k_2 l_2] is given by

$$\cos(\alpha) = \frac{h_1 h_2 + k_1 k_2 + l_1 l_2}{\sqrt{h_1^2 + k_1^2 + l_1^2}\sqrt{h_2^2 + k_2^2 + l_2^2}} \tag{3.1}$$

For example, the angle between the [111] and [110] directions in Figure 3.10 is $\cos\alpha = 2/(\sqrt{3}\sqrt{2})$ yielding α of 35.26°.

For cubic lattices where the direction [h k l] is the normal vector to the (h k l) plane, the spacing, d, between (h k l) planes is given by

$$d = \frac{a_0}{\sqrt{h^2 + k^2 + l^2}} \tag{3.2}$$

where a_0 is the length of one side of the simple cubic face. For example, the spacing between (1 0 0) planes in a simple cubic lattice is $d = \frac{a_0}{\sqrt{1+0+0}} = a_0$. For fcc and bcc cubic structures, note that an extra set of atoms exists in the cube at a distance of $a_0/2$ in the [1 0 0] direction. This corresponds to $h=2$ rather than $h=1$, which correctly yields the interplanar spacing in Equation 3.2. Some care needs to be used in applying this equation to fcc and bcc lattices. In Chapter 12, we will discuss restrictions on the values of h, k, and l for different lattices related to diffraction. These same conditions are useful here. For fcc lattices, the values of h, k, and l must all be either even or odd numbers. For bcc lattices, $h+k+l$=even number. Thus, the three planes from Figure 3.9 for a simple cubic structure are (1 0 0), (1 1 0), and (1 1 1). The corresponding planes in an fcc structure would be (2 0 0), (2 2 0), and (1 1 1) so that the values are all even or all odd. For a bcc structure, these three planes would be (2 0 0), (1 1 0), and (2 2 2).

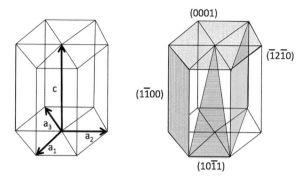

FIGURE 3.11
Hexagonal primitive vectors and common planes.

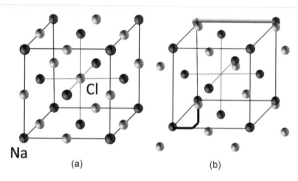

FIGURE 3.12
Crystal structures of (a) NaCl (the two labeled atoms are part of the same basis) and (b) Si with the Si atoms on the (¼, ¼, ¼) lattice shown in lighter shading for clarity (an example of two Si atoms in a basis are connected by a heavy lined path).

In discussing lattices we have focused on cubic systems. Hexagonal lattices are often described by the addition of a fourth index ($h\ k\ i\ l$) where $i=-(h+k)$. These indices are referred to as the Miller-Bravais indices. The purpose of this is to make planes that have similar structures more obvious. For instance, if we use a three-index system, the rectangular sides of a hexagonal structure would include (1 0 0) and (1 $\bar{1}$ 0). These are equivalent planes crystalographically, but in the three-index system, the notations look very different. In the four-index system, these are (1 0 $\bar{1}$ 0) and (1 $\bar{1}$ 0 0), which are clearly related since they just rearrange the same integer values. The unit vectors and common planes in a hexagonal system are represented in Figure 3.11.

With lattices defined, we are ready to create crystals by considering what basis of atoms is located at each lattice point. The simplest basis is a single atom located at each lattice point. This is a common basis in metals such as Cr, which has a bcc lattice and a single Cr atom as the basis with $a_0=2.88$ Å. Another example is Cu, which has an fcc lattice and a single Cu atom basis with $a_0=3.615$ Å. Many materials have a basis of two atoms. The crystal will have the two-atom basis repeating at each lattice point. One example is salt, NaCl, which consists of an fcc lattice and a basis consisting of a Na atom at (0, 0, 0) and a Cl atom at (½, ½, ½) as shown in Figure 3.12a. Another common two-atom basis is Si, which is also an fcc lattice but with a basis of a Si atom at (0, 0, 0) and another Si atom at (¼, ¼, ¼) as shown in Figure 3.12b. This is also the structure of a diamond.

3.2 Reciprocal Lattice

In describing the behavior of waves, it is common to define a wave vector, $\vec{k}$, in the direction of propagation and related to the wavelength, λ, such that

$$k = \frac{2\pi}{\lambda} \tag{3.3}$$

Note that the units of k are just 1/length. This leads us, rather than considering only position space as we have so far, to consider a space of k vectors known as "k space" or "reciprocal space". While this might seem a bit abstract, it is a very powerful concept and particularly of use in diffraction experiments.

In Chapter 12, we will explore in more detail diffraction techniques, but it is useful to note here that crystals act as a diffraction grating for X-rays, electrons, and other particles. In diffraction, we get constructive interference, and therefore a strong signal when Bragg's Law, $2d\sin\theta = n\lambda$, is satisfied where d is the distance between atomic planes perpendicular to the surface, θ is the angle of incidence of the radiation of wavelength, λ, and n is an integer. Notice that $\sin\theta$, which will determine where we will detect output intensity, varies as $1/d$. As a result, diffraction experiments map out reciprocal space.

Figure 3.13 shows a simple, two-dimensional, example of a real space crystal lattice and the reciprocal space lattice that would be associated with it. For crystals with orthogonal axes (cubic, tetragonal, orthorhombic lattices) where $\vec{a}_1 = a\hat{x}$, $\vec{a}_2 = b\hat{y}$, and $\vec{a}_3 = c\hat{z}$, then $\vec{b}_1 = \frac{2\pi}{a}\hat{x}$, $\vec{b}_2 = \frac{2\pi}{b}\hat{y}$, and $\vec{b}_3 = \frac{2\pi}{c}\hat{z}$. The notation $\hat{x}$, $\hat{y}$, and $\hat{z}$ indicates unit vectors in the x, y, and z directions. Notice that long spatial dimensions become short dimensions in reciprocal space.

We can create primitive cells in reciprocal space as well. Using the Wigner-Seitz primitive cell construction discussed earlier in this chapter, the smallest cell that is created is typically referred to as the First Brillouin Zone. Extending this construction technique further, the next larger cell is the Second Brillouin Zone. We could thus construct a whole series of Brillouin zones. A two-dimensional example of Brillouin zones that would correspond to the Wigner-Seitz cell presented in Figure 3.5 is shown in Figure 3.14.

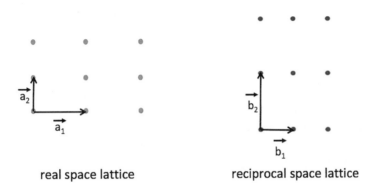

real space lattice   reciprocal space lattice

FIGURE 3.13
Real space lattice and corresponding reciprocal space lattice.

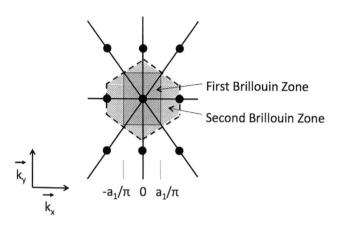

FIGURE 3.14

First and second Brillouin zones for a rectangular lattice

3.3 Bonds in Crystals

So far we have described the positions of atoms in crystals but have not discussed why the atoms occupy those positions. To understand crystal structure we need to explore the bonds between atoms. The process of bonding between atoms will lower the total energy of a system of atoms leading to a solid being a more stable configuration than a group of free atoms. The bonds arise from a balance of attractive and repulsive forces that typically arise from electrostatic and quantum mechanical sources. Bonds between atoms are based on mixtures of four basic types of bonds:

1. Van der Waals
2. ionic
3. covalent
4. metallic

Van der Waals bonds are present in all crystals but are often neglected since they are rather weak. A typical bond energy would be on the order of 0.2 eV/atom. Note that even this relatively weak bond would correspond in thermal energy to a temperature of 2320 K, which explains why solids do not fall apart (melt) until high temperatures. Van der Waals bonds are important in inert gas crystals and some organic materials since these materials do not experience the higher energy bond types. The bonds form from a balance between an attractive force arising from the net dipole interaction between the electron clouds around atoms and a repulsive force from the quantum mechanical Pauli exclusion principle when the electron clouds from two adjacent atoms begin to overlap.

The potential energy can be described as the sum of a positive (repulsive) and negative (attractive) term. One of the most common expressions to use for the total interaction potential energy is the Lennard-Jones potential.

$$U(r) = 4\varepsilon \left[\left(\frac{\sigma}{r} \right)^{12} - \left(\frac{\sigma}{r} \right)^{6} \right]$$

(3.4)

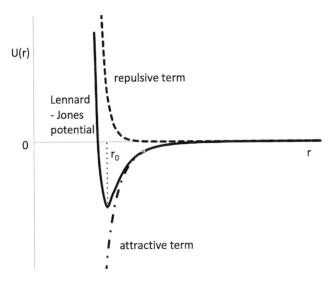

FIGURE 3.15
Interaction potential energy (Lennard-Jones potential) and repulsive and attractive components.

where ε and σ are constants that vary for each material and r is the distance between the atoms. This potential is pictured in Figure 3.15. Energy is minimized by making a bond with length r_0.

A second type of bond is the ionic bond. These are strong bonds with energies typically about 2–10 eV/atom. In ionic bonds, charge is transferred from one atom to the other converting one atom into a negative ion and the other into a positive ion. The resulting bonds again arise from the balance of attractive and repulsive forces. The attractive force comes from the Coulomb attraction between the ions with opposite net charges. The repulsive force arises from the Coulomb repulsion between like charges within each ion. Materials that display ionic bonding include NaCl, MgF_2, and ZnS. The charge transfer from one atom to the other will typically result in ions that have closed electron shells. For instance, Na has one electron beyond a closed shell (a 3s electron) and Cl has 5 3p electrons but needs one more to complete a closed shell. When forming an ionic crystal, both of these atoms end up with closed shells. This will result in an approximately spherically symmetric electron charge distribution around each atom. Thus, the resulting ionic bonds are not directional.

Covalent bonds are another strong bond with energies of 2–10 eV/atom. These are very directional bonds that forms when two electrons of opposite spin (one from each atom) are shared by the two atoms and are localized between the two atoms in overlapping orbitals. Both electrons are able to lower their energy by residing in these overlapping orbitals. The atoms involved in forming a covalent bond must each have at least one half-filled orbital to form a strong bond. These bonds are very hard to deform and are found in non-metallic materials involving elements such as N, O, and C. Semiconductor bonds in Si, Ge, and GaAs are also partly covalent.

Metallic bonds are somewhat weaker bonds with energies typically around 1–2 eV/atom. They arise from a complicated combination of effects including Coulomb forces, Van der Waals interactions, the Pauli exclusion principle, electron correlations, and the Heisenberg uncertainty principle. Metallic bonds are not directional and so can be easily deformed.

Examples of metallic bonding, as the name implies, are most metal elements. In some metals, a Lennard-Jones potential (Equation 3.4) is an appropriate model.

By exploring the details of the repulsive and attractive components of the bond potentials such as in the Lennard-Jones potential of Figure 3.15, we can gain a better understanding of the crystal properties. The lattice parameters and binding energy can be determined by examining where the potential energy has a minimum. The shape of the energy well around this minimum provides information about the elastic modulus of the material and the asymmetry in the shape of the well can provide information about the thermal expansion properties of the material.

3.4 Defects in Crystals

Non-ideal crystalline materials have many defects that cause their structure (and thus the related properties) to differ from a perfect crystal. Most crystalline materials have substantial numbers of defects although some (like Si) can be grown with a minimum number of defects. Defects can be characterized by considering their geometry and breaking them up into three categories: planar defects, line defects, and point defects.

First, consider planar defects. A surface, of course, is a defect in an infinite crystalline material since the symmetry is broken at the surface. We begin exploring the nature of surfaces in the next section. Materials are often polycrystalline consisting of different regions each of which may have the same crystal structure, but are not oriented in the same way. Interfaces between two single crystal regions of different orientations result in a planar defect called a grain boundary as shown in Figure 3.16. Grain boundaries may have a lower density of atoms in their vicinity leading them to resemble surfaces and have similar properties including more weakly bonded atoms at the grain boundary. This difference in bonding may result in the atoms being more reactive and so corrosion and other chemical processes may occur preferentially along the grain boundaries. The open structure may also provide a path for faster atomic diffusion (as discussed in Chapter 4). Typical sizes of crystalline grains are in the 0.01–100 μm range.

To get an idea of the importance of grain boundaries in materials, we could use a very simple model to explore how many atoms in a solid are at grain boundaries. Consider cubic grains with sides of length, l, containing atoms separated by a distance a_0. Let N be

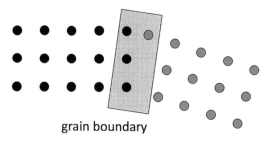

grain boundary

FIGURE 3.16
Grain boundary between two crystalline grains.

the number of atoms in the length l such than $l=Na_0$. Since atoms on the surface of the cubic grains would be on grain boundaries, we can calculate the ratio of the number of atoms on the six sides of the cube to the number of atoms in the bulk.

$$\frac{\text{number of atoms at grain boundaries}}{\text{number of atoms in bulk}} = \frac{6(Na_0)^2/a_0^2}{(Na_0)^3/a_0^3} = \frac{6N^2}{N^3} = \frac{6}{N} = 6\frac{a_0}{l} \qquad (3.5)$$

For a grain size of $l=0.1\ \mu m = 10^3\ \text{Å}$ and a lattice spacing $a_0=5\ \text{Å}$, the ratio would be 0.03. A somewhat better argument using truncated octahedra leads to the fraction being about 3.35 (a_0/l). In later chapters, we will observe in more detail how the grain size in thin films depends on the deposition rate, film thickness, and substrate temperature during deposition.

Dislocations are line defects. The two main types of dislocations are edge and screw dislocations. An example of an edge dislocation is shown in Figure 3.17. In an edge dislocation, we insert an extra row of atoms in the upper part of the crystal that is not present in the lower part. This distorts the lattice and creates stresses in the crystal with the upper part of the crystal being in compression and the lower part in tension. Dislocations are very common and films frequently have 100–10,000 dislocations/μm^2. Dislocations in thin films may arise from dislocations that exist in the substrate and continue into the film, contamination on the substrate, or from the film growth process. Screw dislocation consists of atomic planes which spiral around the linear defect.

Several types of point defects are common in crystalline materials:

1. Self-interstitials are an extra atom of the crystalline element inserted into the crystal but not on a lattice site.
2. Vacancies are missing atoms in the lattice.

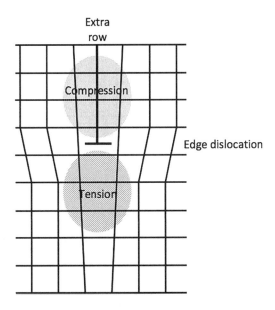

FIGURE 3.17
Schematic representation of an edge dislocation showing an extra row of atoms added in the top section.

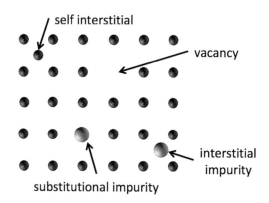

FIGURE 3.18
Common point defects.

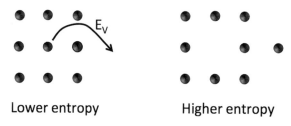

FIGURE 3.19
Vacancies will always exist in a crystal since they increase entropy.

3. Substitutional impurities are foreign atoms that replace a crystalline atom on a lattice site.

4. Interstitial impurities are extra foreign atoms that are inserted into the crystal but not on lattice sites.

These point defects are demonstrated in Figure 3.18.

All of these point defects could be eliminated except vacancies. Vacancies are a normal part of real crystals that arise from thermodynamic arguments regarding entropy. Figure 3.19 shows how the existence of a vacancy produces a configuration that has higher entropy (less order). This will be thermodynamically favored but there is an energy, E_v, required to form a vacancy.

Using our Boltzmann factor argument from Chapter 1, we can determine the probability of forming a vacancy, which is also the ratio of vacancies to atoms in a crystalline solid.

$$f_v = e^{-E_v/k_B T} \tag{3.6}$$

E_v is about 1 eV at room temperature leading to a typical fraction of about 10^{-17}. In thin films, point defects often arise from high deposition rates and/or low substrate temperatures.

3.5 Ideal Surfaces

Our use of hard spheres to model atoms in a solid is, of course, not accurate. Atoms can be better described as consisting of an inner ionic core surrounded by a cloud of electrons. In this case, where is the surface? Is it at the edge of the electron cloud or at the center of the ionic cores or somewhere else? Furthermore, atoms from the surface are expected, from thermodynamics, to be constantly leaving the surface and returning creating a concentration of atoms in the vapor phase above the surface. A useful definition of the location of a surface is presented in Figure 3.20 based on the change in the mass density, which is high in the solid and low in the vapor. The surface is defined by drawing a line creating two equal areas bounded by the limiting densities and the actual density distribution curve.

Returning to our hard sphere approximation, we will frequently use the three cubic structures, pictured in Figure 3.6, since cubic structures are easier to describe mathematically and are commonly found in nature. By cutting the lattice to create a surface with one of the common, low Miller index planes exposed, we can examine the atomic positions to see how these surfaces differ from one another. Figure 3.21 shows some common surfaces for fcc and bcc structures with the atoms in the first, second, and third (where visible) planes of atoms indicated in the figure. Notice how some planes have a very densely packed arrangement of atoms and others are much more open structures. These different structures will lead to different properties of surfaces.

The (1 1 1) face of the fcc lattice demonstrates the highest possible density of atoms and is referred to as a close-packed structure. In three dimensions, two possible ways exist to pack identical atoms into a minimum volume. One is the fcc structure and the other is a hexagonal close-packed (hcp) structure. These structures have identical surface layers and second layers of atoms. In the hcp structure, the third layer of atoms is identical to the surface layer resulting in an ABABAB repeat pattern as you go layer by layer through the crystal. In the fcc structure, the third layer atoms occupy sites that are different from both the surface and second layers resulting in an ABCABC repeat pattern. These arrangements are shown in Figure 3.22.

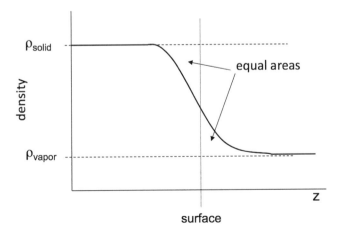

FIGURE 3.20
Defining the surface location through variation in the mass density.

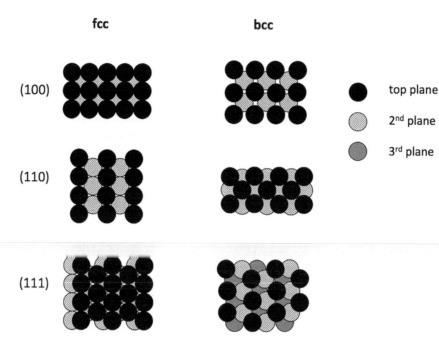

FIGURE 3.21
Simple surfaces of fcc and bcc crystals.

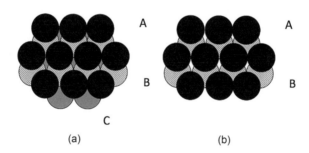

FIGURE 3.22
Top view of close-packed structures (a) hexagonal close-packed (hcp) (b) face-centered cubic (fcc).

3.6 Surface Reconstructions

In creating surfaces we have assumed that the positions of the atoms in the crystal do not change from the bulk. As introduced in Chapter 1, atoms at a surface exist in a different bonding environment since they have no atoms above them (outside the crystal). Considering our simple spring model for bonds from Chapter 1, the absence of an atom above the surface means that there is nothing pulling up on the surface atom. This would suggest that the atom might relax slightly down toward the second layer. Such a reduction in the lattice spacing of the surface atom is observed on many metals with about a 1%–10% reduction in the lattice spacing normal to the surface. A few metals (Be and Al) have an

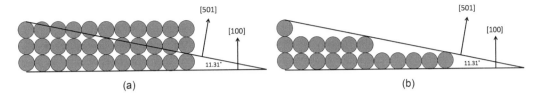

FIGURE 3.23
(a) (5 0 1) Plane through a simple cubic crystal. (b) Resulting surface steps and (1 0 0) terraces.

outward expansion of the surface plane. The extent of this relaxation in lattice parameters is found to vary depending on the material and the particular surface which is exposed. In some materials, changes in lattice constants between additional layers below the surface are also observed that can be expansions or contractions.

Experimentally, we observe that some crystals, especially when cut along low Miller index planes such as (1 0 0), (1 1 0), or (1 1 1), will maintain the structure of the crystal at the surface. Other crystalline materials might show variation in the lattice parameter perpendicular to the surface but still maintain bulk lattice parameters in the plane of the surface. However, even the low index surfaces of some materials may show dramatic reconstructions of the surface resulting in a structural arrangement of the atoms that looks quite different from the corresponding plane of the crystal. If we cut a crystal along a plane that is not simple, a (5 8 3) plane, for instance, we observe that the atoms will typically rearrange to allow terraces of low index planes to be exposed and separated by steps. We consider a simple example in Figure 3.23 where we cut a simple cubic crystal at a small angle from a (1 0 0) surface. This new surface will now intersect two of the axes of the crystal (call them x and z) and so only one Miller index will still be 0. Suppose we cut along the (5 0 1) surface. Equation 3.1 tells us that the angle between the new (5 0 1) surface and the (1 0 0) plane will be 11.31°. Figure 3.23a shows a side view of a simple cubic lattice with the (5 0 1) plane cut into it. If we remove any atoms that extend above this plane, we get a series of steps and (1 0 0) terraces that represent an ideal reconstruction of this surface as indicated in Figure 3.23b. Notice that the terrace is five atoms across and then a 1 atom step. The tangent of 11.31° is 1/5 and so our description is consistent with our equation. We will return to this issue in Chapter 5 where a simple model of surface energy will lead to the same conclusion about steps and terraces.

All surfaces will have defects. Bulk defects such as grain boundaries and dislocations may terminate at surfaces. Terraces and steps are defects in the sense that they are not present in a simple hard sphere model of cutting a crystal. The steps, however, may have additional defects within them if they have kinks that cause them to not continue in a straight line along a simple low index crystal direction. Vacancies will always be present (from thermodynamic arguments) on any surface. Atoms from the material may also protrude up as atoms sitting above the surface plane. Protruding atoms may be present as atoms adsorbed from the vapor (adatoms) and may group together into islands that will be important in thin film formation. Examples of common defects are demonstrated in Figure 3.24. These surface defects may alter the reactivity of surfaces. We will see in Chapter 8 that steps and kinks can be nucleation sites for the formation of thin films.

More complicated reconstructions of surfaces are also observed. For example, the (1 1 0) surfaces of fcc metals often show a reconstruction in which alternate rows along the surface are missing as in Figure 3.25. The Si(1 0 0) and Si(1 1 1)surfaces reconstruct with both vertical and horizontal motion of surface atoms as well as the movement of atoms in lower layers.

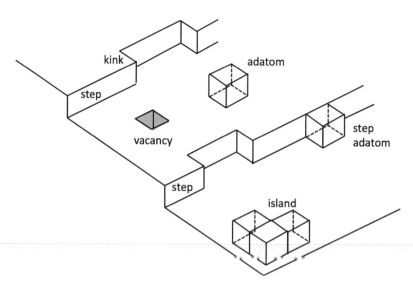

FIGURE 3.24
Common defects on surfaces.

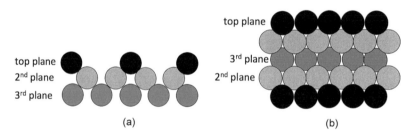

FIGURE 3.25
Missing row reconstruction on fcc (1 1 0) surface: (a) side view and (b) top view.

3.7 Minimizing Energy

Our experimental observations reveal a remarkable diversity of surface structures that are not the same as the ideal surfaces formed by simply terminating the bulk crystal lattice. The origin of these effects typically arises from trying to minimize the energy associated with the creation of a surface. This energy is always positive and so there is always an energy cost in creating a surface. The energy also varies by crystal face so it may be energetically more favorable (lower energy) for a crystal to have one face exposed rather than another. A rearrangement of the atoms on a surface may allow a crystal to minimize the surface energy. We will explore the concept of surface energy in more detail in Chapter 5, but we provide a couple of examples here of calculations that seek to explain atomic positions by examining the energy of a crystal.

Chadi (1978) developed a conceptually simple approach using computer modeling of crystal surface structures for ionic and covalent materials. The total energy of a 14-layer crystal was modeled using three interactions: electron-electron, electron-ion, and ion-ion. These can be expressed in terms of a bulk band-structure term and a contribution from the difference between the ion-ion and electron-electron energies. The band-structure term is essentially constant and the interaction term can be modeled as a short-range interaction. Atoms were placed on the ideal lattice and then allowed to move in order to minimize the energy of the system. This calculation, an example of a tight-binding method, makes use of the electrical properties of solids and surfaces that we will discuss in more detail in Chapter 6.

Shuttleworth (1948) developed a lattice sum technique for examining inert gas and ionic crystals. While the results do not always agree with the experiment, the concepts involved are useful to explore. His calculations assume $T=0$ K so that the atoms are not vibrating. He assumed that the bonding forces (and potential energies) were central and independent of one another. This allowed them to be simply added together. He further assumed that no dipoles were present on the lattice sites.

Shuttleworth considered an infinite solid and then broke it in half and separated the two halves to infinity, allowing him to explore several important energies associated with the resulting surface. The surface energy per unit area, U_a, that we want to minimize, is just the difference of the potential energy of interaction per unit area of the atoms across the plane before it is split, U'_a, and the change in the potential energy per unit area, U''_a when the atoms relax from their ideal positions into positions that minimize U_a. The U'_a term can be found by assuming a Lennard-Jones potential interaction between the atoms (Shuttleworth was actually more general in his calculation) and then summing all the interactions between the atoms. This requires describing the positions of the atoms. For an fcc lattice, for example, one can describe the atoms as being at positions (m_1a_0, m_2a_0, m_3a_0) where m_1, m_2, and m_3 are integers such that $m_1+m_2+m_3$ is even leaving alternate sites empty.

The relaxation energy, U''_a, can be determined by noting that the repulsive component of interatomic forces is a much shorter range than the attractive component. The separation of the two halves of the crystal results in an imbalance between the attractive and repulsive forces. The crystal plane would be expected to move out slightly until this balance is restored. This effect would be predicted to be stronger for less dense planes. Calculations typically show that $U''_a < 0.01\ U'_a$ and so the contribution of this term will be small.

Shuttleworth's model correctly predicts that the fcc (1 1 1) surface should have the lowest energy but is not useful for exact calculations. It does suggest, however, how we can use relatively simple concepts to confirm the ideas of surface reconstruction.

3.8 Surface Roughness

Statistical mechanics does not allow us to create a perfectly smooth surface. There will always be some vacancies of surface atoms as well as adatoms on the surface. In many cases, there will be steps and terraces. If the surface is part of a thin film, this is equivalent to saying that films always have some statistical distribution of thickness across the film. If we apply a simple statistical argument using a Poisson distribution of atoms sticking with no lateral mobility, we would expect that the mean thickness variation of a film, Δd, would increase with film thickness, d, according to

$$\Delta d \propto \sqrt{d} \qquad\qquad (3.7)$$

This is the worst-case model for roughness and typically we observe less roughness. A better understanding of roughness on surfaces requires knowledge of atomic motion through diffusion. This will be developed in Chapter 4, and so we defer further discussion of surface roughness to Section 4.3.3.

3.9 Non-Crystalline (Amorphous) Solids

We have focused on crystalline materials in this chapter, but we should acknowledge that non-crystalline (amorphous) solids, such as glasses, are commonly found. These are materials where the atomic positions do not show long-range order. Short-range order is still governed by the same interatomic interactions found in crystalline solids. While the amorphous state is not stable for most pure metals, amorphous metals can be formed by rapidly (10^6 K/second) cooling pure metals from a melt. They can also be readily formed from many metal alloys, semiconductors, and oxides, especially at low temperatures. Amorphous materials are typically less dense than crystalline materials. Crystalline defects are not defined in these materials since they do have any crystalline structure.

Amorphous materials can still be classified based on their short-range order. Three general classes are typically identified. (1) Continuous random networks, typical of covalently bonded materials such as glasses (oxides) and semiconductors, in which each atom is bonded in good short-range order to nearest neighbors but the next set of atoms will be rotated randomly in space. (2) Random coils are most commonly found in polymer systems in which long polymer chains are intertwined. (3) Random close-packed materials, typical of metallic glasses, in which atoms are closely packed although not achieving the densities found in crystalline close-packed materials. In three dimensions, random close-packed spheres can fill 64% of space compared to 74% for crystalline close-packed systems.

The development of nano-materials represents an interesting case between crystalline and amorphous structures. A crystalline structure that is only a few atoms across will not exhibit long-range order because it is simply too small. The atoms at the surface of a nano-crystal may also exhibit reconstructions that further reduce the order in the nano-crystal. This makes these materials difficult to label as crystalline or amorphous and also difficult to characterize using structural techniques that are based on the existence of long-range order such as most diffraction techniques.

3.10 Characterization of Structure

The structure of surfaces can be observed using a variety of imaging and characterization techniques described in Chapters 11 and 12. Surfaces can be imaged at various scales using optical microscopy, scanning electron microscopy, transmission electron microscopy, low-energy electron microscopy (for steps, terraces, and islands), and various scanning probe microscopies (such as scanning tunneling microscopy and atomic force microscopy). Details are available in Section 11.2. Scattering techniques, such as low-energy ion scattering and Rutherford backscattering, can also provide useful surface structure information.

Diffraction techniques such as low-energy electron diffraction, reflection high energy electron diffraction, and grazing incidence or in-plane X-ray diffraction are sensitive to surface structure. Diffraction and scattering techniques are discussed in Chapter 12. Various chemical techniques (described in Chapter 13) can also provide structural information.

References

Blakemore, J.S. 1985. *Solid State Physics.* 2nd ed. Cambridge: Cambridge University Press.

Chadi, D.J. 1978. Surface states of GaAs: Sensitivity of electronic structure to surface structure. *Phys. Rev. B* 18: 1800–1812.

Hudson, J. 1998. *Surface Science: An Introduction.* New York: John Wiley & Sons, Inc.

Ibach, H. 2006. *Physics of Surfaces and Interfaces.* Berlin: Springer.

Kittel, C. 1996. *Introduction to Solid State Physics.* 7th ed. New York: John Wiley & Sons, Inc.

Luth, H. 2015. *Solid Surfaces, Interfaces and Thin Films.* 6th ed. Berlin: Springer.

Moffatt, W.G., G.W. Pearsall, and J. Wulff. 1964. *The Structure and Properties of Materials, Volume 1: Structure.* New York: John Wiley & Sons, Inc.

Prutton, M. 1994. *Introduction to Surface Physics.* Oxford: Clarendon Press.

Shuttleworth, R. 1948. The surface energies of inert-gas and ionic crystals. *Proc. Phys. Soc.* A62: 167–179.

Somorjai, G.A. 2010. *Introduction to Surface Chemistry and Catalysis.* 2nd ed. New York: John Wiley & Sons, Inc.

Zangwill, A. 1988. *Physics at Surfaces.* Cambridge: Cambridge University Press.

Problems

Problem 3.1 Show that the three areas defined by the primitive cells in Figure 3.4 are the same.

Problem 3.2 What are the Miller indices for a plane that intersects the crystal axes at $x_0=1$, $y_0=2$ and $z_0=4$?

Problem 3.3 What is the angle between the [1 1 1] and [1 0 0] directions in a cubic crystal?

Problem 3.4 What is the distance between planes in fcc Cu for planes parallel to (a) (1 0 0) planes, (b) (1 1 0) planes, (c) (1 1 1) planes? (Note the discussion around Equation 3.2 for how to treat fcc lattices.)

Problem 3.5 Calculate the surface density of atoms (atoms/cm^2) for the (1 0 0) and (1 1 1) planes of an ideal Cu crystal. (Cu has fcc structure with a lattice constant $a_0=3.61$ Å.)

Problem 3.6 From the Lennard-Jones potential (Equation 3.4) determine an equation for the atomic separation, r_0, at the minimum energy in terms of the constant, σ.

Problem 3.7 By analogy with the cubic discussion in Section 3.4, determine the ratio of the number of atoms in the surface of a spherical grain of diameter, l, to the number of atoms in the volume of the spherical grain.

Problem 3.8 It is often useful in surface science to know the number density of atoms on a surface. In two dimensions that is the number of atoms per unit area on the surface. Using a little geometry and trigonometry, calculate the surface density of atoms (atoms/cm²) for the (1 0 0), (1 1 0), and (1 1 1) planes of a Cu crystal. Cu has a face-centered cubic structure and the cube side has a length of 3.61 Å. Since we are on a two-dimensional surface, the atoms can be treated as circles rather than spheres.

Problem 3.9 The potential energy in ionic bonds can be described in terms of an attractive and repulsive term as $U(r) = \dfrac{-\alpha q_e^2}{4\pi\epsilon_0}\dfrac{1}{r} + \lambda e^{-r/\rho}$ where α, λ, and ρ are constants, q_e is the charge on an electron, r is the distance between atoms, and the $4\pi\epsilon_0$ tells us that this is all in SI (mks) units with ϵ_0 being the permittivity of free space.

a. Find the force associated with this energy. Show that the force can be expressed in the form $F(r) = \dfrac{\alpha q_e^2}{4\pi\epsilon_0}\dfrac{1}{r_0^2}\left[-\dfrac{r_0^2}{r^2} + e^{\frac{(r-r_0)}{\rho}} \right]$ where r_0 is the equilibrium position.

b. If the atoms never move very far from the equilibrium position, r_0, we can approximate their positions as $r = r_0 + \delta r$. It is very common to deal with small deviations from equilibrium using a Taylor's series expansion about the equilibrium point. Perform a Taylor's series expansion about $r = r_0$. You only need to keep terms in δr since higher order terms will be very small. Show that the result is $F(r) = -C\,\delta r$ and show the form of the constant C. This is just Hooke's Law and brings us back to the simple mass on spring model discussed in Chapter 1.

Problem 3.10 The interaction potential $U(r)$ plotted in Figure 3.15 for two atoms as a function of separation distance can be more generally expressed as

$$U(r) = \frac{A}{r^n} - \frac{B}{r^m}$$

where A, B, m, and n are positive constants.

a. Show that the binding energy can be given by $U_{bind}(r_0) = \dfrac{B}{r_0^m}\left[\dfrac{m}{n} - 1\right]$ where r_0 is the equilibrium separation distance.

b. The binding energy must be negative for a stable bond to form. What condition does this place on m and n?

c. Recall that force is given by $F = -dU/dr$. If we apply an external force that exceeds the maximum interatomic force, we can break the bond. Show that a molecule formed by two atoms will break up once the atoms are pulled apart to a distance $r_b = r_0\left[\dfrac{n+1}{m+1}\right]^{1/(n-m)}$.

Problem 3.11 A cubic film is deposited onto a substrate such that the (1 0 0) plane in the film makes an angle of 71.56° with the plane of the substrate. What are the Miller indices of the film plane that is parallel to the substrate surface?

Problem 3.12 The lattice parameter can be estimated from the density and crystal structure of a pure material.

    a. Consider Cu, which has a fcc crystal structure and a density of 8.96 g/cm³. Cu atoms have an atomic weight of 63.55. Calculate the lattice parameter for Cu from this information.

    b. Consider Fe, which has a bcc crystal structure and a density of 7.87 g/cm³. Fe atoms have an atomic weight of 55.85. Calculate the lattice parameter for Fe from this information.

Problem 3.13 Assuming a hard sphere model for each atom and a lattice parameter of a_0, what is the distance between the closest atoms (in terms of a_0) in a

    a. fcc crystal

    b. bcc crystal

4

Atomic Motion: Vibrations, Waves, and Diffusion in Solids

Chapter 3 examined the structure of crystals but did not allow for vibrational or significant translational motions of atoms in the crystal beyond surface reconstruction. In Chapter 1, however, we introduced the simple model of atoms in a solid as interacting as though they were connected by springs so that the atoms are not stationary. In this chapter, we explore the concept of atoms in motion. Typically, we find three types of atomic motion in a solid lattice:

1. Random vibrational motion that arises from thermal energy, $k_B T$.
2. Collective motion that can arise from elastic waves that propagate through the solid lattice.
3. Diffusion in which atoms leave their lattice locations and move through the lattice.

We first examine motion within the lattice and then consider atomic diffusion. In each section, we begin with a discussion of the bulk/film properties and then explore how these are modified near the surface. Further details can be found in the literature (Brophy, Rose, and Wulff 1964, Ibach 2006, Kittel 1996, Luth 2015).

4.1 Thermal Vibrations

In Chapter 1, we developed a model of atoms in a crystal being connected by springs that experience a simple force, $F=-C\Delta x$, that always attempts to bring the atom back to the equilibrium position. We saw how the harmonic oscillator potential energy associated with this force is a good approximation to more general potential energy functions when we are near the equilibrium position. In a bulk solid, we can thus model each atom as an independent three-dimensional harmonic oscillator. These atoms will experience random fluctuations in position around their equilibrium lattice position due to the thermal energy in the solid. Figure 4.1 demonstrates that each atom can now be thought of as connected by springs rather than rigid rods to the other atoms. This allows each atom to vibrate in three dimensions.

In one dimension, the average harmonic oscillator potential energy (Equation 1.2) can be set equal to the thermal energy:

$$\left\langle \frac{1}{2}C(\Delta x)^2 \right\rangle = \frac{1}{2}k_B T \tag{4.1}$$

DOI: 10.1201/9780429194542-5

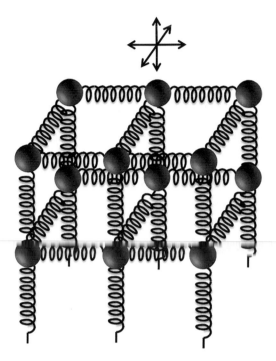

FIGURE 4.1
Model of a solid where atoms vibrate in three dimensions.

From this we see that the amplitude of the random vibrations of the atoms increases with temperature.

$$\left\langle (\Delta x)^2 \right\rangle \propto T \tag{4.2}$$

Near a surface, the bonding environment in the direction perpendicular to the surface can be different from the bulk leading to spring constants, C, that are typically smaller near the surface than in the bulk. This allows for larger amplitude vibrations in the direction perpendicular to the surface, which can be 50%–100% greater than the bulk amplitudes. This lack of symmetry may require the use of an anharmonic oscillator model.

This greater amplitude of vibration has the interesting consequence of surfaces melting before the bulk. If the random fluctuations in position are large enough, the ordering of the crystal structure may be lost and the crystal melts. (This could be described as an order-disorder phase transition). A common way of determining when this occurs is the Lindemann criterion, which describes a crystal as melting when $\left\langle (\Delta x)^2 \right\rangle^{1/2} > 0.25 a_0$. Some researchers replace the 0.25 value with values that are specific to the material being examined. We expect the surface layer to have the lowest melting point with the next layer down having a slightly higher melting point. This increase in melting point as we move away from the surface continues until we reach the bulk melting point. Different surfaces of the same material can also have different melting points. Figure 4.2 shows molecular dynamics calculations of a Cu(1 1 0) surface. The bulk melting temperature of Cu is 1358 K, but the Cu(1 1 0) surface shows evidence of melting below 1100 K.

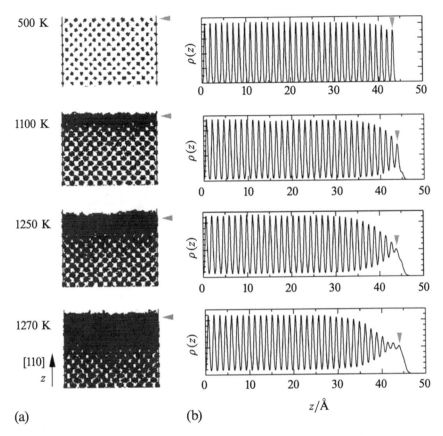

FIGURE 4.2
Surface melting of Cu(1 1 0). (a) Trajectories of atoms near surface for 3 ps. (b) Distributions of atomic number density for (1 1 0) surface model at several temperatures. The arrow represents the position of the initial surface layer. (Kojima and Susa 2002)

4.2 Elastic Waves and Phonons

4.2.1 Elastic Waves

Before focusing on collective motions near surfaces, we introduce elastic waves in crystal-line solids. This involves the displacement of atoms caused by a wave moving through a solid. The direction of the wave is given by a wave vector, $\vec{k}$, whose magnitude is $2\pi/\lambda$. We assume that the wavelength is long when compared to the atomic spacing and that waves only propagate in simple crystallographic directions such as [1 0 0], [1 1 0], or [1 1 1]. We also use the mass and spring model, described in Chapter 1, where the force on each mass is $F = -C\Delta x$. We begin with the simple case where there is just one atom in the basis and so each lattice point has a single atom located at it.

We consider two types of waves. Longitudinal waves have an entire plane of atoms displaced in the direction of the wave vector as indicated in Figure 4.3. This is similar to sound waves propagating through air. Transverse waves, presented in Figure 4.4, have atoms displaced within the plane perpendicular to the wave vector as in simple water waves.

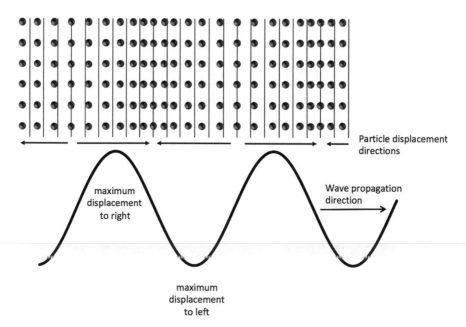

FIGURE 4.3
Longitudinal wave moving to the right. Equilibrium atom positions are indicated by the vertical lines.

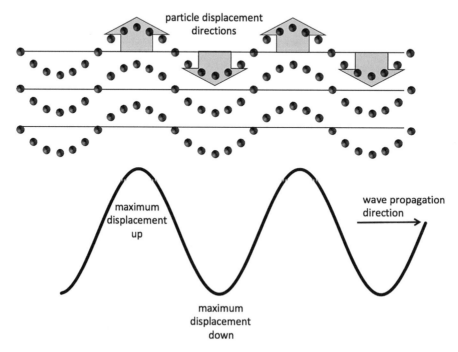

FIGURE 4.4
Transverse wave moving to the right. Equilibrium positions are indicated by horizontal lines.

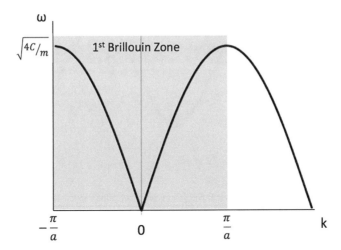

FIGURE 4.5
Dispersion relation for a one-dimensional crystal with a one-atom basis.

If we consider a traveling wave with frequency, ω, and wave vector magnitude, k, propagating through a lattice with atoms of mass, m, separated by a distance, a, we can establish a relationship, known as the dispersion relation, between ω and k. For the single atom basis that we are currently considering, we can set Newton's 2nd law equal to the net force from springs on either side of an atom. This assumption is the nearest-neighbor interaction approximation. Assuming that all displacements have the same time dependence ($e^{-i\omega t}$), we can find the dispersion relation:

$$\omega^2 = \frac{2C}{m}\left(1 - \cos(ka)\right) = \frac{4C}{m}\sin^2\left(\frac{1}{2}ka\right) \tag{4.3}$$

The dispersion relation is graphed in Figure 4.5. Of course, the distance between atoms, a, will vary depending on what direction through the crystal the wave is traveling.

Since we are dealing with the displacements of atoms in a solid lattice, we are only interested in the wave at locations in the solid where atoms exist. Consider a wave with wavelength equal to twice the distance between atoms. At the position of an atom, this wave would produce exactly the same effect as a wave with a wavelength equal to 2/3 the distance between atoms. The shorter wavelength wave would have additional peaks at positions between the atoms, but, since there are no atoms there, this provides us with no new information about atomic displacements. As a result, we do not need to consider any standing waves with wavelength less than twice the distance between atoms since they will not give us any new information. These waves are demonstrated in Figure 4.6.

The condition $\lambda \geq 2a$ is equivalent to $-\frac{\pi}{a} \leq k \leq \frac{\pi}{a}$. Since $\vec{k}$ is a vector in reciprocal space, we see that these limits are the boundaries of the first Brillouin Zone, as discussed in Chapter 3. As a result, we only need to consider wave vectors in the first Brillouin Zone. If we have a wave with a wave vector, $\vec{k}$, that is outside the first Brillouin Zone, we can reduce it to a wave that is inside the zone and contains the same information about atomic positions. If we find a wave vector $k' = k - n\,(2\pi/a)$, where n is an integer, that lies in the first Brillouin Zone, this wave will result in the same atomic displacements as the original wave.

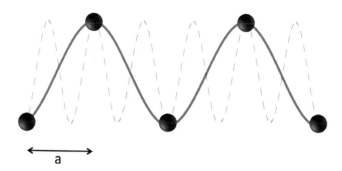

FIGURE 4.6
The dashed wave ($\lambda=2a/3$) contains no new information when compared to the solid wave ($\lambda=2a$).

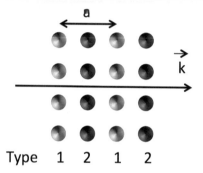

FIGURE 4.7
Two-atom basis with a wave traveling perpendicular to alternating planes of atoms of each type in the basis.

We can examine a somewhat more complicated situation by considering a two-atom basis but restricting ourselves to waves propagating perpendicular to the alternating planes of the two types of atoms as shown in Figure 4.7. We assume that the two types of atoms have different masses, and atoms of the same type are separated by the distance, a.

Two types of motion are now possible when a wave with wavelength large compared to the interatomic spacing propagates through the crystal. These are described as optical and acoustical waves. Figure 4.8 shows the motions of a single row of atoms as a transverse wave propagates to the right along the row of alternating atom types. Figure 4.9 shows the same situation for a longitudinal wave propagating in the same direction.

The dispersion relation, in this case, has two solutions, referred to as the optical branch and acoustical branch, corresponding to the two types of motion.

$$\omega^2_{\text{optical}} = C\left[\left(\frac{1}{m_1}+\frac{1}{m_2}\right)+\sqrt{\left(\frac{1}{m_1}+\frac{1}{m_2}\right)^2-\frac{4}{m_1 m_2}\sin^2\frac{ka}{2}}\right] \qquad (4.4)$$

$$\omega^2_{\text{acoustical}} = C\left[\left(\frac{1}{m_1}+\frac{1}{m_2}\right)-\sqrt{\left(\frac{1}{m_1}+\frac{1}{m_2}\right)^2-\frac{4}{m_1 m_2}\sin^2\frac{ka}{2}}\right] \qquad (4.5)$$

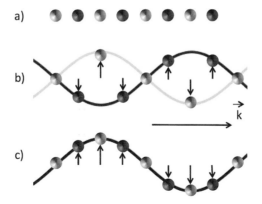

FIGURE 4.8
(a) Undisturbed row in a diatomic linear crystal. (b) Optical mode of a transverse wave traveling to the right. (c) Acoustical mode of a transverse wave traveling to the right.

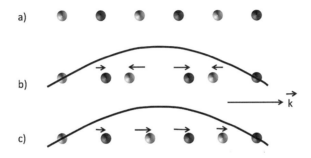

FIGURE 4.9
(a) Undisturbed row in a diatomic linear crystal. (b) Optical mode of a longitudinal wave traveling to the right. (c) Acoustical mode of a longitudinal wave traveling to the right.

These are shown in Figure 4.10. Notice that not all combinations of ω and k are allowed. There is a gap between the two modes indicating that no wave can be generated with ω in that gap.

4.2.2 Phonons

Historically, classical wave models such as those discussed here failed to explain certain phenomena. The classical wave model of light could not describe blackbody radiation. The classical wave model of solids could not describe heat capacity. In each case, the solution was to quantize the wave and treat these quanta as particles. In light, the resulting particle is a photon and in lattice waves, the particle is a phonon. An elastic wave becomes a stream of phonons.

The photon model for light and phonon model for lattice waves have interesting similarities and differences summarized in Table 4.1.

Photons and phonons can interact with one another. For instance, phonons can be created or absorbed when light (photons) reflects from a solid surface. An incident photon with wave number, k, can strike a surface and create a phonon in the solid. The phonon takes some energy away from the photon and so the reflected photon has a different wave

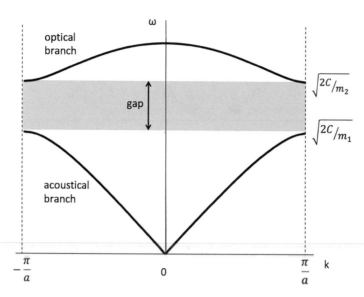

FIGURE 4.10
Dispersion relation for a one-dimensional crystal with a two-atom basis.

TABLE 4.1

Comparison of Photons and Phonons

Photons	Phonons
$E = n\hbar\omega$	$E = \left(n + \dfrac{1}{2}\right)\hbar\omega$
$v = c$	$v = v_{group} = d\omega/dk = v_{sound}$
$\vec{p} = \hbar\vec{k}$	$\vec{p} = \hbar\vec{k}$

number, k'. Similarly, if an existing phonon interacts with an incident photon of wave number, k, the phonon can be absorbed and lose all of its energy to the photon. The resulting photon will be reflected with more energy than it had and will have a new wave number, k'. Two phonons can also collide with one another to produce a third phonon and phonons traveling through a crystal with defects can scatter from those defects or from surfaces.

The dispersion relation in Figure 4.11 demonstrates the origin of the names of the two branches in Figure 4.10. The linear dispersion relation of the photon (moving at the speed of light) will only intersect the optical phonon branch. The acoustical phonons and photons both start at the origin, but the acoustical phonons can never have slope as great as the photons and so the two will not interact.

4.2.3 Surface Waves and Phonons

Lattice vibrational modes that are not allowed in the bulk can be found at the surface. These have a wave vector parallel to the surface and an amplitude of vibration that decays exponentially into the bulk. These can be modeled from a macroscopic elastic continuum approach or from a microscopic model.

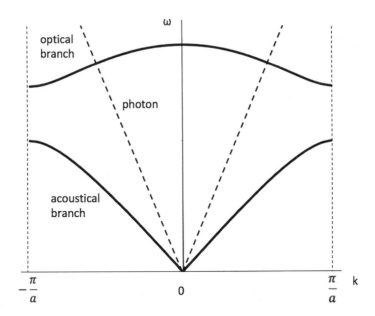

FIGURE 4.11
The dispersion curve for the photon intersects the phonon optical branch but not the acoustical branch.

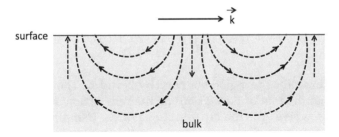

FIGURE 4.12
Displacements of atoms at a fixed time in a Rayleigh wave showing both vertical and horizontal displacement.

Isotropic elastic continuum theory gives insight into long-wavelength surface waves. The presence of the surface generates an additional boundary condition that there can be no forces directed into the vacuum above the surface. The resulting surface modes are confined to the plane perpendicular to the surface and in the direction of $\vec{k}$. These are known as sagittal modes. These acoustical surface waves are often called Rayleigh waves. They are a blend of transverse and longitudinal waves as indicated by the displacements shown in Figure 4.12. Surface acoustical waves have found broad industrial applications.

Microscopic theory provides insight into short wavelength surface vibrations where both optical and acoustical vibrations are allowed at the surface. Shear horizontal modes are also allowed with displacements in the plane of the surface.

Modeling of these modes depends on the type of bonding in the crystal. Ionic crystals are fairly straightforward to model using two-body Coulomb interactions. Semiconductors are harder to model since covalent bonds are more directional and often require three-body interactions to be considered. Metals are also harder to model due to the need to consider the role of the electrons (often considered as an electron gas) in the solid.

4.3 Diffusion

4.3.1 Bulk Diffusion

Atomic motions are not restricted to vibrations around the equilibrium lattice positions in a crystal. Atoms may move through the crystal, along defects, or along surfaces through diffusion. A driving force for diffusion is the thermodynamic goal of having a random distribution of all elements through a material. This will create a tendency for elements to diffuse from regions of high concentration to regions of lower concentration. In one dimension, this can be expressed mathematically through Fick's 1st law.

$$J(x,t) = -D\frac{\partial C(x,t)}{\partial x} \tag{4.6}$$

where J is the flux of diffusing atoms (atoms/m²s), D is the diffusion coefficient (m²/s), C is the concentration of the diffusing elements (atoms/m³) at the position x and time t. The negative sign indicates that the flow of atoms is from regions of high concentration to regions of low concentration.

We obtain greater insight into this process by considering the flux of atoms into and out of a particular volume in the solid. Since both $J(x, t)$ and $C(x, t)$ will vary with position, we relate the flux at position x_2 to the flux at a nearby position x_1 by

$$J(x_2,t) = J(x_1,t) + \frac{\partial J(x,t)}{\partial x}\Delta x \tag{4.7}$$

In Figure 4.13, we examine the region bounded by x_1 and x_2. The rate of increase of atoms in the region is just the rate of flow in minus the rate of flow out. Mathematically this is $[J(x_1, t) - J(x_2, t)]A$ where A is the cross-sectional area. Using Equation 4.7, we express this rate of increase of atoms in the region as $-\frac{\partial J(x,t)}{\partial x}A\Delta x$. Equally valid would be to

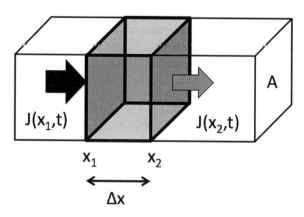

FIGURE 4.13
Flow of atoms into and out of a region bounded by x_1 and x_2.

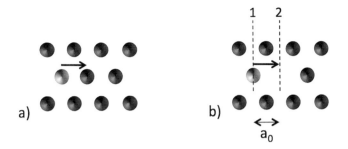

FIGURE 4.14
(a) Diffusion into a space occupied by another atom is difficult. (b) Diffusion into a space occupied by a vacancy is easier.

relate this rate of increase of atoms to the change in concentration of atoms in the volume $\frac{\partial C(x,t)}{\partial t} A \Delta x$. Since these two expressions are both describing the same increase,

$$-\frac{\partial J(x,t)}{\partial x} = \frac{\partial C(x,t)}{\partial t} \tag{4.8}$$

On the left side of the equation, we substitute in for J from Equation 4.6 to obtain Fick's 2nd Law:

$$\frac{\partial C(x,t)}{\partial t} = D \frac{\partial^2 C(x,t)}{\partial x^2} \tag{4.9}$$

Using appropriate boundary conditions, this partial differential equation can be solved to determine the concentration at different positions and times.

With this macroscopic view of diffusion in place, we now seek to understand the process on an atomic scale. For simplicity, we will consider a simple cubic lattice. If all of the lattice positions are filled with atoms, as in Figure 4.14a, it is very difficult for diffusion to occur since the diffusing atom has nowhere to go. Much faster diffusion can be achieved if a vacancy is present as in Figure 4.14b. In Chapter 3, we observed that vacancies are always present in crystals. We restrict our attention now to diffusion into vacancies. In the case of diffusion into an adjacent lattice position, the diffusing atom will move a distance, a_0. There will be some energy barrier, E_j, that must be overcome to jump into the vacancy.

The diffusion from position 1 to position 2 in Figure 4.14b will depend on how often the diffusing atom attempts to jump in that direction, the probability that a vacancy exists in the adjacent location, the probability that the diffusing atom has enough energy to overcome the barrier and diffuse, and the total number of atoms of the diffusing element available per unit area. Mathematically, this can be written out (with the terms in the same order):

$$J_{1 \to 2} = \left[\frac{1}{6} v \right] \left[e^{-E_V/k_B T} \right] \left[e^{-E_j/k_B T} \right] [C a_0] \tag{4.10}$$

where v is the vibrational frequency of the diffusing atom and E_V is the energy to create a vacancy as defined in Chapter 3.

If the diffusing atom that moves to position 2 were to immediately return to position 1, then there would be no net flux. The concentration of diffusing atoms, however, is different across the crystal and so we can write the diffusion from position 2 to position 1 as

$$J_{2\to1} = \left[\frac{1}{6}v\right]\left[e^{-E_V/k_BT}\right]\left[e^{-E_j/k_BT}\right]\left[\left(C+\frac{\partial C}{\partial x}a_0\right)a_0\right] \tag{4.11}$$

The net flux to the right is then

$$J_{net} = J_{1\to2} - J_{2\to1} = -\left[\frac{1}{6}a_0^2v\right]\left[e^{-(E_V+E_j)/k_BT}\right]\frac{\partial C}{\partial x} \tag{4.12}$$

Comparing this to Fick's 1st law, Equation 4.6, we see that the diffusion coefficient can now be expressed in terms of fundamental properties

$$D = D_0 e^{-E_D/k_BT} \tag{4.13}$$

where $D_0 = \dfrac{1}{6}a_0^2 v$ and $E_D = E_V + E_j$ is the diffusion energy per atom. Note that the diffusion constant increases with temperature. This will be very important in many applications. The diffusion constant also depends on the structure and bonding of the material in which diffusion is taking place as well as the specific elements that are diffusing. Some situations are so common that is worth specifying a notation for diffusion coefficients related to certain diffusing species and environments.

- D_A self-diffusion (element A in A)
- D_V diffusion of vacancies
- D_{AB} chemical diffusion (element A in B)
- D_{gb} diffusion along grain boundaries
- D_s diffusion along surfaces

The diffusion coefficient may vary with position in the crystal and with time since there may be temperature gradients and the jump frequency depends on the local atomic arrangement and defects in real crystals. Different crystal directions may also have different diffusion coefficients. Concentrations, defects, and temperature may also vary with time making the diffusion coefficient time-dependent.

The diffusion energy, as other barrier or activation energies, can often be determined experimentally from measurements of the diffusion coefficient at various temperatures.

Taking the natural logarithm of both sides of Equation 4.13

$$\ln(D) = \ln(D_0) - \frac{E_D}{k_BT} \tag{4.14}$$

We see that a plot of $\ln(D)$ vs $1/T$ will have a slope that is just $-E_D/k_B$. This Arrhenius plot is shown in Figure 4.15.

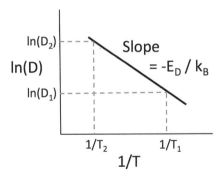

FIGURE 4.15
Arrhenius plot used to determine the diffusion energy.

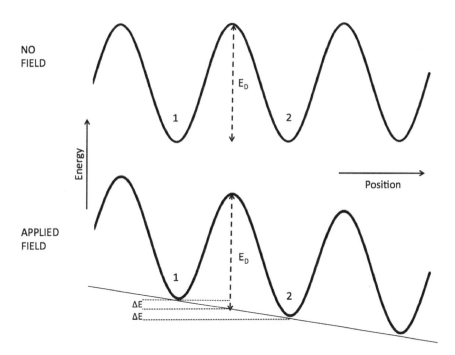

FIGURE 4.16
Diffusion in the absence and presence of an applied field.

As an example of calculating the diffusion energy, consider that the values indicated in Figure 4.14 are: $T_1=1000$ K, $D_1=2.2\times10^{-15}\,\text{m}^2/\text{s}$, $T_2=1200$ K, and $D_2=2.7\times10^{-13}\,\text{m}^2/\text{s}$. The slope$=\dfrac{\ln(D_2)-\ln(D_1)}{1/T_2-1/T_1}=\dfrac{\ln(2.7\times10^{-13})-\ln(2.2\times10^{-15})}{1/1200-1/1000}=-28,800$. Solving for the diffusion energy, $E_D = 28,800\times8.617\times10^{-5}\,\text{eV}\quad K = 2.48\,\text{eV}$.

Diffusion can also be changed by the application of stress, electric fields, or any other external influences that cause an energy gradient in the crystal. We demonstrate the effects of applying a field that results in an energy gradient in Figure 4.16.

From Figure 4.16 we see that the energy barrier for jumping from position 1 to 2 has been lowered by ΔE and that the energy barrier for jumping from position 2 to 1 has been increased by ΔE due to the application of a field. Since forces are defined by $F = -dU/dx$, an energy gradient will produce a force that will change the rate of diffusion.

The fluxes to the right and left can now be expressed as:

$$J_{1 \rightarrow 2} = \left[\frac{1}{6} v \right] \left[e^{-(E_D - \Delta E)/k_B T} \right] [C a_0]$$
(4.15)

$$J_{2 \rightarrow 1} = \left[\frac{1}{6} v \right] \left[e^{-(E_D + \Delta E)/k_B T} \right] \left[\left(C + \frac{\partial C}{\partial x} a_0 \right) a_0 \right]$$
(4.16)

The net flux to the right is then

$$J_{\text{net}} = D \left[\frac{C}{a_0} e^{\Delta E/k_B T} - \frac{C}{a_0} e^{-\Delta E/k_B T} - e^{-\Delta E/k_B T} \frac{\partial C}{\partial x} \right]$$
(4.17)

Noting that $\sinh(x) = (e^x - e^{-x})/2$, we can express this as:

$$J_{\text{net}} = D \left[2 \frac{C}{a_0} \sinh\left(\frac{\Delta E}{k_B T} \right) - e^{-\Delta E/k_B T} \frac{\partial C}{\partial x} \right]$$
(4.18)

For small x, $\sinh(x) \approx x$ and $e^x \approx 1 + x$. In most cases, $\Delta E \ll k_B T$, and so it is a good approximation to rewrite this as

$$J_{\text{net}} = D \left[\frac{2C\Delta E}{a_0 k_B T} - \left(1 - \frac{\Delta E}{k_B T} \right) \frac{\partial C}{\partial x} \right]$$
(4.19)

$$J_{\text{net}} = D \left[\frac{\Delta E}{k_B T} \left(\frac{2C}{a_0} + \frac{\partial C}{\partial x} \right) - \frac{\partial C}{\partial x} \right]$$
(4.20)

In addition to the original diffusion term, we now have a term arising from the applied field creating a drift of atoms through the crystal. If the concentration gradient is sufficiently small,

$$J_{\text{net-field}} \approx D \frac{2C\Delta E}{a_0 k_B T} = CD \left(\frac{2\Delta E}{a_0} \right) \frac{1}{k_B T}$$
(4.21)

Noting that the speed of the atoms from the applied field $v_d = J_{\text{net-field}}/C$ and that the quantity $(2\Delta E/a_0)$ is just the applied force from that field:

$$v_d = \frac{DF}{k_B T}$$
(4.22)

which is the Nernst-Einstein relation.

4.3.2 Surface and Interface Diffusion

In real crystals, fast diffusion paths are typically found at surfaces, grain boundaries and in three-dimensional networks of dislocations. These are more open structures with higher jump frequencies and lower energy barriers. As a result, D_0 and E_D are different along these paths compared to the bulk. We focus our attention on surfaces although grain boundaries are similar. Typically $D_S > D_{gb} > D$ although this may not be true at very high temperatures.

For diffusion along a surface or grain boundary, we are restricted to diffusion in two dimensions. If all directions along that surface are equally likely, the atom can be treated as performing a random walk on the surface. The displacement, R, of the atom from its original position can then be related to the diffusion coefficient and time

$$\left\langle R^2 \right\rangle^{1/2} \approx \sqrt{Dt} \tag{4.23}$$

If all directions are not equally likely, then a numerical coefficient will be introduced into the equation. For instance, on a cubic (1 0 0) surface, we might expect that atoms would jump in the x or y directions, but no other direction. This leads to a factor of 2 on the right side of Equation 4.24. The jumps that the diffusing atom takes need not always be to the adjacent atom position. Longer jumps are allowed and are observed.

Figure 4.17 shows a representation of the diffusion of an adatom on a cubic (1 0 0) surface. Atoms vibrate in a lattice position without diffusing and diffusion is limited to the [1 0 0] type lattice directions. While most diffusion steps are single lattice spacings, longer steps are also observed.

As we have seen the diffusion coefficient depends on the specific mechanism of diffusion. Some examples of types of surface diffusion are demonstrated in Figure 4.18. Figure 4.18a shows the diffusion of an atom on top of a surface. Figure 4.18b shows a process

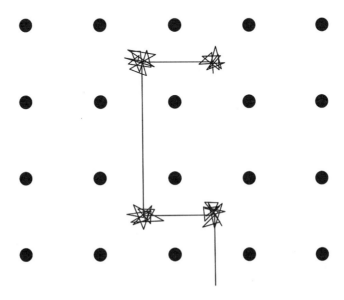

FIGURE 4.17
Representation of surface diffusion of an adatom on a (1 0 0) surface.

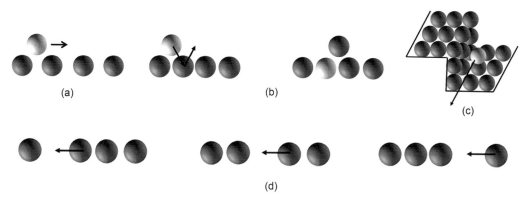

FIGURE 4.18
Surface diffusion mechanisms. (a) On-top diffusion. (b) Exchange diffusion. (c) Step diffusion. (d) Vacancy diffusion.

where an atom on top of the surface can diffuse by moving into the surface and pushing a surface atom up onto the next lattice position. We know that real surfaces have steps and terraces and Figure 4.18c shows diffusion along a step. Figure 4.18d shows the diffusion of a vacancy through the top layer of atoms. Other possibilities exist and which mechanism may dominate often depends on the surface temperature. Different crystalline surface planes will also exhibit different diffusion properties.

4.3.3 Surface Roughness

Surface roughness was briefly introduced in Section 3.8. Roughness implies the existence of multiple layers exposed at the surface. In addition to the surface diffusion processes that we have discussed, which were assumed to be along a flat surface on a single layer (intralayer diffusion), the existence of surfaces of multiple layers separated by steps opens up the possibility of interlayer diffusion as well. Diffusion across a step from one layer to another will typically have a higher energy barrier than surface diffusion within a single layer (Schwoebel and Shipsey 1966). This extra energy barrier is referred to as the Ehrlich-Schwoebel barrier, E_{ES}. The relevant energies are shown in Figure 4.19.

An atom approaching a step from the upper layer experiences a barrier energy that is the sum of the surface diffusion energy, E_D, and the Ehrlich-Schwoebel barrier energy. An atom approaching from below a step, however, experiences a very large barrier consisting of the sum of the surface diffusion energy, the Ehrlich–Schwoebel barrier energy and a large barrier energy, E_S, arising from a large number of nearest neighbors available at the bottom of the step. This large number of nearest neighbors will make this location a particularly stable place for an atom to remain. The difference in these energy barriers means that, given enough energy to overcome the Ehrlich-Schwoebel energy barrier, atoms will tend to preferentially diffuse from higher layers to lower layers tending to smooth out roughness. A common approach to smoothing roughness is annealing a film at high temperatures to encourage interlayer diffusion.

The process of smoothing a surface can be discussed in a macroscopic continuum sense by considering the chemical potential dependence on surface shape. We will discuss chemical potential more in the next chapter where thermodynamics will be the primary focus, but we note here that the chemical potential of a convex surface (hill) is higher than

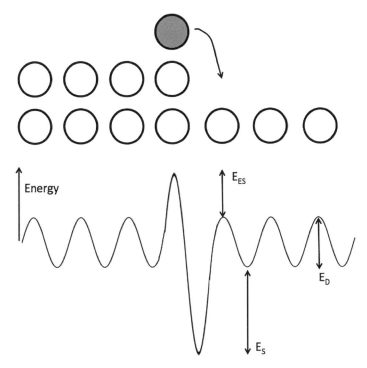

FIGURE 4.19
Interlayer diffusion across a step with relevant energies.

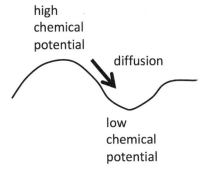

FIGURE 4.20
Chemical potential difference drives diffusion from convex to concave regions on the surface.

the chemical potential of a concave surface (valley). As a result, there is a thermodynamic driving force to move mass from hills to valleys resulting in a smoothing of the surface as suggested in Figure 4.20. A more detailed analysis of this process shows that roughness on a short length scale should be able to be smoothed by heating for relatively short times, but large length scale roughness is very difficult to smooth in reasonable times.

The roughness of a surface can also be impacted by diffusion in the surface. For instance, kinks and corners along steps can change the diffusion along the step in ways very similar to the Ehrlich-Schwoebel barriers for interlayer diffusion. This results in a change in the structure of steps on the surface.

When thin films are grown on a substrate, the surface of the film changes in time during the growth process. This kinetic process leads to a variety of kinetic surface roughening mechanisms. Generally, the roughness of a growing surface is found to be greater than on a static equilibrium surface. In Chapter 3, a very simple roughness model for growing thin films was presented. Kinetic roughening models typically examine scaling properties in both space and time to model the surface of the growing film.

When growing thin films, the mismatch between the lattice parameters of the substrate and the film can lead to strain in the film. This can cause roughening of the film surface. If surface diffusion is very limited and the film is uniformly strained, the film may grow with the substrate lattice parameters until it reaches a critical thickness at which dislocations are introduced into the film to allow it to grow further with the lattice parameters typical of the film. If surface diffusion is significant, then the film may not grow uniformly but rather can form islands of film that effectively roughen the surface of the growing film. We will explore the growth processes of thin films in more detail later in this book.

4.4 Characterization of Atomic Motion

A wide variety of techniques can be used to characterize atomic motion. Imaging techniques such as scanning electron microscopy or scanning probe microscopies are discussed in Chapter 11. Electron diffraction techniques (discussed in Chapter 12) can detect atomic vibrations and surface melting. Phonons can be detected using a variety of techniques including Brillouin light scattering, electron energy loss spectroscopy, and others described in Chapter 13. Surface roughness can be measured using scanning probe microscopies (Chapter 11), or several structural techniques discussed in Chapter 12.

References

Brophy, J.H., R.M. Rose and J. Wulff. 1964. *The Structure and Properties of Materials Volume II: Thermodynamics of Structure*. New York: John Wiley & Sons, Inc.
Ibach, H. 2006. *Physics of Surfaces and Interfaces*. Berlin: Springer.
Kittel, C. 1996. *Introduction to Solid State Physics*. 7th ed. New York: John Wiley & Sons, Inc.
Kojima, K. and M. Susa. 2002. Surface melting of copper with (100), (110) and (111) orientations in terms of molecular dynamics simulation. *High Temp. High Pressur.* 34: 639–648.
Luth, H. 2015. *Solid Surfaces, Interfaces and Thin Films*. 6th ed. Berlin: Springer.
Schwoebel, R.L. and E.J. Shipsey. 1966. Step motion on crystal surfaces. *J. Appl. Phys.* 37: 3682–3686.

Problems

Problem 4.1 We will derive the dispersion relation for a simple one-dimensional system with a one-atom basis assuming nearest-neighbor interactions only. We assume a simple mass and spring model with spring constant C. The position of the nth atom is x_n and the $(n+1)^{st}$ atom is at x_{n+1} etc. The equilibrium distance between atoms is a.

a. Find the total force on the nth atom, which is just the vector sum of the forces on both sides of the atom.

b. Use Newton's 2nd law to show that the motion of the atoms is described by

$$m\frac{d^2x_n}{dt^2} = -C(2x_n - x_{n-1} - x_{n+1}).$$

c. Assume a traveling wave solution of the form $x_n(t) = Ae^{i(kna-\omega t)}$. By direct substitution into the differential equation, derive the dispersion relation (Equation 4.3).

Problem 4.2 We are often interested in the long wavelength limit where $\lambda \gg a$, which is equivalent to ka being small. In the following calculations, it may be helpful to recall the Taylor's series expansions for small q: $\sqrt{1-q} \approx 1 - \frac{1}{2}q$, $\sin(q) \approx q$, and $\cos(q) \approx 1 - \frac{1}{2}q^2$.

a. Show that Equation 4.3 reduces to $\omega = \sqrt{\frac{ca^2}{m}}k$.

b. Show that Equation 4.5 reduces to $\omega = \sqrt{\frac{ca^2}{2(m_1 + m_2)}}k$.

Problem 4.3 The model of a one-dimensional infinite chain discussed in Problems 4.1 and 4.2 can be extended to consider surface vibrations by making the chain semi-infinite. The surface being defined as the end of the chain which terminates. Far from this surface, we expect the solutions for the infinite chain to still be valid (bulk waves). Near the surface we seek solutions that have amplitudes which decay exponentially as you move into the solid (surface waves). One way to model this is to assume that the wave vector has an imaginary part, $k=k_R+ik_I$. We examine the behavior of Equation 4.3 using the form with the $cos(ka)$ with k complex.

a. Using a complex k, show that the $cos(ka)$ term can be written as $\cos(k_Ra)\cosh(k_Ia) - i\sin(k_Ra)\sinh(k_Ia)$. The trigonometric identities $\cos(a+b)=\cos(a)\cos(b)-\sin(a)\sin(b)$ and $\cos(ix)=\cosh(x)$, $\sin(ix)=i\sinh(x)$ will be helpful.

b. While k is complex, ω must be real. This requires the imaginary part of $cos(ka)$ to be zero. Since we seek surface wave solutions, we are not allowed to set $k_I=0$ and so this requirement restricts the values of k_R. Determine the conditions on k_R for surface waves to exist. Which conditions correspond to solutions in the first Brillouin Zone?

c. The two solutions in the first Brillouin Zone yield $\omega^2 = \frac{2C}{m}[1-\cosh(k_Ia)]$ for the $k_R=0$ case and $\omega^2 = \frac{2C}{m}[1+\cosh(k_Ia)]$ for the $k_R=\pi/a$ case. Our requirement that ω is real, however, excludes the $k_R=0$ solution for this one-atom basis. (The solution does exist for the two-atom basis.) The $k_R=\pi/a$ solution exists at the boundary of the first Brillouin Zone. Determine any upper or lower limits on ω for this solution and note where these surface wave solutions would lie on Figure 4.5.

Problem 4.4 The solution of Fick's 2nd law (Equation 4.9) is mathematically messy for even simple boundary conditions. It can be shown that releasing a mass per cross-sectional area of material M at $x=0$ and $t=0$ in an infinite solid, will result in a concentration profile $C(x,t) = \dfrac{M}{\sqrt{4\pi Dt}} e^{-x^2/4Dt}$. As expected the concentration at $x=0$ goes down as time goes on and the concentration spreads out in both $\pm x$ directions.

a. Determine the time as a function of x and D at which the concentration will be a maximum for any given position in the sample.

b. What is the value of this maximum concentration (in terms of M and x)?

Problem 4.5 Consider a surface at $x=0$ and a semi-infinite solid extending to $x=+\infty$. Imagine that we put an infinite supply of material at $x=0$. Such that the boundary condition at $t=0$, is $C(0,0)=C_0$. We also assume that the diffusing material never reaches $x=\infty$ and so $C(\infty, t)=0$. These two boundary conditions allow us to solve the 2nd order differential equation by using a superposition of the solutions described in Problem 4.4. The final solution can be written in the form $C(x, t)=C_0$

$\left[1 - \mathrm{erf}\left(\dfrac{x}{\sqrt{4Dt}}\right)\right]$ where the error function is a standard function with the form

$\mathrm{erf}(x) = \dfrac{2}{\sqrt{\pi}} \displaystyle\int_0^x e^{-q^2} dq$. Note that the error function depends only on the upper limit of the integral. The error function varies from $\mathrm{erf}(0)=0$ to $\mathrm{erf}(\infty)=1$.

a. Suppose we deposit a thick film of material A on a surface of material B such that we can treat it as an infinite source for the time periods of interest. Assume the diffusion coefficient for material A in B is $D=10^{-11}\,\mathrm{cm^2/s}$ at the temperature used for the diffusion and the initial concentration of material A is 10^{20} atoms/$\mathrm{cm^3}$. What will be the concentration of material A at a depth of 1 µm after 2 hours?

b. At what depth will the concentration be 10^{18} atoms/$\mathrm{cm^3}$?

c. What total time would we need to allow diffusion to last to reach twice the the concentration $(2\times10^{18}$ atoms/$\mathrm{cm^3})$ at the depth in part (b)?

Problem 4.6 Diffusion is an activated process. We described the diffusion coefficient using an exponential term involving a diffusion energy. Below is approximate data for diffusion in Ag. From the graph, determine the diffusion energy for the cases of volume diffusion, grain boundary diffusion, and surface diffusion. Notice that the horizontal axis has been flipped so the small values of $1/T$ (higher T) is on the right.

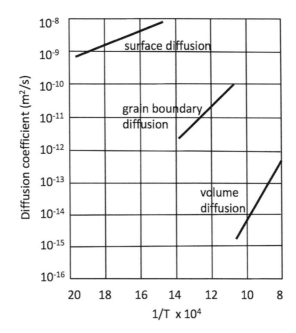

Problem 4.7 For a particular diffusing system at 500 K we find that the diffusion coefficient $D=6\times10^{-11}\,m^2/s$. Assume that the diffusion energy $E_D=1$ eV/atom and is constant. What is the diffusion coefficient at 600 K?

Problem 4.8 Suppose that we deposit a thick film of Cu on Si. We diffuse the Cu into the Si by heating the sample for 8 hours at 600 K where the diffusion coefficient is $8\times10^{-11}\,m^2/s$. How much time would it take for diffusion at 500 K where the diffusion coefficient is $9\times10^{-12}\,m^2/s$ to produce the same diffusion profile? (If you assume a random walk, you can avoid nasty calculations.)

Problem 4.9 Many films are polycrystalline and so diffusion through the film will be a combination of volume (D_V) and grain boundary (D_{GB}) diffusion. Consider a simple case of cubic grains which are as high as the film is thick. What grain width (d) would be needed to have equal amounts of material transported by grain boundary and volume diffusion? Derive an expression for d in terms of D_V, D_{GB}, and the width (b) of a grain boundary. Then calculate a value for Ag diffusing in Ag at 800 K assuming the grain boundaries are $b=0.5\,nm$ wide and $D_{GB}=2.5\times10^{-13}\,cm^2/s$ and $D_{vol}=1.8\times10^{-14}\,cm^2/s$.

5

Thermodynamics

Equilibrium thermodynamics is a powerful tool for understanding processes involving surfaces and thin films. It is important to recall, however, that while thermodynamics provides restrictions on what is allowed to happen, it does not address how quickly it will happen. We begin by examining the thermodynamics of solids, which is important in understanding thin films. We then see how this is modified in the presence of a surface. A deeper discussion of these topics is available in the literature (Blakely 1973, Brophy, Rose, and Wulff 1964, Herring 1951, Hudson 1998, Ibach 2006, Meier 2014, Ohring 2002, Smith, Van Ness, and Abbott 2005, Somorjai 2010, Tsao 1993, Zangwill 1988). The thermodynamics of atoms adsorbed on surfaces will be discussed in Chapter 7 as we transition from surface science to thin film science.

We examine the relationships among several quantities:

Extensive variables:

U = internal energy

G = Gibbs free energy

H = enthalpy

S = entropy

V = volume

N = number of particles

Intensive variables:

T = temperature

P = pressure

μ = chemical potential

Intensive variables do not scale when we combine systems while extensive variables do scale. Consider combining two systems each of which has volume V and temperature T. The resulting volume is $2V$, but the resulting temperature is still T. Mathematically, we can define some function $f(x)$. If we scale the variable x by a factor λ, $f(\lambda x) = \lambda^m f(x)$ where $m = 0$ for intensive variables and $m = 1$ for extensive variables.

DOI: 10.1201/9780429194542-6

5.1 Thermodynamics in Solids

We start with equilibrium thermodynamics of a one-component, bulk, system where the internal energy can be described as a function of S, V, and N.

$$U = U(S, V, N) \tag{5.1}$$

From the definition of an exact differential involving partial derivatives of the internal energy, we define the intensive variables temperature, T, pressure, P, and chemical potential, μ, of the system.

$$dU = \left(\frac{\partial U}{\partial S}\right)_{V,N} dS + \left(\frac{\partial U}{\partial V}\right)_{N,S} dV + \left(\frac{\partial U}{\partial N}\right)_{S,V} dN \tag{5.2}$$

$$dU = TdS - PdV + \mu dN \tag{5.3}$$

where we have defined:

$$T = \left(\frac{\partial U}{\partial S}\right)_{V,N} \tag{5.4}$$

$$P = -\left(\frac{\partial U}{\partial V}\right)_{N,S} \tag{5.5}$$

$$\mu = \left(\frac{\partial U}{\partial N}\right)_{S,V} \tag{5.6}$$

The subscripts indicate which variables are being held constant.

Using $U(S, V, N)$ and the properties of extensive and intensive variables, we can express the Euler equation:

$$U = TS - PV + \mu N \tag{5.7}$$

Differentiating with intensive variables T, P, and μ as our independent variables, and using Equation 5.3 to eliminate the dU term, yields the Gibbs–Duhem equation:

$$0 = SdT - VdP + Nd\mu \tag{5.8}$$

If we choose T, P, and N as our independent variables (often appropriate in surface and thin film science), we define the Gibbs free energy.

$$G(T, P, N) = U + PV - TS = H - TS \tag{5.9}$$

and

$$dG = VdP - SdT + \mu dN \qquad (5.10)$$

where again we have used Equation 5.3 to eliminate the dU term.

Note that when operating under conditions of constant temperature ($dT = 0$) and pressure ($dP = 0$), which are common in many experiments, this expression simplifies to $dG = \mu dN$ which identifies the Gibbs free energy as the chemical potential per particle. We will return to these arguments in Section 5.3 when we add surface terms in order to explore the thermodynamics of surfaces.

For many applications, the change in the Gibbs free energy, ΔG, and how it is related to the other variables is of particular interest. We relate ΔG to enthalpy, entropy, and temperature.

$$\Delta G = G_{\text{final}} - G_{\text{initial}} = \Delta H - T\Delta S \qquad (5.11)$$

The sign of ΔG provides us with restrictions on which processes are allowed or forbidden:

- $\Delta G < 0$ process is allowed
- $\Delta G > 0$ process is forbidden
- $\Delta G = 0$ equilibrium

These inequalities are equivalent to the statement that nature will minimize the free energy of a system. For constant T, P, and N, the total entropy change of a system in contact with the environment can be shown to be

$$dS_{\text{total}} = -\frac{1}{T} dG_{\text{system}} \qquad (5.12)$$

The entropy is increased (consistent with the Second Law of Thermodynamics) be reducing the Gibbs free energy of a system.

In this book, we focus on the Gibbs free energy which is sufficient to understand a wide range of surface and thin film thermodynamics. The Helmholtz free energy, $F = U - TS$, which is the energy that is provided as work to create a system from nothing, is often used to mathematically describe the thermodynamics of interfaces.

A common application of these ideas is in chemical reactions where we consider a reversible reaction involving the chemical species A, B, C, and D.

$$A + B \leftrightarrow C + D$$

We relate the free energy change for this reaction to the concentration, C_i, of each chemical species or to the partial pressure of each chemical species if in the gas phase.

It can be shown:

$$\Delta G = \Delta G^0 + k_B T \, \ln \left[\frac{C_C C_D}{C_A C_B} \right] \qquad (5.13)$$

ΔG^0 is determined in the standard state (gases at pressures of 10^5 Pa (=1 bar = 0.987 atm) and aqueous solution concentrations at 1 M). In many laboratory situations, $\Delta G \approx \Delta G^0$. Frequently standard state data is reported at approximately room temperature.

A useful plot arising from Equation 5.13 is the Ellingham diagram of ΔG^0 vs. temperature. Figure 5.1 shows such a diagram for the formation of several oxides. The chemical reactions have all been normalized to having one unit of O_2. This allows direct comparison of reactions to determine that oxides which are lower on the diagram are more stable.

Since the partial pressure of oxygen at which a solid metal and its solid oxide coexist can also be determined from Equation 5.13, the dissociation pressure of an oxide can be measured directly from an Ellingham diagram using

$$P_{\text{dissoc}} = P_0 e^{\Delta G_{\text{oxide}}/k_B T} \tag{5.14}$$

where P_0 is the pressure in the standard state in the desired units.

We see that the oxidation of Al to Al_2O_3 involves a more negative change in free energy than the oxidation of Si to SiO_2. Thermodynamics tells us that if Al and Al_2O_3 are in contact with Si and SiO_2, the oxygen will prefer to move from the SiO_2 to the Al to form Al_2O_3 lowering the free energy of the system. We expect then that the SiO_2 will be reduced to Si, and the Al will be oxidized to Al_2O_3. Recall that no information about the rate at which this will happen is provided by these thermodynamic arguments.

Similarly, if we expose a Ni-Al alloy to oxygen, the Al will preferentially oxidize in order to minimize the energy of the system. As a result, the alloy may have an Al_2O_3 film on top when exposed to oxygen. This will draw some of the Al out of the upper region of the alloy

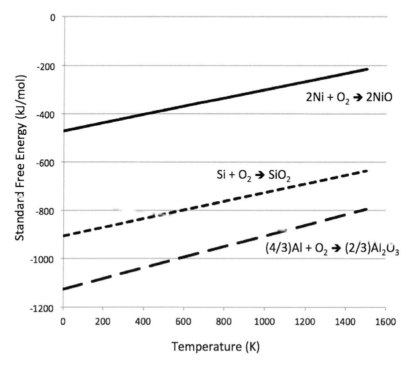

FIGURE 5.1
Ellingham diagram for the oxidation of Ni, Si, and Al.

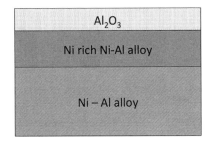

FIGURE 5.2
Preferential oxidation of Al in a Ni-Al alloy.

leaving a nickel-rich layer of alloy below the surface before reaching bulk alloy concentrations deeper in the material. This is demonstrated in Figure 5.2.

Example: Consider the deposition of a thin Ni film onto a SiO_2 surface at a temperature of 600 K. Will the Ni film reduce the SiO_2 and form NiO? We can read the values for ΔG^0 from Figure 5.1.

$$2Ni + O_2 \Leftrightarrow 2NiO \quad \Delta G^0{}_{NiO}(600\,K) = -380\,kJ/mol$$

$$Si + O_2 \Leftrightarrow SiO_2 \quad \Delta G^0{}_{SiO_2}(600\,K) = -800\,kJ/mol$$

We can eliminate O_2 from these reactions by reversing one reaction and then combining them.

$$2Ni + O_2 + SiO_2 \Leftrightarrow 2NiO + Si + O_2$$

$$2Ni + SiO_2 \Leftrightarrow 2NiO + Si$$

The free energy change for this reaction at 600 K is determined by subtracting the SiO_2 energy from the NiO energy (since we reversed the SiO_2 reaction).

$$\Delta G^0(600\,K) = -380 - (-800) = +420\,kJ/mol$$

Since this is a positive value, the reaction of Ni reducing SiO_2 is not allowed and the Ni will remain in a metallic state.

5.2 Phase Diagrams

Another useful application of thermodynamics is understanding different phases and the transitions between them using phase diagrams arising from minimizing the Gibbs free energy for each phase and seeing which is favored under various conditions of pressure, temperature, and composition.

5.2.1 One-Component Systems

Consider first a simple one-component system where the only variables are pressure and temperature. Figure 5.3 shows schematically part of the phase diagram of carbon.

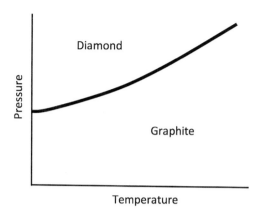

FIGURE 5.3
One-component phase diagram of carbon.

Two phases (diamond and graphite) are observed with diamond being favored at higher pressures. The two phases are equally favorable energetically on the line that separates them and so both phases coexist on that line.

Another example, Fe, is shown in Figure 5.4. Three different solid phases are shown along with the liquid and vapor phases. Three phases can coexist at very well-defined points known as the triple points. The body-centered cubic α phase of Fe is stable at room temperature and atmospheric pressure and over a considerable range of pressures and temperatures around that point.

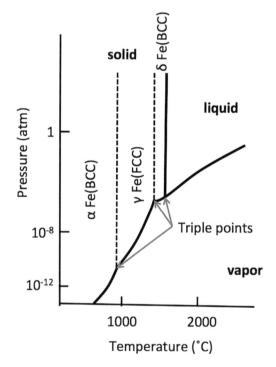

FIGURE 5.4
One-component phase diagram of iron.

5.2.2 Two-Component Systems

Two-component systems, such as GaAs, NiCr, or WSi, have three-dimensional phase diagrams since pressure, temperature, and composition are all variables. Since many experiments are done at atmospheric pressure, we consider here the two-dimensional phase diagrams of temperature and composition at a fixed pressure of 1 atm. Several common types of two-component phase diagrams are observed and we examine each in sequence.

5.2.3 Binary Solid Solutions

Binary solid solutions have both elements completely soluble in liquid and solid phases at all compositions. An example, SiGe, is shown in Figure 5.5. Note how the liquid and solid phases are separated by a region where the two phases coexist. This region is bounded on top by a curve known as the liquidus and on the bottom by a curve known as the solidus. To demonstrate the use of the phase diagram considers points at several temperatures with a composition C_0. At the highest temperature (point I in Figure 5.5), we have a pure liquid phase with a composition C_0 that is about 30% Si and 70% Ge. At point II, we are on the liquidus separating the two-phase region (liquid and solid) from the liquid region. Both liquid and solid phases coexist at this point. The composition of the liquid phase is still C_0. The composition of the solid phase can be obtained by projecting a horizontal line from point II to where it intersects the solidus and then reading the composition at this point (labeled $C_S(II)$ in the figure). At point III, we are in the middle of the region where the liquid and solid phases coexist. The compositions of each phase are obtained by considering a horizontal line through point III. Where it intersects the liquidus, we can read off the composition of the liquid phase $C_L(III)$ and where it intersects the solidus, we can read off the composition of the solid phase $C_S(III)$. At lower temperatures, point IV is in the region where only the solid phase exists with a composition of C_0.

The amount of each phase present at a point on the phase diagram can also be determined. The atomic (or mole) fraction of material in the solid (f_{solid}) and liquid (f_{liquid}) is given by

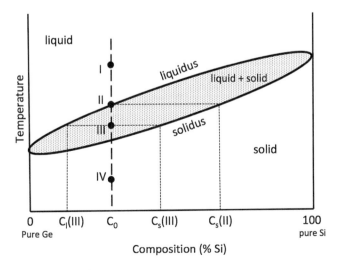

FIGURE 5.5
Binary solid solution phase diagram of Ge-Si alloy.

$$f_{\text{solid}} = \frac{C_0 - C_L}{C_S - C_L} \qquad (5.15)$$

$$f_{\text{liquid}} = \frac{C_S - C_0}{C_S - C_L} \qquad (5.16)$$

In order to form a binary solid solution, the elemental components must have the same crystal structure. Some other examples are Cu-Ni, Pt-Rh, and NiO-MgO.

5.2.4 Binary Eutectics

Components that have only limited solubility in the solid state may form phases described by a binary eutectic phase diagram. An example, Pb-Sn, is shown in Figure 5.6. In this case, we have two solid phases, labeled α and β. Note the eutectic point where the liquid, α and β phases all coexist. In this example, the α phase is an fcc substitutional solid of Sn in Pb and the β phase is a tetragonal substitutional solid of Pb in Sn. Note also that adding a little Sn to pure Pb or Pb to pure Sn will lower the melting point from the pure material value.

We again explore the phases and compositions found at several points on the phase diagram. At point I, we are in the liquid phase with a composition of C_0 that is about 35% Sn and 65% Pb. At point II, the liquid can now coexist with the α solid phase. The liquid composition is still C_0. The α solid composition is obtained by drawing a horizontal line and looking at where it intersects the boundary with the pure α solid phase region. The composition at this point is $C_\alpha(\text{II})$. There is no β solid. At point III, we are into the region where the liquid and α solid coexist. The solid composition is again obtained from a horizontal line intersecting the boundary with the α solid region. This yields a composition of $C_\alpha(\text{III})$. The liquid composition is determined by the same horizontal line intersecting the boundary with the liquid region. The composition at that point is $C_L(\text{III})$. There is no β

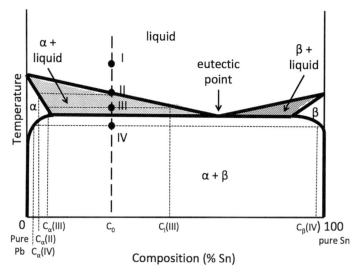

FIGURE 5.6
Binary eutectic phase diagram of Pb-Sn alloy.

solid under these conditions. Point IV is into the region where the two solid phases α and β coexist. We obtain the composition of the α phase by projecting a horizontal line to the intersection with the α phase region yielding a composition $C_\alpha(IV)$. The intersection of this line with the β region provides the composition of the β phase as $C_\beta(IV)$.

In the mixed phase regions, the atomic fraction of each phase can be determined using the same equations as for the binary solid solution, Equations 5.15 and 5.16. Bi-Cd, Al-Si, and CaO-MgO are all examples of other systems having binary eutectic phase diagrams.

5.2.5 Other Considerations

The phase diagrams presented here are equilibrium phase diagrams. Thin films, however, are often grown under non-equilibrium conditions. Often the temperature is too low or the deposition rate is too high to obtain equilibrium thermodynamics. As a result, other structures may be observed. Thermodynamics indicates that these meta-stable structures will convert to stable equilibrium structures to minimize energy but the time for this to happen may be very long.

Although many experiments are conducted at atmospheric pressure, films may be grown in a vacuum environment where the pressure is far from atmospheric pressure. Under these conditions, the phase diagram may be very different from the atmospheric pressure phase diagram. Figure 5.7 shows an example of several pressures for the Si-Ge system.

5.3 Surface Thermodynamics

5.3.1 Surface Energy, Tension, and Stress

The thermodynamics of surfaces is based on standard thermodynamics as discussed in the first sections of this chapter but assigning separate contributions to thermodynamic quantities from additional surface terms. The surface free energy, g_S, is the excess energy per unit area associated with the existence of a surface. This term is positive, so creating surfaces always comes with an energy cost. The surface tension, γ, is the reversible work done in creating a unit area of a new surface at constant T, V, and N. The units are often expressed as force/length, which is the same as energy/area. As we will see soon, the surface free energy and surface tension are often equal. The energy and tension are scalars that depend on the crystal plane under consideration. The surface stress, σ_{ij}, is the work done per unit area in reversibly deforming a surface. It is a second-rank tensor that depends on the crystal plane and the direction along that plane.

We can think of the surface stress by considering the simple surface reconstructions discussed in Chapter 3. Surface atoms are missing the nearest-neighbor atoms directly above the surface, which can lead to a change in the surface lattice spacing. The forces experienced by the surface can be described as a surface stress that will be positive (tensile stress) if it minimizes the surface and negative (compressive stress) if it increases the surface. This change in the bonding environment can also lead to a surface energy associated with these surface atoms.

To represent a surface, we add a surface energy term to the First Law of Thermodynamics (Equation 5.7).

$$U = TS - PV + \mu N + \gamma A \qquad (5.17)$$

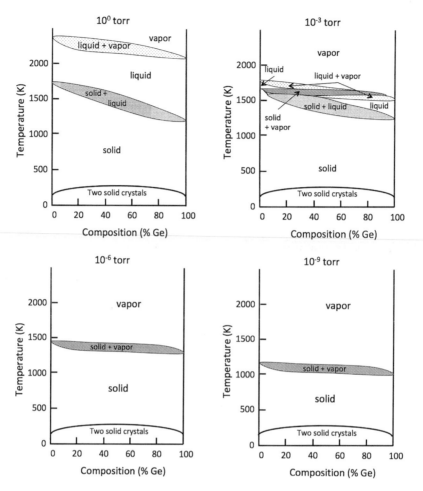

FIGURE 5.7
Pressure dependence of phase diagram for Si-Ge alloy.

where A is the area of the surface created.

Experimentally, we typically choose temperature, pressure, and area as our independent variables leading to a Gibbs free energy change, $dG(P, T, A)$, defined by

$$dG = -SdT + VdP + \gamma dA \qquad (5.18)$$

We often operate under conditions of constant T and P such as might be found on a table-top. Under these conditions, the differentials dT and dP are zero and our expression simplifies to

$$dG_{T,P} = \gamma dA \qquad (5.19)$$

Equivalently, we could define the Gibbs free energy of a system as coming from the Gibbs free energy of the individual atoms in the system, G_0, and a term from the existence of the surface, Ag_S

$$G = NG_0 + Ag_S \qquad (5.20)$$

where N is the number of atoms in the material and A is the area of the surface. If we create a new surface only by moving existing atoms without changing their properties, then $dG = 0 + d(Ag_S)$ and, for our specific case,

$$dG_{T,P} = d\left(g_S A\right) \qquad (5.21)$$

There are two ways to get more surface area using existing atoms. One is to move atoms to the surface around the edges of the surface. If we do this,

$$d\left(g_S A\right) = g_S dA \qquad (5.22)$$

since we are only changing the area without changing any properties of the atoms themselves. Comparing the result to Equation 5.19, we see that

$$\gamma dA = g_S dA \qquad (5.23)$$

which allows us to identify $\gamma = g_S$ under these conditions for a one-component system.

A second way to increase the surface area is to keep the number of surface atoms fixed but stretch them out to increase the surface area. The surface will now be strained, and we need stress terms in our differential expression. In a liquid, the stress terms are isotropic, and the surface stress can be shown to be equal to the surface tension, $\sigma = \gamma$. In a solid, the tensor nature of the stress typically needs to be considered. If the surface stress is isotropic, it can be related to the surface tension.

$$\sigma = \gamma + A \frac{d\gamma}{dA} \qquad (5.24)$$

for a solid with isotropic stress.

The temperature behavior of the surface tension can be shown to depend on the surface entropy per unit area, s_S.

$$\left(\frac{\partial \gamma}{\partial T} \right)_{V,A} = -s_S \qquad (5.25)$$

Since the surface entropy is typically positive, the surface tension is expected to decrease with increasing temperature.

5.3.2 Minimizing Energy

In equilibrium, a solid crystal will try to minimize the total energy. Since the contribution from the surface energy is positive, the crystal will minimize the surface energy. γ is found to depend on which crystal face is exposed at the surface. This suggests that certain crystal surfaces may be unstable and are expected to break into other surfaces in order to lower the total energy. This may actually increase the area of the surface even as it minimizes the total energy.

We explored this situation in Chapter 3 where Figure 3.23 showed a hard sphere model for a cubic crystal cut at an angle to form a (5 0 1) surface. Consider a more general case cut at some angle θ. A simple model of surface tension examines the number of bonds broken in forming this surface. If we examine the one-dimensional case where the spacing between atoms is *a*, we can count the number of vertical and horizontal bonds broken per unit length.

$$n_{\text{vertical}} = \frac{1}{a}.$$ (5.26a)

$$n_{\text{horizontal}} = \frac{\tan\theta}{a}$$ (5.26b)

and find the total number of broken bonds per unit length

$$n_{\text{total}} = \frac{1}{a}(1+\tan\theta)$$ (5.27)

The total number of bonds broken is then

$$N_{\text{total}} = n_{\text{total}}a = (1+\tan\theta)$$ (5.28)

and the length along the surface is $a/\cos\theta$. We can then describe the surface tension in this one-dimensional example in terms of the bond energy, E_b.

$$\gamma(\theta) = \frac{1}{2}\frac{(1+\tan\theta)E_b}{a/\cos\theta}$$ (5.29)

where the factor of ½ arises since two surfaces were created. Some mathematical manipulation yields:

$$\gamma(\theta) = \frac{E_b}{a\sqrt{2}}\cos\left(\theta-\frac{\pi}{4}\right) \text{ where } 0<\theta<(\pi/2)$$ (5.30)

which is the equation of a circle passing through the origin and centered in the first quadrant. Symmetry requires additional circles in each of the other three quadrants resulting in Figure 5.8, which is a simple example of a Wulff plot. A Wulff plot can be used to approximate the

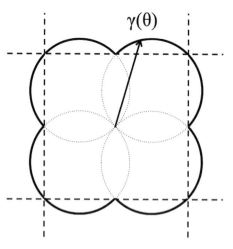

FIGURE 5.8
Wulff plot demonstrating the equilibrium shape of a cubic crystal based on minimization of the total energy at $T=0$ K.

equilibrium shape of a crystal at $T = 0$. If a crystal is cut on a plane that is not a low-energy plane, it is predicted to facet to expose planes that will minimize the surface energy. Figure 5.8 shows a typical polar plot used to explore the faceting of a cubic crystal. The outer curved figure (solid line) is a plot of the magnitude of the surface tension in that direction $\gamma(\theta)$ for each of the quadrants. Points that are closest to the center represent lower energy configurations. For this case, the cusps in the vertical and horizontal directions represent the lowest energy directions. A crystal cut along any other direction will break up into terraces and steps exposing these two surfaces in order to minimize energy. In this case, only two crystal orientations, at right angles to one another are predicted to be observed (dashed lines). This was displayed in Figure 3.23b where only horizontal and vertical surfaces are exposed. If a more complicated model of bonding is used, which includes more than just nearest-neighbor bonds, the resultant Wulff plot will have more structure to it with more cusps.

At finite temperatures, there will also be a thermodynamic contribution from entropy that will be maximized. This will create more disorder on the surface by roughening. The competition between minimizing energy and maximizing entropy results in a phase transition above a roughening temperature, T_r. Above this temperature, the surface will exhibit much longer scale variations in height (hills and valleys) rather than short-scale, random variations as demonstrated in Figure 5.9. On the Wulff plot, this would be indicated by a rounding of the cusp-like features leading to a reduction in the size of the facets.

5.3.3 Segregation in Two-Component Systems

In a two-component system, the solid is made up of materials α and β. These might be two elements in an alloy or could be impurities in a one-element solid. The Euler equation (Equation 5.7) now needs terms with the chemical potentials of each component and the number of particles of each component.

$$U = TS - PV + \mu_a N_a + \mu_b N_b + \gamma A \tag{5.31}$$

Solving for γ'

$$\gamma = (1/A)[U - TS + PV - \mu_a N_a - \mu_b N_b] \tag{5.32}$$

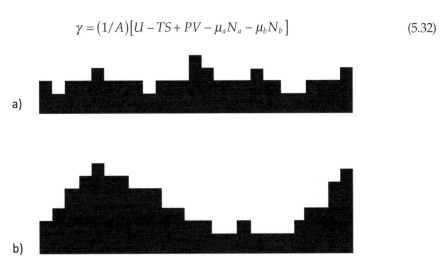

FIGURE 5.9
Schematic representation of surface at (a) $T < T_r$ showing random height variations and (b) $T > T_r$ showing long-scale height variations.

Differentiating this expression with T, P, and μ as the independent variables,

$$dy = (1/A)\left[-SdT + VdP - N_a d\mu_a - N_b d\mu_b\right] \tag{5.33}$$

The energy of the system can be minimized by having a surface composition of the two components that differs from the bulk composition. This difference, known as segregation, is commonly observed.

We examine the conditions under which the β component will segregate to the surface at constant pressure by considering how γ changes when T or μ change.

$$d\gamma = \frac{\partial\gamma}{\partial T}dT + \frac{\partial\gamma}{\partial\mu_\beta}d\mu_\beta \tag{5.34}$$

We define a surface excess of β

$$\Gamma_\beta = -\frac{\partial\gamma}{\partial\mu_\beta} \tag{5.35}$$

If we restrict our consideration to the common experimental conditions of constant T and β being dilute in the binary alloy, it is found that

$$\mu_\beta = \mu_{\beta0} + k_B T \ln\left(a_\beta C_\beta\right) \tag{5.36}$$

where C_β is the concentration of component β, a_β is a constant when β is dilute and $\mu_{\beta0}$ is the chemical potential of the β component in the standard state. Using this to solve for the surface excess we find

$$\Gamma_\beta = -\frac{1}{k_B T}\left(\frac{\partial\gamma}{\partial\ln C_\beta}\right)_T \tag{5.37}$$

If the surface energy decreases with the increasing concentration of component β, then the surface excess will be positive and β will segregate to the surface. Note that it is not just the lower surface energy material that segregates. It depends on the change in surface energy with increased concentration. Segregation will depend on the temperature, the crystallographic plane of the surface and the presence of any defects at the surface as well.

5.4 Characterization of Thermodynamics

Characterization of thermodynamic properties of surfaces is often connected with other characterization issues. The presence of different phases may best be determined by looking at structural information from diffraction techniques such as grazing incidence X-ray diffraction, low-energy electron diffraction, or reflection high-energy electron diffraction. These structural techniques are discussed in Chapter 12. Oxidation and reduction processes as well as segregation to the surface of one component in a multi-component system may be examined using surface-sensitive chemical characterization techniques. Auger electron diffraction, X-ray photoelectron spectroscopy, and secondary ion mass spectrometry, for example, might all be useful in studying surface segregation or oxidation and reduction. Chemical characterization is described in Chapter 13. Direct measurement of thermodynamic properties such as surface tension is possible and is discussed in Chapter 14.

References

Blakely, J.M. 1973. *Introduction to the Properties of Crystal Surfaces*. Oxford: Pergamon Press.

Brophy, J.H., R.M. Rose and J. Wulff. 1964. *The Structure and Properties of Materials: Volume II Thermodynamics of Structure*. New York: John Wiley & Sons, Inc.

Herring, C. 1951. Some theorems on free energies of crystal surfaces. *Phys. Rev.* 82: 87–93.

Hudson, J. 1998. *Surface Science: An Introduction*. New York: John Wiley & Sons, Inc.

Ibach, H. 2006. *Physics of Surfaces and Interfaces*. Berlin: Springer.

Meier, G.H. 2014. *Thermodynamics of Surfaces and Interfaces: Concepts in Inorganic Materials*. Cambridge: Cambridge University Press.

Ohring, M. 2002. *Materials Science of Thin Films*. 2nd ed. San Diego, CA: Academic Press.

Smith, J.M., H.C. Van Ness and M.M. Abbott. 2005. *Introduction to Chemical Engineering Thermodynamics*. 7th ed. New York: McGraw Hill.

Somorjai, G.A. 2010. *Introduction to Surface Chemistry and Catalysis*. 2nd ed. New York: John Wiley & Sons, Inc.

Tsao, J.Y. 1993. *Materials Fundamentals of Molecular Beam Epitaxy*. San Francisco, CA: Academic Press Inc.

Zangwill, A. 1988. *Physics at Surfaces*. Cambridge: Cambridge University Press.

Problems

Problem 5.1 Using Equation 5.3, derive Equation 5.8 from Equation 5.7.

Problem 5.2 Using Equation 5.3, derive Equation 5.10 from Equation 5.9.

Problem 5.3 Consider a molten mixture of 70 atomic % Ga and 30 atomic % As at a temperature of 1200°C. We slowly cool the mixture down to 0°C. The phase diagram for this system is provided below. GaAs is a double eutectic phase diagram. This is difficult to see since the two eutectic points are located very close to the edges of the diagram. This system can be treated as two side-by-side eutectic phase diagrams – one for Ga-GaAs and one for GaAs-As. For each of the temperatures: 1300°C, 1000°C, 700°C, 100°C, 30°C, and 29°C indicate which phases are present (solid, liquid) and what is the relative amount of each phase. Also specify the chemical composition of each phase.

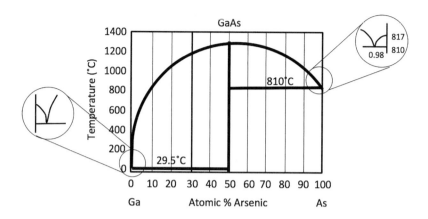

Problem 5.4 The surface energy of a particular surface of a metal can be estimated using a broken bond model where the surface energy is proportional to the fraction of broken bonds, f, and the surface atomic density (atoms/area), σ, $\gamma \propto f\sigma$. Since the proportionality constant depends only on the material, we can use this to identify the low-energy surface of a metal. Consider Ag, which has an fcc structure.

a. Show that the atomic diameter, assuming the atoms are hard spheres that touch one another, is $2.88\,\text{Å}$.

b. Calculate σ, for both the (1 1 1) and (1 0 0) surfaces of Ag.

c. Calculate f for both the (1 1 1) and (1 0 0) surfaces of Ag. Figure 3.21 may be helpful. Count the total number of nearest-neighbor atoms that would be missing from the plane above an atom when a surface is created and divide that into the total number of nearest neighbors that a bulk atom has.

d. Based on your calculations, what would be the ratio of the surface energies for the (1 1 1) to (1 0 0) surfaces of Ag?

Problem 5.5 Consider the Ellingham diagram in Figure 5.1. Calculate the dissociation pressure and determine if a NiO sample in an ultra-high vacuum chamber at a pressure of 10^{-8} Pa would be likely to reduce to Ni metal at a temperature of 600K?

Problem 5.6 In the nearest-neighbor bond model that we used to determine the shape of the Wulff plot, it can be established that the surface energy (or tension) will depend on the atomic density of the surface plane. The higher the density, the more stable the surface is expected to be.

a. Compare the surface density of the (1 0 0), (1 1 0) and (1 1 1) surfaces of an fcc crystal and predict whether the (1 1 0) surface would be stable or would tend to facet.

b. The figure below shows the Wulff plot for a (1 1 0) surface. Explain how this plot supports your prediction from the surface densities.

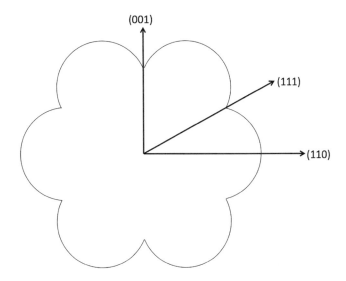

Problem 5.7 The equations of the straight line relationships on an Ellingham diagram for several oxidation reactions are:

$2Cu + O_2 = 2CuO$ $\Delta G^0 = -305 + 0.17T$

$2Ni + O_2 = 2NiO$ $\Delta G^0 = -471 + 0.17T$

$(3/2)Co + O_2 = \frac{1}{2} Co_3O_4$ $\Delta G^0 = -479 + 0.23T$

$2Co + O_2 = 2CoO$ $\Delta G^0 = -491 + 0.16T$

All energies are in kJ/mol.

a. At 600 K, which of these oxides would be the most stable and which would be the least stable?

b. Will Co_3O_4 ever be the more stable form of Co oxide? If so, calculate the temperature at which it becomes the more stable form. If not, then explain why it might be observed.

Problem 5.8 Based on Figure 5.1, which material Al or Ni would be appropriate to evaporate from a SiO_2 crucible? Explain your answer.

Problem 5.9 Construct the Wulff plot for a two-dimensional hexagonal close packed crystal (see Figure 3.22) using the nearest-neighbor bonding model. Derive an equation for $\gamma(\theta)$ and sketch the resulting Wulff plot showing the expected equilibrium crystal shape. The trigonometric identity $\cos(u - v) = \sin(u)\sin(v) + \cos(u)\cos(v)$ may be of help to put your result into the form of the polar equation of a circle.

Problem 5.10 Derive Equation 5.26 from Equation 5.24.

Problem 5.11 Wulff plot: Sketch the equilibrium shape of a small crystallite of a material having a surface tension plot like that shown in the figure below.

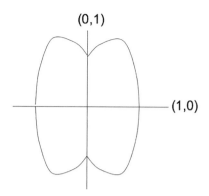

Problem 5.12 Figure 5.4 demonstrates that, at one atmosphere of pressure, Fe undergoes a phase transition at 910 C from α Fe, which has BCC structure, to γ Fe, which has FCC structure. When Cr or Mo is added to Fe, the phase transition is pushed to higher temperatures (meaning the BCC structure is stable to higher temperatures). When Ni is added to Fe, the transition temperature is reduced. Considering only very basic properties of these metals, why does this happen?

Problem 5.13 The figure below is the phase diagram for a binary metallic alloy made of elements A and B. Use this phase diagram to answer the following questions.

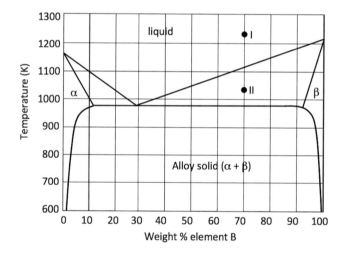

a. What is the temperature and alloy composition of the eutectic point?

b. What is the composition and phase (liquid, solid, mixed) at point I? (If there is more than one phase, describe each phase and give the composition of each phase.)

c. What is the composition and phase (liquid, solid, mixed) at point II? (If there is more than one phase, describe each phase and give the composition of each phase.)

6

Electrical, Magnetic, Optical, and Thermal Properties

The preceding chapters have focused on "atoms", which typically means the position of the nucleus and those electrons which are strongly bound to it (often called the ion core). Now we expand this view to consider the more weakly bound electrons, which will allow us to understand more about the electrical, magnetic, optical, and thermal properties of our materials. As in previous chapters, we begin by exploring the bulk properties which are relevant in understanding thin films and then focus our attention on properties at surfaces. These concepts will also be useful in understanding some of the characterization techniques that we describe in the final section of this book. The bulk properties are described in various books (Patterson and Bailey 2007, Rose, Shepard and Wulff 1964). Other books (Gross 2009, Ibach 2006, Zangwill 1988) focus on the surface properties.

6.1 Electrical Properties

We begin by considering electrons that are free to move through the lattice with no interactions with other electrons or with the ion cores. We then increase the interaction between the electrons and the ion cores to gain an understanding of a broad range of materials. The descriptions here are brief introductions. Further detail is available in books on solid-state physics (Kittel 1996, Omar 1975).

6.1.1 Free Electron Gas Model

The free electron gas model is particularly good for describing the behavior of electrons in some metals. Outer shell electrons are assumed not to be bound at all to the ion core and are able to move freely through the lattice with no interaction with each other or with the ion cores. This model forms the simplest approximation for examining electrical properties, but it is useful in many cases.

Quantum mechanics allows us to describe electrons as waves that obey the Schrödinger wave equation. In one dimension, the solution to the time-independent wave equation is a wave function, ψ_n, that has a corresponding energy eigenvalue, ε_n. For an electron confined in a one-dimensional region of length, L,

$$\psi_n = A \sin\left(\frac{n\pi}{L} x\right) \tag{6.1}$$

$$\varepsilon_n = \frac{\hbar^2}{2m_e}\left(\frac{n\pi}{L}\right)^2 \tag{6.2}$$

DOI: 10.1201/9780429194542-7

where A is the amplitude of the wave, n is an integer quantum number, L is a characteristic dimension of the sample, $\hbar$ is Planck's constant divided by 2π, and m_e is the mass of the electron.

The quantum number n describes different states for the system. We are allowed to place two electrons (since they are fermions) in each state. One with intrinsic spin up and one with spin down. At $T=0$, we can place two electrons in each state until N total electrons occupy the lowest energy $N/2$ states. This fills the quantum states up to the Fermi number $n_F=N/2$. The energy of the highest filled state is called the Fermi energy, ε_F, which is the energy corresponding to $n=n_F$.

The energy eigenvalue contains a term $(n\pi/L)$ that corresponds to a wave vector, $k=\pm n\pi/L$. We can define our energy now in terms of k to obtain a view of the free electron model in reciprocal space (defined in Section 3.2).

$$\varepsilon = \hbar\omega = \frac{\hbar^2}{2m_e}k^2 \qquad (6.3)$$

where $k=0, \pm 2\pi/L, \pm 4\pi/L \ldots$ This includes only even terms in order to satisfy the boundary conditions on the wave function.

In three dimensions, we can develop similar expressions in the y and z directions, leading to k being a radial vector in reciprocal space. We fill states in reciprocal space as in Figure 6.1a out to some radius, k_F, corresponding to the Fermi energy

$$\varepsilon_F = \frac{\hbar^2}{2m}k_F^2 \qquad (6.4)$$

From Figure 6.1b we see that in any given energy range $(d\varepsilon)$ there exist a finite number of available electron energy states (dN). We define the density of states, $D(\varepsilon)$, as the number of electron states available in a range of energy

$$D(\varepsilon) = dN/d\varepsilon \qquad (6.5)$$

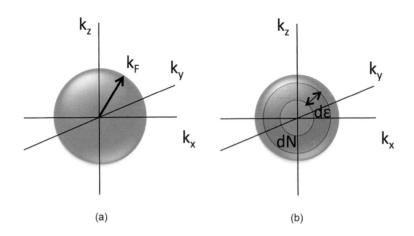

(a)   (b)

FIGURE 6.1
(a) Filled states in reciprocal space. (b) A spherical shell region of filled states in reciprocal space containing dN states in an energy range $d\varepsilon$.

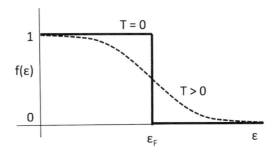

FIGURE 6.2
Fraction of energy states available as a function of energy for $T=0$ and at a finite temperature.

The fraction of the energy states, $f(\varepsilon)$, available at a given energy is given by Fermi-Dirac statistics yielding

$$f(\varepsilon) = \frac{1}{e^{(\varepsilon - \mu)/k_B T} + 1} \tag{6.6}$$

where μ is the chemical potential. At $T=0$, we note that $\mu = \varepsilon_F$. The fraction of energy states is plotted in Figure 6.2 for both $T=0$ and finite T. Even at room temperature, most materials are reasonably well modeled using the $T=0$ case.

This simple free electron metal model has made important contributions to our understanding of the heat capacity of solids, electrical conductivity, the effects of magnetic fields, thermal conductivity, and many other subjects involving bulk solids. For surfaces, the model can be used to consider how electrons can escape from solid surfaces through the process of thermionic emission, which will be discussed later in this chapter.

6.1.2 Jellium and Nearly Free Electron Model

The presence of the positive ions can be accounted for in various ways. The jellium model (homogeneous electron gas model) treats the ion cores as a uniform positive charge density that interacts with the mobile electrons. Electrons are also allowed to interact with one another. The Pauli exclusion principle restricts the distance between electrons (which are fermions) and thus the electron density. Electrons with parallel spins will keep further apart to minimize their exchange energy. Electrons with anti-parallel spins can lower their correlation energy by keeping further apart and reducing the Coulomb repulsion. The electrons move in a constant, attractive, potential energy from the ion cores given by the coulombic image potential energy

$$U(z) = -\frac{1}{4\pi\varepsilon_0} \frac{e^2}{4(z - z_0)} \tag{6.7}$$

where z_0 is the distance between the boundary of the uniform positive charge and the image plane. This model has only one adjustable parameter, which is the density of the uniform positive charge.

The nearly free electron models allow interaction with discrete ion cores by including an attractive potential energy $U(x)$ at each ion core. Electrons do not interact with one another and move through the lattice, which now includes a periodic potential from the ion cores (each separated by distance a) as indicated in Figure 6.3.

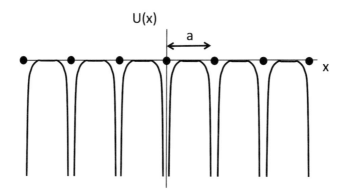

FIGURE 6.3
Periodic potential wells at each ion core position.

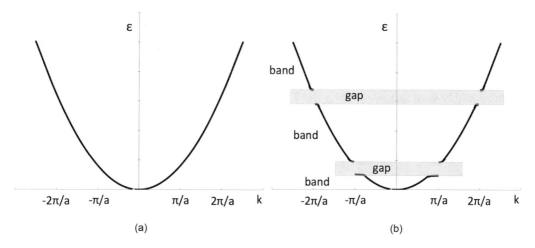

FIGURE 6.4
(a) Free electron dispersion relation. (b) Nearly free electron dispersion relation.

The relationship between the energy, ε, or angular frequency, ω, and the wave vector, k, is called the dispersion relation as discussed in Section 4.2 for elastic waves. For free electrons, Equation 6.3 gives the dispersion relation, which is plotted in Figure 6.4a. In the nearly free electron model, electrons that are not near ion cores will obey this dispersion relation.

When the electrons are near ion cores, however, the dispersion relation is modified since the electron wave will be reflected by the periodic structure of the ion cores. As a result, there will not be a traveling electron wave for certain wave vectors and thus there will not be any allowed value of k for those energy values. The condition for reflection from a periodic structure is $k=\pm n\pi/a$. We sketch the resulting dispersion relation in Figure 6.4b. Notice that there are now gaps in energy where no electron states are allowed with allowed bands of energy on either side of the gaps. The width of these gaps (in energy) depends on the depth of the periodic potential wells. Deeper wells lead to wider energy gaps.

This model allows us to address the difference between conductors and insulators. In conducting materials, some electrons are bound in the ion cores while other electrons behave as nearly free electrons. Conductors have partly filled energy bands. In other

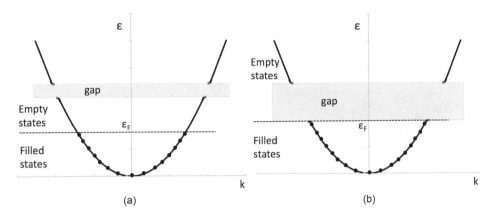

FIGURE 6.5
(a) Dispersion relation for a conductor. (b) Dispersion relation for an insulator.

words, the Fermi energy lies in the middle of a band of allowed energies. This means that there exist empty electron states that are allowed and are accessible to the electrons. Figure 6.5a shows the dispersion relation for a conductor. If an electric field is applied to the material, higher energy allowed states are accessible for the electrons to occupy and a current is allowed to flow in the material.

Insulators, in contrast, have filled energy bands and large (>5 eV) energy gaps between the filled and empty states. This is represented in Figure 6.5b. When an electric field is applied to an insulator, no accessible allowed states are available to the electrons and so no current is able to flow unless the electric field is very large.

We have examined dispersion relations in what is typically called the extended zone scheme. Just as for phonons discussed in Chapter 4, we do not need to consider all possible values of the wave vector, k. We only need to consider those k values in the First Brillouin Zone. We can redraw the dispersion relation using a reduced zone scheme by translating all k values by $2\pi n/a$ (where n is an integer) until they lie in the First Brillouin Zone. Figure 6.6a shows this process for the nearly free electron dispersion relation resulting in Figure 6.6b, which is the reduced zone scheme representation of the nearly free electron dispersion relation.

6.1.3 Other Electronic Models

Many computational models of electronic structures use the density functional method, which assumes that the total electron density distribution $n(\vec{r}, t)$ of the ground state of any multi-electron system contains all of the information about that system. In other words, all electronic properties are functionals (functions of a function) of the ground state electron density which is a function of position and time. This assumption significantly simplifies computational calculations. The process starts by assuming that the positions of the nuclei (much more massive than the electrons) are fixed at the positions that minimize the energy of the system. These fixed nuclei produce a potential to which is added potential terms from electron-electron interactions and any applied external potential. The specific form of these potentials depends on the problem being addressed. For instance, the jellium potential can be used in the density functional method. The total potential is then used in the Schrödinger equation.

Other models allow for greater interaction between the electrons and the ion cores than the nearly free electron model. A variety of tight binding models use periodic ion core

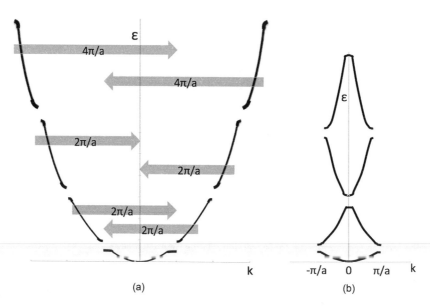

FIGURE 6.6
(a) Translations to move from extended zone scheme to reduced zone scheme. (b) Resulting in reduced zone scheme dispersion relation for the nearly free electron model.

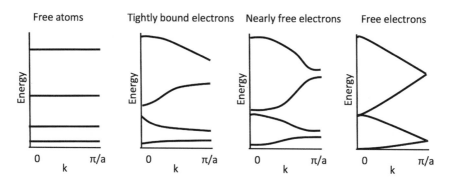

FIGURE 6.7
Energy diagrams for different models of solids.

potentials (modeled as pseudopotentials) and have much stronger interactions between the electrons and the ion cores. A common choice of wave functions in these tight binding models is a linear combination of atomic orbitals (LCAO). These tight binding methods are particularly appropriate for examining the band structure of low-lying bands where the electron shell radius is much smaller than the spacing between atoms.

We can compare the models by considering the reduced band scheme dispersion relations for several models as presented in Figure 6.7. If we start with a group of identical free atoms that are infinitely separated, they have discrete atomic energy levels that are well defined by atomic physics. If we gather those atoms into a solid, but use a tight binding model where the ion cores are very atom-like and we do not have "free" electrons, then we begin to see the development of a band structure. Moving to a nearly free electron model, the bands become broader in energy. Finally, if some electrons are free to move unhindered by the ion cores, all energies are allowed.

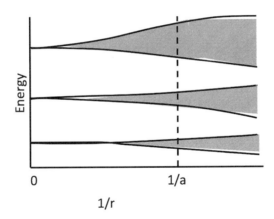

FIGURE 6.8
Building a solid from N infinitely separated atoms.

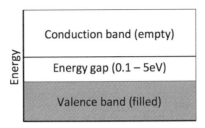

FIGURE 6.9
Energy levels in semiconductors at $T=0$ K.

The process of building a solid containing N atoms from infinitely separated atoms is pictured in Figure 6.8. At infinite separation, we have 2N atomic states possible for each atomic energy level. As the atoms are brought closer and closer together, bands develop that are typical of solids rather than the individual levels of the free atoms. This leads to 2N orbitals in each band.

6.1.4 Semiconductors

Before examining surface electrical properties, we briefly examine the properties of semiconductors. These materials have an energy gap of 0.1–5 eV leading to a variety of different electrical behaviors depending on temperature and applied electric field. At low temperatures and low applied fields, these materials behave as insulators. At high temperatures or high applied fields, however, electrons can enter the conduction band and move through the solid. This is pictured in Figure 6.9. From this simple picture, the energy gap, E_G, is clearly the energy of the bottom of the conduction band minus the energy of the top of the valence band.

A more detailed examination of the dispersion relations in semiconductors indicates that they can be divided into two types depending on the k values corresponding to the minimum energy of the conduction band and the maximum energy of the valence band. These two types are indicated in Figure 6.10. If the bottom edge of the conduction band and the top edge of the valence band occur at the same value of k, then the semiconductor has a direct gap (like GaAs). If these two energies occur at different values of k, then the semiconductor has an indirect gap (like Si).

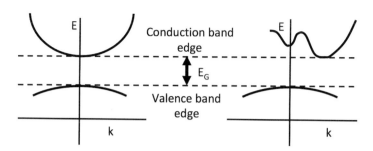

FIGURE 6.10
Dispersion relations for direct gap and indirect gap semiconductors.

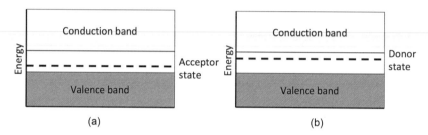

FIGURE 6.11
Semiconductor doping creates (a) acceptor or (b) donor states in the band gap.

The electrical properties of semiconductors can also be changed by adding small quantities of other elements (doping) that create additional allowed electronic states lying in the gap between the valence and conduction bands. For instance, Si or Ge are often doped using elements from the adjacent columns in the periodic table. Boron, for instance, has one fewer electron than Si. It is an electron acceptor, which means that it can take an electron from the valence band resulting in p-type doped Si. Phosphorus has one more electron than Si. It is an electron donor and can give up an electron to the solid crystal resulting in n-type Si. The doping creates allowed acceptor or donor states in the energy gap making it easier for electrons to cross the gap at lower temperature or applied fields by a two-step process of excitation from the valence band into the acceptor or donor state and then from this state into the conduction band. This is demonstrated in Figure 6.11.

6.1.5 Resistivity and Conductivity

As suggested in Chapter 1, a simple model of electrical transport properties involves looking at the motion of charge carriers and how frequently they scatter in their motion. Electrical current in a material depends on the average velocity of the electrons, which will decrease with collision frequency. Resistivity (measured in ohm-m) of a material will increase as the mean free path decreases (more collisions). The inverse of resistivity is conductivity (measured in mho/m). Both of these are material-dependent parameters unlike resistance or conductance that depend on both the material and the geometry.

Resistivity (and conductivity) will depend on lattice vibrations. Since higher temperatures result in larger vibration amplitudes (as shown in Chapter 4), greater scattering will occur and the resistivity is expected to increase with temperature. Scattering can also occur from impurities and defects in the crystalline structure leading to an additional term

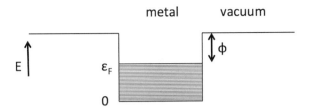

FIGURE 6.12
Finite energy well for a metal with filled energy states up to ε_F.

known as the residual resistivity. Higher impurity or defect concentrations will increase the resistivity. Figure 6.5 explored the connection between the size of the gap between the valence and conduction bands and the electrical conductivity of a material. We return to these concepts in Chapter 14 where we explore electrical characterization techniques.

6.1.6 Surfaces

As is always the case with surfaces, we break the symmetry that was inherent in a discussion of infinite solids by creating a surface. We have previously discussed how this will cause atoms at the surface to change positions to minimize the surface energy. These changes will also change the electrical properties at a surface.

The presence of a surface means that electrons could escape from the solid. Using the free electron model, if we treat a metal as a finite energy well, as in Figure 6.12, then electrons with kinetic energies greater than $\varepsilon_F + \phi$ will be able to escape from the solid if they are moving in the correct direction (+z). ϕ is the work function of the surface material.

The condition for an electron with a z component of momentum, p_z, to escape is

$$\frac{p_z^2}{2m_e} \geq \varepsilon_F + \phi \tag{6.8}$$

A complete analysis yields a current density, J, of escaping electrons given by the Richardson – Dushman equation,

$$J = A_0 (1-r) T^2 e^{-\phi/k_B T} \tag{6.9}$$

where A_0 is a constant and r is a reflection coefficient to indicate electrons that are reflected back into the material. This current density is very sensitive to temperature and work function.

In metals, a free electron model does not allow for any directionality in the bonds between metal atoms. The electron charge density spreads out at the surface and so does not extend as far above the ion cores as in the bulk. As a result, the top ion core layer is pushed back into the solid. This changes the lattice spacing perpendicular to the surface for the top layer or two. This is an example of a surface reconstruction as discussed in Chapter 3 and is pictured in Figure 6.13.

In the "jellium" model, the mobile electrons interact with one another and we model interactions between the electrons and the ion cores by assuming the existence of a uniform positive charge distribution inside the material. We create a surface by having $\rho_+ = \overline{\rho_+}$ inside the solid and $\rho_+ = 0$ outside the solid. This creates some interesting effects at the surface as indicated in Figure 6.14.

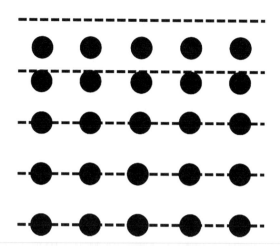

FIGURE 6.13
Vertical surface reconstruction in metals.

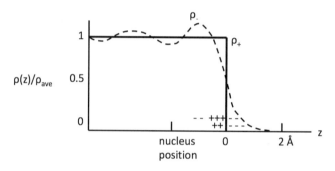

FIGURE 6.14
Jellium model of a surface showing the positive (solid line) and negative (dashed line) charge densities and the creation of a surface dipole.

The electron density smears out on the surface in order to screen the positive charge distribution. This leads to the formation of a double layer on the surface with dipoles that have a negative charge on the outside (furthest from the material) and a positive charge on the inside. This model also leads to Friedel oscillations in the negative charge density $\rho_-(z)$ arising from reflected electron waves at the surface resulting from the sharp step in the positive charge density distribution.

A more accurate model of the ion cores would be to assume a periodic ion core potential. This periodic potential leads to strong variations in the electron densities within the solid. These effects however are smoothed out at the surface where the periodicity is terminated. The formation of a double layer on the surface can be described within this periodic potential model. Consider the two-dimensional model in Figure 6.15. Each dot represents a positive ion core. Each square encloses enough negative charge so that the net charge in the square is zero. The surface (at the top of the figure) is electrically neutral.

Now consider two competing effects. Spreading of the electrons at the surface will create a negative double layer at the surface (dipoles with the negative side out). The absence of the next row of ion cores allows the electrons to spread outward in order to reduce the

Surface

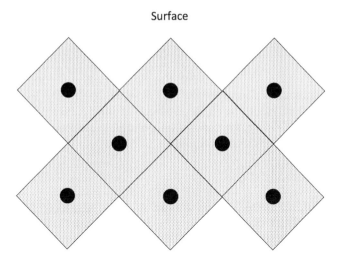

FIGURE 6.15
Simple model of positive ion cores (circles) and surrounding electron charge density (squares).

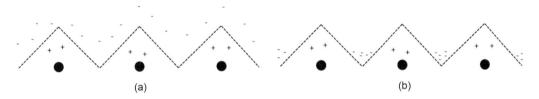

(a)   (b)

FIGURE 6.16
(a) Spreading of electrons creating a negative surface double layer. (b) Smoothing of electrons creating a positive surface double layer.

energy of the system. This is pictured in Figure 6.16a. The loss of electrons from within the square surrounding each ion core results in a net positive charge inside the square. Smoothing of the surface can also occur. The energy of the electron cloud at the surface is reduced if the electrons are bounded by a flatter surface. This will cause the electrons to "flow" into the valleys between the positive cores as shown in Figure 6.16b. The result is a positive double layer at the surface (dipoles with the positive side out). Competition between these two opposing effects makes the double layer sensitive to which crystal face is exposed and also to the presence of adsorbed atoms on the surface.

This surface double layer will modify the work function at the surface. The work function now depends on three terms. The first term is the attractive well created from the ion core as discussed above. The second term comes from a positive image charge that is created in the solid when a negative charge is outside the surface. The third term comes from the net surface double layer.

$$\phi = E_{\text{Fermi}} - eV_{\text{image}} + eV_{\text{double}} \qquad (6.10)$$

The work function will tend to be lower for lower atomic density surfaces. Surfaces that are atomically "rough" will tend to have more smoothing, which will also lower the work function. Similarly, atomic steps on surfaces can reduce the work function.

Semiconductors have directional covalent bonds and the process of creating a surface will leave some bonds open. These are referred to as "dangling bonds". The surface will reconstruct to try to minimize the total energy. This may distort the surface both laterally and vertically.

We can increase the interaction between the electrons and the ion cores even further using tight binding models. The electrons are now bound to the ion core. The solid is treated as a perturbation where the electron clouds around each ion core overlap with the electron clouds of neighboring ion cores. The potentials are now much more atom-like. One approach to this model is to use a LCAO.

6.1.7 Surface States

The possibility of surface states in semiconductors was introduced in an earlier section. Surface states are electronic states of the crystalline solid that are localized at the surface. In order to be localized at the surface, these states must decay both into the vacuum and into the bulk. They may be able to be described by a wave parallel to the surface. These conditions require that surface states have energies and wave vectors that are different from bulk states. These surface states are located in the band gap of the material which results in a change in the electronic properties of the material near the surface.

The surface states arise from changes in the potential of the bulk. The simplest change is the presence of a surface itself which represents the termination of the bulk potential. This can then be refined by relaxation and reconstruction of the crystal structure at surfaces. Defects in the crystal structure at the surface and the presence of adsorbed impurity atoms at the surface can also change the nature of surface states.

An example of surface states arises from the reconstruction on semiconductor surfaces which can lead to the formation of additional allowed electron states in the energy gap. These surface states, represented in Figure 6.17, are only found near the physical surface of the crystal and can lead to the electrical properties near the surface of a semiconductor being different from those in the bulk. The states may be occupied or vacant and can cause bending of the energy bands near the surface.

Consider an n-type semiconductor with excess electrons occupying the surface states. The band bending pictured in Figure 6.18 will cause changes in the properties of the semiconductor near the surface.

More generally, the creation of surface electron states requires three primary conditions:

1. There must be a solution to the Schrödinger equation.
2. The electron must not be able to escape to the vacuum $E_{surface} < E_{vacuum}$.

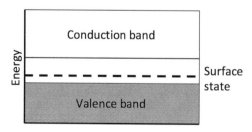

FIGURE 6.17
Surface state in a semiconductor.

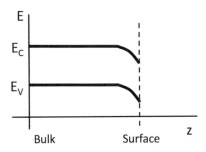

FIGURE 6.18
Band bending from surface states in an n-type semiconductor.

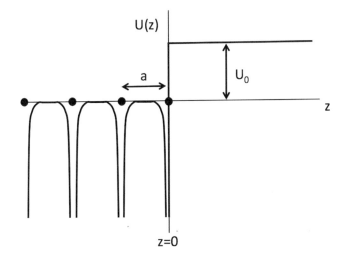

FIGURE 6.19
Potential energy model for surface states showing periodicity in the lattice and a surface barrier outside.

3. The electron must not be able to escape to the bulk. This means that no bulk states exist with the same energy and parallel component of the wave vector as the surface state.

We can use the same models that we have applied earlier in this chapter to examine surface states. The free electron and jellium models do not have surface states since they lack atomic structure. The nearly free electron model gives rise to Shockley states in which the electron is localized at the surface with energy in the band gap. The electron wave function amplitude decreases exponentially with position in the crystal.

Solving the Schrödinger equation for this nearly free electron model is relatively straightforward. The standard, one-dimensional Schrödinger equation is

$$-\frac{\hbar^2}{2m_e}\frac{d^2\psi}{dz^2}+U(z)\psi(z)=E\psi(z) \qquad (6.11)$$

The first step in solving this is to identify the potential energy $U(z)$ appropriate for the problem. Figure 6.19 shows a potential that meets all of the requirements. It contains a surface barrier, $U(z)=U_0$ for $z>0$, to keep the electrons in the solid. The potential in the

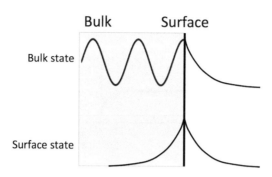

FIGURE 6.20
Wave function solutions at surfaces.

solid, $z<0$, must be periodic with a general form $U(z)=U(z+na)$ where n is an integer. The Schrödinger equation can then be solved for each region ($z<0$, $z>0$). The resulting wave functions, pictured in Figure 6.20, will have one solution that is periodic in the bulk and decays exponentially into the vacuum. These are bulk states. Another solution will exist, however, that decays exponentially both into the vacuum and into the solid. These solutions are Shockley surface states. In an infinite crystal, the surface states do not exist since we require the wave function to be finite as z goes to infinity. The presence of a surface at $z=0$ allows these states to exist.

At the other extreme for the interaction between electrons and ion cores, the tight binding model gives rise to Tamm states. The electron is now localized to a surface atom and the surface state has energy in the band gap. The Schrödinger equation involves a potential that can be treated as a sum of atomic potentials for atoms located on the lattice points of the crystal. The solutions will include states that only have significant amplitude near the surface atoms.

The bulk density of states (number of states per unit energy, dN/dE) can often be treated as uniform and so does not need to be a function of position. We write this as $n_{st}(E)$ which only depends on the energy. Surfaces, however, will have a spatial dependence in the density of states. This is referred to as a local density of states (LDOS) and is written $n_{st}(E, \vec{r})$. The surface density of states should be different from the bulk due to the presence of additional surface states and possibly atoms or molecules adsorbed on the surface. Typically by about the third layer of atoms down into the solid, the surface and bulk densities of states are the same.

6.1.8 Excitons and Plasmons

Collective behavior can impact electron transport in materials. This can lead to excitons and plasmons. Excitons occur when an electron is excited into an unoccupied state leaving behind an electron vacancy (hole). The electron and hole are bound together by a Coulomb attractive force and form an exciton (neutral quasiparticle) in insulators or semiconductors with a binding energy typically less than 1 eV as shown in Figure 6.21. The physical separation of the electron and hole can be relatively large (greater than a lattice spacing) and the energy of the exciton can be in the band gap often just below the conduction band edge. The exciton can move through the crystal until the electron drops back down and fills the vacancy. Thus the exciton transports energy but not charge. Typical lifetimes for excitons are in the nanosecond range. Excitons have applications in solar cells since they can change the way a material absorbs light. A more mathematical introduction to excitons is available (Reynolds 1981) as well as extensive research journal literature.

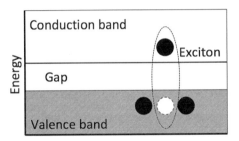

FIGURE 6.21
Excitons consist of an excited electron and a hole.

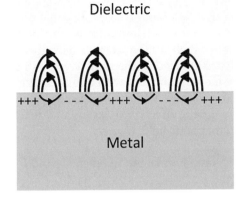

FIGURE 6.22
Surface plasmon charge density fluctuation and the resulting electric field.

Plasmons arise in the free electron model as a collective electron density fluctuation in metals. This can be treated as waves of electron density. In quantum mechanics, these waves can be quantized and treated as a particle (a plasmon), just as photons and phonons. Bulk plasmons have typical energies of a few eV. Surface plasmons, confined to two dimensions, have lower energies.

$$E_{\text{bulk}} = \hbar \omega_p \tag{6.12}$$

$$E_{\text{surface}} = \hbar \omega_p / \sqrt{2} \tag{6.13}$$

where ω_p is the plasma frequency of the free electron gas defined in Equation 2.19. Surface plasmons, in the jellium model, will be located (have a maximum charge density) at a position outside the edge of the positive charge density. Figure 6.22 shows how the surface plasmon electric field extends further into the dielectric material (air, vacuum, glass, etc.) than into the metal. The field of plasmonics encompasses a wide range of possible applications including high-frequency transmission of information, molecular sensors, solar cells, and light-emitting diodes. Further details are available in the literature (Helsey 2011, Sarid and Challener 2010).

6.2 Magnetic Properties

We briefly introduce the basic elements of magnetic applications which are important in thin film materials and surfaces. Magnetic thin films are discussed in more detail elsewhere (Volkerts 2011) as is the subject of surface magnetism (Diep 2014, Kaneyoshi 1991).

6.2.1 Magnetic Materials

A simplified picture of an atom is presented in Figure 6.23. An electron is shown orbiting in a circular orbit in a counter-clockwise direction. This results in an angular momentum vector, $\vec{L}$, whose direction is given by the right-hand rule. Since current is the flow of positive charge, a negatively charged electron moving counter-clockwise is equivalent to a current flowing in the clockwise direction. This results in a magnetic dipole moment, $\vec{\mu}_m$, that will point downward using the right-hand rule. More generally, the total magnetic dipole moment of the atom will also have a contribution from the intrinsic spin of the electron.

If electrons are filling energy shells of the atom, as described earlier in this chapter, the up and down spins, which are allowed in the same level, will cancel one another to give no permanent magnetic moment in an atom with a filled energy shell and in the absence of an applied magnetic field.

For materials that show magnetic effects, we classify magnetic materials by their behavior in an externally applied magnetic field, $\vec{H}_{ext}$, as

1. diamagnetic materials have atoms with filled shells and $\vec{\mu}_m$ anti-parallel to $\vec{H}_{ext}$.
2. paramagnetic materials have atoms with partially filled shells and $\vec{\mu}_m$ parallel to $\vec{H}_{ext}$.
3. ferromagnetic materials have atoms with partially filled shells and can have a large permanent magnetic moment even in the absence of an applied field.

Diamagnetism arises from the cyclotron motion of negatively charged electrons in an applied magnetic field. The magnetic force will cause the electrons to circle around the applied magnetic field lines in a direction that will create a dipole moment opposite to the applied field as in Figure 6.24. The diamagnetic effects are very weak and typically can be neglected except in atoms with filled shells. Cu, Ag, and Au are examples of materials that display diamagnetism.

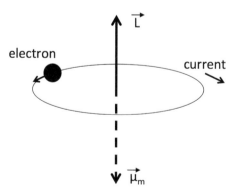

FIGURE 6.23
Relationship of electron motion, angular momentum vector and magnetic dipole moment vector.

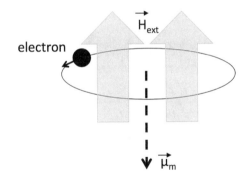

FIGURE 6.24
Diamagnetic response in the presence of an applied magnetic field induces a magnetic dipole moment opposite to the applied field.

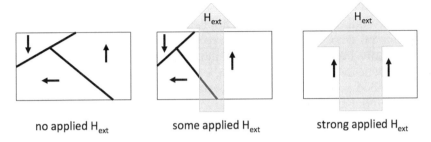

no applied H$_{ext}$ some applied H$_{ext}$ strong applied H$_{ext}$

FIGURE 6.25
Domains oriented with magnetic moments parallel to the applied magnetic field grow as the strength of the field increases.

The behavior of paramagnetic materials (Al, W, …) is dominated by the conduction electrons in the partially filled shells. They can contribute in two ways. Spin paramagnetism arises from the electrons wanting to align their magnetic dipole moment with the external field to minimize energy, but the Pauli exclusion principle tries to prevent spins from flipping. As a result, only a few electrons near the Fermi energy will flip. The second contribution is a diamagnetic tendency to align the dipole moment anti-parallel to the external field due to the electron cyclotron motion.

Ferromagnetic materials (Fe, Ni, Co …) have many interesting applications (besides refrigerator magnets) and we will emphasize an understanding of ferromagnetic materials. The magnetization of a material, $\vec{M}$, is the total magnetic moment per unit volume.

$$\vec{M} = \frac{1}{V}\sum_i \vec{\mu_i} \qquad (6.14)$$

It will have a preferred direction in a crystal that will create a minimum energy situation arising from the interaction of the electron spins with the periodic electric field of the crystal lattice. This direction is called the "easy axis". For Fe, the easy axis is [1 0 0] while for Ni it is [1 1 1].

Ferromagnetic materials have regions, called domains, in which all the magnetic moments are aligned. Adjacent domains are randomly oriented with respect to one another as indicated in Figure 6.25. These domains do not necessarily correspond to the crystalline

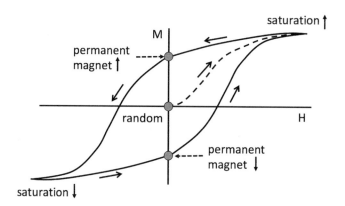

FIGURE 6.26
Hysteresis curve for a ferromagnetic material.

grains in polycrystalline materials. In an applied magnetic field, the domain walls will move with domains whose magnetization is parallel with the applied field growing at the expense of domains oriented in other directions. At higher applied magnetic fields, rotation of the magnetic moments to align with the field will also occur.

When the applied magnetic field is removed some of the domains lose their orientation, but the material does not return all the way to a random configuration. As a result, it retains some magnetic properties; it has become a permanent magnet. The magnetic properties of these materials can be described by plotting a hysteresis loop for the magnetization, $\vec{M}$, of the material as a function of the applied magnetic field, $\vec{H}_{ext}$. A typical plot is shown in Figure 6.26.

In Figure 6.26, the material starts out at the origin with grains randomly aligned producing no magnetization in no applied field. As the magnetic field is increased the domain walls move and the magnetization increases aligned with the field. Eventually, all of the domains have magnetizations aligned with the field and the magnetization saturates. If we now decrease the field the magnetization will slowly decrease as some of the domains lose their orientation with the field. When zero field is reached, the material still has some residual magnetization and has become a permanent magnet. If the direction of the field is now reversed and increased in the opposite direction, the domain walls will again move to favor domains aligned with the new reversed field. This will change the magnetization down through zero and to a saturated state in the opposite direction. As this applied field is reduced to zero, some grains will again lose their orientation with the field, but a net magnetization will remain. The material is now a permanent magnet but in the opposite direction from before. Reversing the field back to the direction we started with allows us to again saturate the material in the first direction by increasing the field.

Ferromagnetic materials that saturate at low applied magnetic fields are referred to as "soft" materials. These correspond to materials in which the domain walls can easily be moved. Those materials that require a large external magnetic field to reach saturation are known as "hard" magnetic materials and have domain walls that are difficult to move. Defects in a material will often increase the magnetic hardness of the material.

Ferromagnetic materials have a magnetic moment in the absence of an applied magnetic field. This suggests the existence of some type of interaction between magnetic moments that tends to align them parallel to one another which is known as the exchange field. This tendency to align is opposed by a thermally induced tendency toward randomness. At sufficiently high temperatures, the spontaneous magnetization will be lost and the material will undergo a phase transition from the ferromagnetic state at lower temperatures to a paramagnetic state at higher temperatures. The critical temperature at which this transition occurs is known as the Curie Temperature. For $T < T_{Curie}$ the material exhibits an ordered ferromagnetic phase. For $T > T_{Curie}$ a disordered paramagnetic phase exists. Iron, for example, has $T_{Curie} = 1043$ K. This phase transition is demonstrated in Figure 6.27, which shows the behavior of the magnetic dipole moment as a function of temperature.

6.2.2 Ferromagnetic Energies

We examine the energy terms involved in the ferromagnetic materials. The exchange energy depends on the strength of coupling of adjacent magnetic moments. This energy term is minimized by having the magnetic moments parallel to one another as suggested in Figure 6.28. The exchange energy will tend to resist changes in the direction of the magnetization.

The anisotropy energy, E_A, involves the orientation of the magnetization relative to the easy axis, which will depend on the crystal structure as well as induced stress in the material. It is minimized when the magnetization is aligned with the easy axis. The external field contributes an energy term, E_H, that takes on a minimum energy when the magnetization aligns with the external field.

$$E_H = -\vec{H}_{ext} \cdot \vec{M} \tag{6.15}$$

A demagnetizing field will be created at surfaces where the boundary conditions require that $\vec{\nabla} \cdot \vec{M} = 0$ outside the material. This condition creates dipoles leading to a demagnetizing

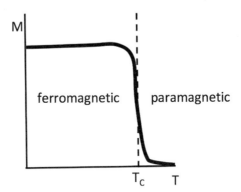

FIGURE 6.27
Magnetic dipole moment as a function of temperature showing the phase transition from ferromagnetic to paramagnetic behavior.

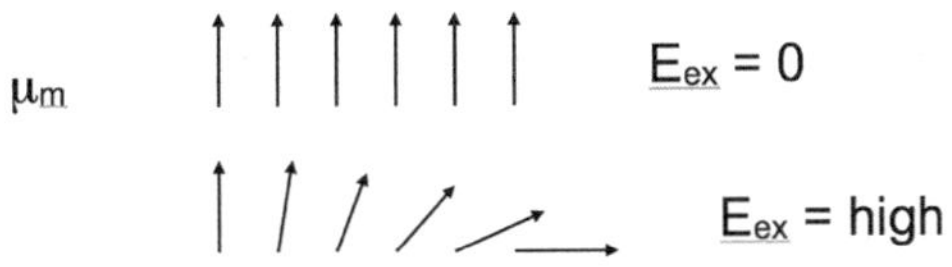

$$\mu_m \quad \uparrow\uparrow\uparrow\uparrow\uparrow\uparrow \quad E_{ex} = 0$$

$$\uparrow / / / \nearrow \quad E_{ex} = \text{high}$$

FIGURE 6.28
Exchange energy and orientation of magnetic moments.

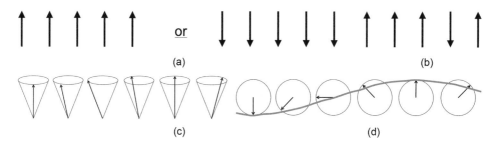

FIGURE 6.29
(a) Ground state configuration of magnetic dipoles. (b) Excited state. (c) Spin wave excited state. (d) Top view of spin wave showing wave pattern. Horizontal axis represents adjacent lateral positions in the material.

field inside the material and a stray field outside the material. This reduces the total magnetic moment in the material and leads to the formation of ferromagnetic domains.

With energies defined, we can consider the energy states of a system of magnetic dipole moments in the absence of an applied magnetic field. The lowest energy state (ground state) will have all of the magnetic dipoles aligned with one another as shown in Figure 6.29a. A simple excited state might consist of one of those dipole moments being inverted as demonstrated in Figure 6.29b. A more complicated excited state that is of low energy is a spin wave, Figure 6.29c. In this case, the dipole moments all precess around the direction of being completely aligned. Each adjacent dipole moment is slightly advanced in phase by a constant angle over the previous dipole moment. Each dipole moment is separated by a lattice spacing. The wave nature of this spin wave is demonstrated in Figure 6.29d.

Just as waves of light were quantized as photons and elastic waves were quantized as phonons, we can quantize these spin waves and call them magnons. The interactions between photons, phonons, and magnons can then be treated as particle interactions.

6.2.3 Surface Magnetism

We have seen that the LDOS at the surface varies from that in the bulk and so we expect new magnetic states at the surface. Two types of surface spin waves are observed.

Very long wavelength (small k) magnetostatic waves are observed similar to the Rayleigh waves in phonons discussed in Section 4.2. These propagate in one direction on the surface but NOT in the reverse direction. If we compare the top surface to the bottom surface, the directions are reversed but are always perpendicular to the direction of the magnetization vector in the material. These are demonstrated in Figure 6.30.

Short wavelength surface spin waves are also observed that demonstrate optical and acoustic modes as observed for elastic waves. The orientation of the spin vector is indicated in Figure 6.31 for both types of waves. Figure 6.29c and d showed the cone in which the bulk spin vector precesses. Figure 6.31a shows a surface acoustic spin wave where the magnitude decays as you move in from the surface. The direction of the spin, however, is the same as you move into the bulk. In Figure 6.31b, the surface optical spin wave reverses spin direction for each layer as you move into the bulk. The variation along the surface is the same as for the acoustical wave and the wave decays in amplitude as you move away from the surface.

These surface magnetic waves tend to be very sensitive to surface structure (including the crystal face of the surface) and to the presence of defects and impurities. Adsorbed atoms on the surface will tend to reduce the magnetic moments at surfaces.

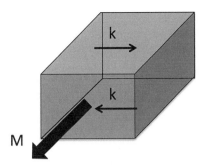

FIGURE 6.30
Long wavelength magnetostatic waves propagate in one direction along the surface.

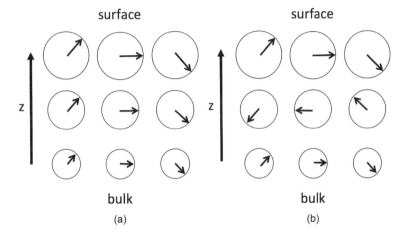

FIGURE 6.31
(a) Top view of spin vector precession for a surface acoustic spin wave. (b) Top view of spin vector precession for a surface optical spin wave.

6.3 Optical Properties

Once again this section provides only a brief introduction to a very extensive topic. Further details on the optical properties of surfaces and thin films (including rough surfaces and discontinuous films) are available in the literature (Aspnes 1982a, b, Bedeaux and Vlieger 2002, Heavens 1991).

6.3.1 Electromagnetic Waves in Materials

Optics classically uses a transverse wave model to describe electromagnetic radiation. In free space, the electric and magnetic field vectors oscillate perpendicular to one another and perpendicular to the direction of propagation of the wave. Introductory treatments of optics focus on the far-field approximation where we assume that the distances involved in our experiments are large compared to the wavelength of light. Visible light has a typical wavelength of around 500 nm. A typical electric field oscillation for a plane wave is shown in Figure 6.32.

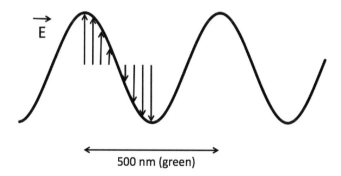

FIGURE 6.32
Electric field variation in plane wave.

We are primarily interested here in the interaction of light with matter, which is the interaction of oscillating electric and magnetic fields with electrons and ion cores. This interaction is described by several inter-related, wavelength-dependent terms. The complex index of refraction, $\tilde{N}$, can be defined in terms of a real index of refraction, n, and an extinction coefficient, k. The absorption index, κ, is also used where $k = n\kappa$.

$$\tilde{N} = n - ik = n(1 - i\kappa) \tag{6.16}$$

where $i = \sqrt{-1}$ indicates the imaginary term in the complex quantity.

An absorption coefficient, α, can also be defined as

$$\alpha = 2\frac{\omega}{c}k = 2\frac{2\pi}{\lambda}k \tag{6.17}$$

Equivalently we can define the complex dielectric constant, ε, and relate it to n and k.

$$\varepsilon = \varepsilon_1 - i\varepsilon_2 \tag{6.18}$$

$$\varepsilon_1 = n^2 - k^2 \tag{6.19}$$

$$\varepsilon_2 = 2nk \tag{6.20}$$

An electromagnetic wave, with wavelength, λ, traveling in the x direction through a material characterized by a complex index of refraction can then be described by looking at the electric field vector, $\vec{E}$.

$$\vec{E}(x) = \vec{E}_0 e^{-i\tilde{N}\frac{2\pi}{\lambda}x} = \vec{E}_0 e^{-k\frac{2\pi}{\lambda}x} e^{-in\frac{2\pi}{\lambda}x} \tag{6.21}$$

The first exponential term is decaying and describes the absorption of light in the material. The second exponential term can be written as a traveling wave and represents the propagation of light through the material.

While n and k are convenient macroscopic coefficients, the interaction of light with matter involves the ability of an oscillating electric field to excite oscillations of the electrons and ion cores in the material. If the frequency of vibration of the light is less than the natural frequency of a vibrational mode in the solid, it can excite that oscillation. We identify three possible oscillations to excite in the solid. The tightly bound, inner shell electrons have a natural frequency of around 10^{19} Hz. The loosely bound valence electrons have a frequency around 10^{16} Hz and ion cores in the lattice have a frequency around 10^{12} Hz. Since visible light has frequencies in the 10^{14}–10^{15} Hz range, it can excite both the inner shell electrons and the valence electrons. The inner shell electrons, however, are so tightly bound that the impact of this excitation is small, and we focus our attention on the valence electrons.

In metals, the valence electrons can be treated as a free electron gas as we described earlier in this chapter. From this model, the dielectric/optical properties of the metal can be described in terms of the angular frequency, ω, of the incident electromagnetic radiation, the electron gas plasma frequency, ω_p, which depends on the density of free electrons, and the Lorentz-Sommerfeld frequency of the electron gas, ω_0, which measures the coupling of the electrons and phonons.

$$\varepsilon_1 = n^2 - k^2 = 1 - \frac{\omega_p^2}{\omega^2 + \omega_0^2} \tag{6.22}$$

$$\varepsilon_2 = 2nk = \frac{\omega_p^2 \omega_0}{\omega\left(\omega^2 + \omega_0^2\right)} \tag{6.23}$$

where ω_p was defined in Equation 2.17 and $\omega_0 = 1/$relaxation time.

For low frequencies (long wavelengths) $\omega \ll \omega_0$ and $\omega_p \approx \omega_0$ so these equations yield

$$n = k = \sqrt{\varepsilon_2/2} \tag{6.24}$$

which is observed in the far infrared.

The free electron model predicts that metals will typically be very reflective at lower frequencies such as the visible and infrared parts of the spectrum and will have high absorption at higher frequencies ($\omega > \omega_0$ and ω_p) in the visible and ultraviolet parts of the spectrum. This explains the observed colors of metals. Ag and Al typically reflect throughout the visible spectrum giving them a whitish appearance. Cu and Au reflect in the long wavelengths (yellow and red) and absorb in the short wavelengths (blue and green) giving those materials a reddish appearance.

If we consider a good conductor, we can define a skin depth, d_{skin}, in which most of the light is absorbed

$$d_{\text{skin}} = \frac{2}{\alpha} \approx \frac{\lambda}{2\pi} \tag{6.25}$$

For visible light of 500 nm wavelength, this would correspond to about an 80 nm skin depth. This provides a quick way of estimating the penetration depth of light into a metal.

Dielectric materials cannot be described using the free electron model. In these materials, the valence electrons are bound with a natural vibration frequency, ω_n. It can be shown

$$\varepsilon_1 = 1 + \omega_p^2 \frac{\omega_n^2 - \omega^2}{\left(\omega_n^2 - \omega^2\right)^2 + \omega_0^2 \omega^2} \tag{6.26}$$

$$\varepsilon_2 = \omega_p^2 \frac{\omega_0 \omega}{\left(\omega_n^2 - \omega^2\right)^2 + \omega_0^2 \omega^2} \tag{6.27}$$

Note that when $\omega=\omega_n$ we get a maximum absorption that is typically in the UV part of the spectrum. This model suggests that dielectric materials will absorb and/or reflect light in the ultraviolet part of the spectrum and transmit light in the visible. At higher wavelengths (shorter frequencies) the infrared part of the spectrum again will have low transmission of light from excitation of lattice vibrations.

This model, if applied to semiconductors, would predict that the transparent region would shift down into the infrared part of the spectrum with low transmission of light in the visible and ultraviolet parts of the spectrum.

6.3.2 Optical Properties at Surfaces

We have already seen how the electrical properties of materials change near surfaces. The importance of the interaction of electrons with light in materials then suggests that optical properties will experience similar changes in the surface region. A more advanced treatment of surface optics is available (Bacsa, Bacsa and Myers 2020).

Our discussion of the optical properties of bulk materials has implicitly assumed that the electric field changes slowly on the atomic scale. Since light has a typical wavelength of about 500 nm and atomic dimensions are around 0.5 nm, this is a reasonable approximation. At an interface, however, the material properties change rather abruptly over less than 1 nm. This sudden change may force a rapid change in the electric field that suggests that the surface must have different optical properties from the bulk.

Most optical modeling relies on a layer model where each layer has a sharp planar boundary and has uniform optical properties. A simple macroscopic layer model, represented in Figure 6.33, takes into account the existence of different optical properties at the surface by defining a surface layer of thickness, d_s, which could be described by some new optical properties, n_s and k_s while keeping the bulk optical properties, n_b and k_b as determined by the methods of the last section.

Exploring this in more detail, we note that classical physics provides a contradiction at the surface. Using the z-direction to be perpendicular to the surface, classical physics assumes that E_z varies slowly. Maxwell's equation $\vec{\nabla}\cdot\vec{D}=0$ where $\vec{D}=\varepsilon\vec{E}$, however, requires that E_z be discontinuous at a surface that has charge present. To work around this contradiction, we keep Maxwell's equation but modify the relationship of $\vec{D}$ to $\vec{E}$.

The results of this model are that nothing is changed for incident frequencies $\omega\leq0.7\omega_p$. For $0.7\omega_p<\omega\leq\omega_p$ electron-hole pairs are created at the surface causing energy to be lost

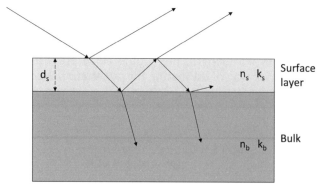

FIGURE 6.33
Reflection and refraction of incident light in a layer structure.

from the light. The surface electric field can couple indirectly with surface plasmons creating a surface plasmon polariton. A polariton is a quantized particle associated with a transverse wave caused by the coupling of phonons and photons. This can occur when ω and the wave vector for both the phonon and photon are approximately equal: $\omega_{photon} \approx \omega_{phonon}$ and $\vec{k}_{photon} \approx \vec{k}_{phonon}$. If $\omega > \omega_p$ the light can excite a bulk plasmon as discussed in Section 6.1.

Non-linear optical effects can be accentuated at surfaces. In optics, we often assume that relationships are linear. For instance, the polarizability is proportional to the electric field.

$$\vec{P} = \chi \vec{E} \tag{6.28}$$

where χ is the susceptibility. This is actually an approximation to the general expression

$$P = \chi_1 E + \chi_2 E^2 + \chi_3 E^3 + \dots \tag{6.29}$$

Typically we find that $\chi_2 \approx 10^{-8} \chi_1$ and $\chi_3 = 10^{-15} \chi_1$ so we will only see the non-linear effects with very intense electric fields such as those generated in powerful lasers.

The leading non-linear term, χ_2, depends on the symmetry of the environment. In a centrosymmetric environment, such as that of a bulk material, $\chi_2 = 0$. A surface, however, breaks the centrosymmetric symmetry and allow for the non-linear effects to be observed. One consequence of the non-linear term is that on reflection of light with frequency, ω, we can generate second harmonic radiation with frequency, 2ω as indicated in Figure 6.34. For instance, we can send in infrared radiation and have both infrared and green light emitted by the surface. For a simple reflection from a Ag surface, the efficiency of producing the second harmonic is about 10^{-15} although higher efficiencies are possible with other systems.

6.3.3 Adsorbates on Surfaces

While adsorbed atoms (atoms on top of the surface rather than in the surface layer) will be treated in detail in Chapter 7, we consider the optical properties of such layers here. Submonolayer coverages of adsorbed atoms will change the optical properties in the surface region. A simple macroscopic approach to this is to describe the adsorbed atoms as a layer with optical properties that differ from the surface and the bulk. Note that the surface optical properties may change in the presence of the adsorbed atoms. Sometimes the surface and adsorbate effects are combined by describing the system as a single adsorbate layer on top of the bulk as indicated in Figure 6.35.

There are two common macroscopic models for describing the optical properties of a layer of adsorbate atoms as the coverage changes. One is to assume that the layer has

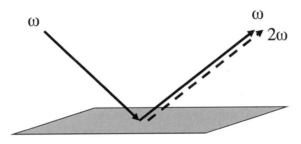

FIGURE 6.34
Near surfaces light can be reflected with both the incident frequency and the second harmonic.

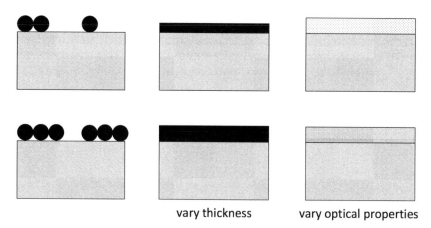

vary thickness   vary optical properties

FIGURE 6.35

Adsorbed atoms on a surface can be treated as a layer with constant optical properties but variable thickness or a layer of constant thickness with variable optical properties.

a constant n and k. We then allow the thickness of the layer to vary with coverage. The second is to assume that the thickness is constant (typically one atomic layer) and allow the optical properties to vary with coverage. Both of these models clearly attempt to use macroscopic quantities in describing atomic level properties.

A microscopic model of sub-monolayer distributions of adatoms (Strachan 1933) treats the atoms as oscillators that interact with light by scattering. Assuming the separation of the adsorbed atoms or molecules is smaller than the wavelength of light, the oscillators can be treated as a continuous distribution. This model gives reasonable agreement if the adsorbed atoms are physisorbed but very poor agreement for chemisorbed layers since it fails to take into account adsorbate-surface interactions. Models employing polarizable dipoles, quadrupoles, or multipole expansions can be used to treat adsorbed atoms on surfaces (Bedeaux and Vlieger 2002).

The presence of adsorbed atoms on the surface also changes the local interaction with the electric field from the incident light. The non-linear susceptibility term χ_2 is found to be proportional to the surface coverage of adsorbed atoms.

6.3.4 Films

The simplest optical modeling of films is again to use a layer model where we assume that the film can be described as a homogeneous layer characterized by thickness and optical properties as indicated in Figure 6.36.

The optical properties of thin films are complicated by the non-ideality of the materials. The surfaces and interfaces may have roughness and contamination. The film itself may have porosity when compared to the bulk material and may also have contamination. Within a layer model, we can attempt to take into account the surface and interfacial effects by adding additional layers to our model. Each layer has a uniform thickness and homogeneous optical properties as demonstrated in Figure 6.37.

A rough surface layer, a porous layer, or a layer that is a composite of more than one material would all be modeled here as a layer of uniform thickness and optical properties. We explore here the use of effective medium approximations (Aspnes 1982a,b) to define the uniform optical properties of a heterogeneous layer. In these approximations,

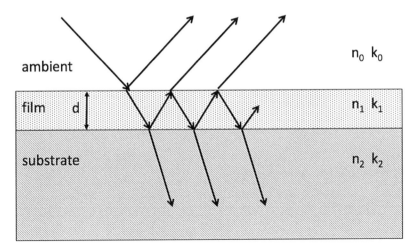

FIGURE 6.36
Reflection and refraction in a thin film layer model.

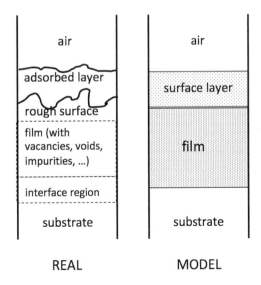

REAL   MODEL

FIGURE 6.37
Comparison of reality with an optical layer model.

a two-component layer is assumed to consist of spherical regions of material A and spherical regions of material B as shown in Figure 6.38. The regions in the layer that are not spheres of one material or the other are described as a host material. The choice of that host material properties will vary among the various effective medium approximations.

We note that a simple attempt to average the dielectric or optical properties of components A and B will not work since these macroscopic properties are not additive. It is atomic level properties that should be added such as the atomic polarizabilities, α, of the components. The Clausius–Mossotti equation relates the complex dielectric coefficient, ε, to the complex atomic polarizability, α, while assuming that the host medium was air with $\varepsilon = 1$.

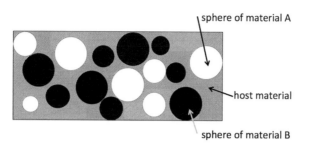

FIGURE 6.38
Effective medium model for heterogeneous materials.

$$c = \cfrac{1 + 2\left(\dfrac{4\pi}{3} n_{\text{dipole}} \alpha\right)}{1 - \left(\dfrac{4\pi}{3} n_{\text{dipole}} \alpha\right)} \tag{6.30}$$

where n_{diplole} is the number density of atomic dipoles. This can be solved for α.

$$\frac{\varepsilon - 1}{\varepsilon + 2} = \frac{4\pi}{3} n_{\text{dipole}} \alpha \tag{6.31}$$

A more general solution for multiple components in a host medium yields

$$\frac{\varepsilon - \varepsilon_h}{\varepsilon + 2\varepsilon_h} = \frac{4\pi}{3} \sum_i n_{\text{dipole}_i} \alpha_i \tag{6.32}$$

where ε_h is the dielectric coefficient of the host medium and we sum over i components. Since the α_i are often unknown, but the dielectric properties of the pure components are known, we substitute back into the expression to eliminate α_i in favor of the dielectric properties of the components, ε_i. This yields the general expression of the effective medium approximation.

$$\frac{\varepsilon - \varepsilon_h}{\varepsilon + 2\varepsilon_h} = \sum_i f_i \frac{\varepsilon_i - \varepsilon_h}{\varepsilon_i + 2\varepsilon_h} \tag{6.33}$$

where f_i is the volume fraction of the ith component.

Several choices are possible for the host medium. The Lorentz-Lorenz model assumes that $\varepsilon_h = 1$ for vacuum or air. The Garnett model assumes that we have spheres of one component imbedded in the other component and so $\varepsilon_h = \varepsilon_A$ or ε_B. The Bruggeman model assumes a self-consistent solution and so equates the host material dielectric properties with the overall dielectric properties of the layer and sets $\varepsilon_h = \varepsilon$.

An effective medium approximation is frequently used to model the optical properties of a composite film. Porosity can be modeled by choosing one of the materials to have $\varepsilon = 1$. A rough surface could be modeled as a surface layer consisting of material and empty space ($\varepsilon = 1$) as represented in Figure 6.39. In this sense, a rough layer and a porous layer would be modeled in the same manner.

Wiener (1912) established absolute limits on the possible combinations of two material components in any geometric arrangement. They correspond to horizontal alternating

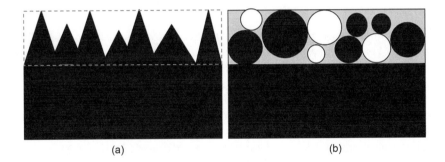

FIGURE 6.39
(a) Rough surface. (b) Bruggeman effective medium approximation showing spheres of the material (black) and void (white) imbedded in a self-consistent host material (gray).

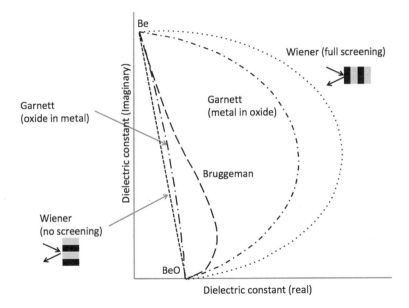

FIGURE 6.40
Effective medium models of the dielectric constant of a heterogeneous mixture of Be and BeO.

layers and vertical alternating layers. Plotting these limits, along with the effective medium approximations, can give some insight into the modeling of films. An example for a mixture of metallic Be and transparent BeO is provided in Figure 6.40. Any mixture of Be and BeO should lie within the Wiener limits. The exact location of experimental data in that region would give insight into the quantity and the physical arrangement of the two materials.

6.4 Thermal Properties

The thermal behavior of materials depends on both the lattice vibrations of the atoms (phonons) as discussed in Chapter 4 and the motion of electrons in materials.

6.4.1 Specific Heat

Specific heat is a material property defined as the heat capacity per unit mass. The specific heat approaches zero at very low temperatures and approaches a constant value of 3R where R is the universal gas constant at room temperature and above for many materials. The specific heat depends primarily on lattice vibrations. The electron contribution is small since only electrons that can be thermally excited into an empty state will contribute. Referring back to Figure 6.2 suggests that only electrons that are in states near the Fermi energy are likely to have empty states that are close enough in energy to be available to them. Since surfaces have lattice vibrations that are different from the bulk, a difference in specific heat near the surface is expected. It may be difficult, however, to measure this difference in the presence of the much more massive bulk of the material.

6.4.2 Thermal Conductivity

The thermal conductivity depends on both phonons and electron motion. Good electrical conductors (such as metals) also tend to be good thermal conductors and electron motion dominates thermal conduction in these materials. Materials with ionic or covalent bonds may have phonon-dominated thermal conduction. The thermal conduction is typically driven by a temperature gradient across the material, dT/dx with heat flowing from regions of high temperature to low temperature. In a crystalline solid, the thermal conductivity depends on the direction in the crystal. In bulk polycrystalline materials, the conductivity is an average of all of the directions. Thin films with oriented crystallites may have different thermal conductivity in the directions perpendicular and parallel to the surface.

If we treat the solid as a gas, then the conductivity should depend on how quickly the electrons or phonons are moving, the mean free path between collisions, and the amount of heat that they are able to carry (specific heat). In good metallic conductors, the mean free path of electrons is much longer than that for phonons, but the phonon-specific heat is much greater. Detailed calculations suggest that electron conduction is more important in these materials. Increasing temperature will increase the velocities but decrease the mean free paths resulting in a thermal conductivity that is relatively constant with temperature.

In Chapter 1 we presented a simple argument that electrons will experience greater scattering (shorter mean free paths) near surfaces. In Chapter 4, we examined differences in lattice vibrations near surfaces. These differences suggest that surface thermal conductivity may be different from bulk thermal conductivity. Again, these measurements would be difficult due to the presence of the bulk material.

6.4.3 Thermal Expansion

In Chapter 4, we saw that an increase in temperature would increase the amplitude of vibration of the atoms in the lattice (Equation 4.2). The potential energy well that the atoms vibrate in can often be described by something like the Lennard-Jones potential (Equation 3.4). The important observation about this potential is that it is not symmetric. As the amplitude of vibration increases, the motion will favor increasing the interatomic spacing. We thus expect all materials to expand when heated.

Metallic bonds result in a more asymmetric potential leading to a greater thermal expansion. Covalently bonded materials have a more symmetric potential that results in a smaller thermal expansion. This connection to bonding potential energy also leads to a crystallographic direction dependence for thermal expansion.

Chapter 3 demonstrated that the bonding environment of surface atoms is different from the bulk in many materials. This should lead to differences in surface thermal expansion when compared to bulk thermal expansion.

6.5 Characterization of Surface and Film Properties

The electrical properties of surfaces and thin films can be probed on an atomic scale using a scanning tunneling microscope. Surface states can be observed and the density of states of a particular surface atom can be mapped. This is described in Chapter 11. Band structure of filled states can be probed with photoelectron spectroscopies and the unoccupied states can be examined with near-edge X-ray absorption fine structure. These techniques are discussed in Chapter 13. The Hall effect can determine the concentration and sign of charge carriers. Electrical resistivity can be measured. These techniques are discussed in Chapter 14.

Magnetic properties of materials can be examined using the Magneto-optic Kerr effect, spin-polarized electron techniques, magnetic force microscopy, and Brillouin light scattering, and various types of magnetometer measurements. These are all included in Chapter 14. Optical properties of surfaces and thin films can be determined using ellipsometry, reflectance measurements, and interferometry as described in Chapter 14. The thermal properties of materials can be explored using micro-thermal microscopy, photothermal analysis, and cross-thermal analysis which are described in Chapter 15.

References

Aspnes, D.E. 1982a. Optical properties of thin films. *Thin Solid Films.* 89:249–262.

Aspnes, D.E. 1982b. Local-field effects and effective-medium theory: A microscopic perspective. *Am. J. Phys.* 50: 704–709.

Bacsa, W., R. Bacsa, and T. Myers. 2020. *Optics near Surfaces and at the Nanometer Scale.* Cham: Springer Nature.

Bedeaux, D. and J. Vlieger. 2002. *Optical Properties of Surfaces.* London: Imperial College Press.

Diep, H.T. 2014. *Theory of Magnetism: Application to Surface Physics.* Singapore: World Scientific Publishing Co.

Gross, A. 2009. *Theoretical Surface Science: A Microscopic Perspective.* 2nd ed. Berlin: Springer Verlag.

Heavens, O.S. 1991. *Optical Properties of Thin Solid Films.* New York: Dover Publications, Inc.

Helsey, K.N. ed. 2011. *Plasmons: Theory and Applications.* Hauppauge, NY: Nova Science Publishers.

Ibach, H. 2006. *Physics of Surfaces and Interfaces.* Berlin: Springer.

Kaneyoshi, T. 1991. *Introduction to Surface Magnetism.* Boca Raton, FL: CRC Press.

Kittel, C. 1996. *Introduction to Solid State Physics.* 7th ed. New York: John Wiley & Sons, Inc.

Omar, M.A. 1975. *Elementary Solid State Physics: Principles and Applications.* Reading, MA: Addison-Wesley Publishing Company.

Patterson, J.D. and B.C. Bailey. 2007. *Solid-State Physics.* Berlin: Springer.

Reynolds, D.C. 1981. *Excitons: Their Properties and Uses.* New York: Academic Press.

Rose, R.M., L.A. Shepard and J. Wulff. 1964. *The Structure and Properties of Materials: Volume IV Electronic Properties.* New York: John Wiley & Sons, Inc.

Sarid, D. and W. Challener. 2010. *Modern Introduction to Surface Plasmons: Theory, Mathematica Modeling and Applications*. Cambridge: Cambridge University Press.

Strachan, C. 1933. The reflection of light at a surface covered by a monomolecular film. *Proc. Cambridge Philos. Soc.* 29: 116–130.

Volkerts, J.P. ed. 2011. *Magnetic Thin Films: Properties, Performance and Applications*. Hauppauge, NY: Nova Science Publishers.

Wiener, O. 1912. Die Theorie des Mischkörpers für das Feld der stationären Strömung. 1. Abjandlung: Die Mittelwersätze für Kraft, Polarisation und Energie. *Abh. Math. Phys. Kl. Königl. Sächs. Ges.* 32: 509–604.

Zangwill, A. 1988. *Physics at Surfaces*. Cambridge: Cambridge University Press.

Problems

Problem 6.1 In the free electron model, it can be shown that the Fermi Energy (in three dimensions at $T=0$) is given by:

$$E_F = \left(\frac{h^2}{8m_e} \right) \left(\frac{3N}{\pi V} \right)^{2/3}$$

where N is the number of free electrons in the sample and V is the volume. As you can see the Fermi Energy (in this model) depends only on the number density of free electrons. Calculate the Fermi Energy for Aluminum ($n=18.1\times10^{22}$ electrons/cm³) and for Potassium ($n=1.4\times10^{22}$ electrons/cm³). Compare these numbers to the thermal energy ($k_B T$) associated with room temperature. Comment on your comparison.

Problem 6.2 One simple model for a solid, which is in the nearly free electron category, is the Kronig-Penney model. This uses a simple quantum mechanical approach. The potential of the lattice is approximated as a series of finite square wells – rather than a more realistic shape as shown in Figure 6.4.

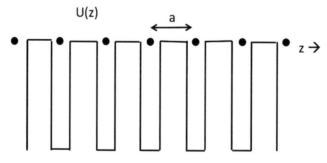

This is the potential that the electrons experience. This potential can be put into Schrödinger's Equation which can then be solved. We will get two types of solutions depending on what region of the crystal we are in. When far from the ion cores, the electrons are essentially free and the wave function will be

$$\psi = Ae^{ikz} + Be^{-ikz}$$

where the wave number, k, is given by $k^2 = 2mE/\hbar^2$. In the regions near the ion cores the electrons are tunneling through the barriers and the wave function is

$$\psi = Ce^{kz} + De^{-kz}$$

with $k^2 = 2m(V_o - E)/\hbar^2$.

By applying the boundary conditions (ψ and $d\psi/dz$ continuous at boundaries) this problem can be solved. (It is rather tedious to do).

One result is the following equation

$$(A' / ka)\sin(ka) + \cos(ka) = \cos(Ka)$$

where A and K are constants, "a" is the lattice spacing and k is the wave number.

Note that the left side of the equation can take on values greater than 1 or less than –1 while the right side can only have values within the range –1 to 1. As a result there will be certain values of k for which the equation cannot be solved.

a. Make a plot of the left side of the equation as a function of ka and show that the model predicts a set of evenly space forbidden zones (or gaps). For simplicity, let $A'=3\pi/2$ and plot ka from 0 to 4π.

b. This same model can be used to describe a crystal which terminates at a surface (in one dimension). Draw the potential as a function of z (similar to (a) in the figure given above). Explain any changes you make in the potential as a result of the presence of a surface. There are several "right" answers to this. Give some thought to how the potential should be impacted by a surface and decide what makes the most sense to you and why.

Problem 6.3 We claimed that the energy of a surface plasmon $E_S = \dfrac{E_B}{\sqrt{2}}$, where E_B is the bulk plasmon energy and $E_B = \hbar\omega_P$ relates the energy to the plasma frequency. This result basically follows from electromagnetic theory and the application of boundary conditions at the surface. We find a solution to Laplace's equation and then apply the boundary condition that the normal component of **D**, the electric displacement vector, must be continuous across the boundary at $z=0$. The solid is located in the $z>0$ region. Vacuum is in the $z<0$ region. The solution to Laplace's equation gives the components of the electric field:

We have used the facts that the parallel component of the electric field must be continuous at the boundary and that $\nabla \cdot E = 0$ in the vacuum to determine the forms of these components.

a. Recall that inside the material $\mathbf{D} = \varepsilon(\omega)\,\mathbf{E}$ and that outside the material $\mathbf{D}' = \mathbf{E}'$ since $\varepsilon = 1$ in vacuum. $\mathbf{D}$ is the electric displacement vector. Now use the boundary condition that the normal component of $\mathbf{D}$ (D_z) must be continuous across the surface ($z = 0$) to show that $\varepsilon(\omega) = -1$ at the surface. (Notice that there are two types of frequencies in this problem. The plasma frequency is a property of the material. The electric field also varies with a frequency, ω. At the surface ($z = 0$) the electric field frequency interacts with the material and so we are interested in $\omega = \omega_s$ the frequency of the surface plasmon.)

b. From free electron gas theory, it can be shown that

$$\varepsilon(\omega) = 1 - (\omega_n / \omega)^2,$$

Use the result from part a along with this to show that $E_s = E_B / \sqrt{2}$.

Problem 6.4 The Richardson-Dushman equation relates the thermionic current density (J) to the temperature.

$J = AT^2 e^{-\Phi/k_B T}$ where $A = 120$ amps/cm² deg²

a. Calculate the temperature required to obtain a thermionic electron emission current density of 10^{-2} amps/cm² from wires of aluminum and tungsten. You may need to look up some physical properties of the metals. (This equation is not easy to solve analytically, so you may want to use a computer, an iterative method, or a graphical method to solve it.)

b. Considering other properties of the materials (melting point, etc.), which of these metals would make a practical themionic emitter and why?

Problem 6.5 In a free electron gas, the plasma frequency is given by $\omega_P = \sqrt{4 n_e e^2 / m_e}$ in cgs units, where n_e is the number density of free electrons.

a. Estimate the number density of valence electrons from the density of Silicon (2.33 g/cm³), the atomic weight of Silicon (28.0855 amu), and the expectation from the periodic table that Silicon should have four valence electrons per atom.

b. Use the result from (a) to calculate the bulk and surface plasmon energies of Si. (Experimental measurements yield a Si bulk plasma energy of 17.2 eV.)

Problem 6.6 Interaction of light with dielectric materials

For techniques like ellipsometry to be useful, we need to relate the properties of light to the properties of solid materials.

Since light is a time varying electric field, it is likely to interact most strongly with the electrons in the solid. Instead of free electron materials, let us consider dielectric materials where the electrons are bound to the ion cores by some sort of force constant, K at an equilibrium distance, r. As the electric field interacts with the electrons, it will move the electrons around the equilibrium position. This results in polarization of the atoms,

$$P = -ner,$$

where n is the number of electrons per unit volume and each electron has charge, $-e$.

The equation of motion of the electrons can be expressed as a damped harmonic oscillator with the damping described by a constant mb:

$$m\frac{d^2r}{dt^2} + mb\frac{dr}{dt} + Kr = -eE$$

a. If the electric field (and thus the electrons) vary harmonically ($e^{-i\omega t}$), show that the equation of motion can be written as

$$\left(-m\omega^2 - imb\omega + K\right)r = -eE$$

(An effective resonant frequency of the bound electrons can be defined as $\omega_0 = \sqrt{K/m}$.

b. Show that the resulting atomic polarization can be expressed as

$$P = \frac{ne^2/m}{\omega_0^2 - \omega^2 - i\omega b}E$$

This polarization term changes the electric field (light) that propagates through or reflects from a solid because the general wave equation contains a source term involving the polarization. If we take the analysis further, we could show that the electric field depends on the optical properties of the solid and involves both a term indicating that some light is absorbed and a term that shifts the phase of the electric field of the light. This would change the polarization of the light which makes ellipsometry work!

7

Adsorbed Atoms on Surfaces

We now consider what happens when atoms or molecules from a gas phase bond to a clean surface. This process, where the atoms are sitting on top of the surface, is adsorption. The adsorbed atoms are called adatoms. In this chapter, we consider only cases where there are relatively few atoms on the surface. Typically, we will explore situations where we have less than one atomic layer (monolayer). This topic is typically regarded as surface science, but it begins the transition into the world of thin films. As more atoms come down onto the surface, we start to create a thin film, which we discuss in Chapter 8. More extensive discussions of adsorbed atoms on surfaces are available in the literature (Bruch, Cole and Zaremba 1997, Gross 2009, Hudson 1998, Ibach 2006, Prutton 1994, Zangwill 1988).

7.1 Thermodynamics of Adsorbed Atoms

We begin by exploring a macroscopic thermodynamic approach to adsorption. We identify two types of bonds with which atoms may be adsorbed. Physical adsorption (often called physisorption) is a weak bond originating from a van der Waals interaction (introduced in Section 3.3) between the adsorbed atom and the surface. No charge is transferred between the atom and surface, but the adsorbed atom is polarized with a negative charge being pulled closer to the surface. This is a relatively weak bond of typically less than 0.25 eV. This bond will be of particular interest later in this chapter because the energy associated with the adsorbate–substrate interaction is comparable to that of the interactions between adsorbate atoms. This allows for the formation of ordered surface phases of adsorbate atoms.

The other type of bond is chemical adsorption (often called chemisorption) where a stronger bond (typically ionic or covalent as described in Section 3.3) is formed that involves charge transfer between the atom and the surface. In this case, the adsorbate–substrate interaction is much stronger than the interactions between the adsorbate atoms, which will hinder the formation of any ordered surface structures. A plot of the potential energies involved in physisorption and chemisorption is shown in Figure 7.1.

A clean surface in the presence of atoms in a gas is not a thermodynamically favorable situation. The surface tension, γ, as described in Chapter 5, can decrease by adsorbing atoms onto the surface. The presence of adatoms may also stimulate a surface reconstruction that would further reduce γ as described in Chapter 3. We can thus assign an adsorption energy, E_a, to this process indicated in Figure 7.1.

We explore a simple thermodynamic model of chemisorption that examines the conditions under which charge transfer may occur. We are interested in what happens to the electrons in the adatom, so we need three terms. (1) The ionization energy of the adatom, I, is the energy needed to remove an electron from the adatom. (2) The electron affinity of the adatom, A, is the energy lost by adding an electron to an adatom. (3) The work function, ϕ,

DOI: 10.1201/9780429194542-8

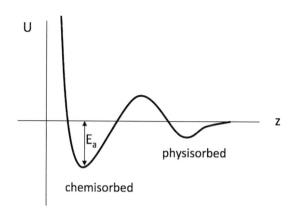

FIGURE 7.1
Adsorption states on a surface

of the surface is the energy attracting an electron to the surface, which is also the energy inhibiting an electron from leaving the surface.

A simple model like this predicts that if $\phi > I$, an electron should transfer to the surface. Similarly, if $A > \phi$, an electron should transfer to the adatom. Finally, if $A < \phi < I$, no charge transfer should take place and chemisorption will not occur. Calculation of these three energies, however, is complicated since atoms near surfaces will cause changes in the energy levels of both the adatoms and surface atoms.

Atoms that are adsorbed to the surface will typically desorb again very quickly. If we consider a desorption energy, E_d, which is the barrier energy for an atom leaving the surface, we can define a typical residence time, τ, that an atom will remain on the surface.

$$\tau = \tau_0 e^{E_d / kT} \tag{7.1}$$

Typical values of $\tau_0 = 10^{-12}$ s and $E_d = 0.25$ eV, suggest that at $T = 300$ K, $\tau = 2 \times 10^{-8}$ s and a surface cooled to $T = 100$ K would have $\tau = 4$ s.

With both adsorption and desorption happening simultaneously, we consider an equilibrium surface coverage, σ, arising from a balance of these two competing processes. The units of surface coverage are atoms/area, which is also often reported as the fraction of a monolayer on the surface. The incident flux of particles, Φ, can often be determined from the kinetic theory of gases as in Equation 2.11. The total number of particles that a surface has been exposed to will depend on this flux and the amount of time that the flux continues. This can be described in terms of an exposure, which is given as the pressure multiplied by the time. This is often reported in units of Langmuirs where 1 Langmuir = 10^{-6} Torr seconds.

It is useful to define a condensation coefficient, α_C, which is the probability that impinging particles will be adsorbed on the surface. With these definitions, we can express the surface coverage as

$$\sigma = \alpha_c \Phi \tau \tag{7.2}$$

Notice that if the flux of atoms is stopped, the coverage goes to zero as adatoms desorb. Maintaining sub-monolayer adsorbed atoms on a surface requires the presence of the desired gas in the chamber. In Chapter 8, we will discuss how adsorbed atoms can be stabilized on a surface to create a thin film.

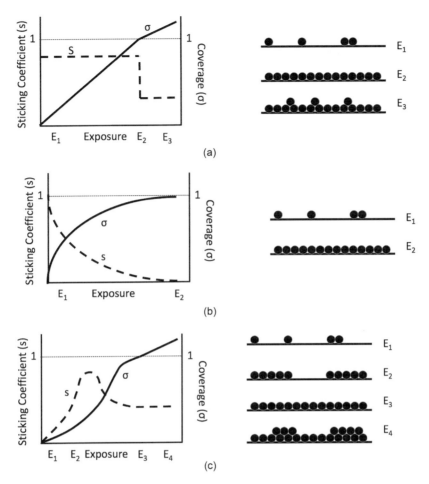

FIGURE 7.2
Coverage and sticking coefficient in several different cases.

Another useful term is the sticking coefficient, s, which is the surface coverage (number of atoms/area on the surface) divided by the total number of atoms/area incident to the surface. While the term "sticking" suggests a permanence on the surface that is not accurate, the concept is useful. We consider several physical situations leading to sticking coefficients that will change as atoms adsorb onto the surface. This will also change the surface coverage.

Figure 7.2 shows several possible scenarios for how sticking coefficient and coverage change as the surface is exposed to additional adatoms. Figure 7.2a considers a case where the adsorbing atom interacts with the substrate differently than with other adsorbed atoms. As long as some surface is exposed, the sticking coefficient has one constant value. Once a full monolayer of adsorbed atoms covers the surface, the sticking coefficient changes to a different (in this example, lower) value. With this model, the surface coverage would be expected to increase linearly up to $\sigma=1$ and then continue to increase but at a lower slope. Figure 7.2b considers the case where the sticking coefficient depends on the amount of surface available for adsorption. If the adsorbed atoms do not stick at all to other adsorbed atoms, then the sticking coefficient would initially be some high value (1 in this example) and decreases as more and more of the surface is covered until reaching a value of 0 at one monolayer of coverage. The surface coverage corresponding to this situation would increase gradually to a value of 1 and then not change after that since no more atoms could be adsorbed. Figure 7.2c considers the case where adsorption occurs only at the edges of islands of adsorbed atoms. The sticking coefficient starts at 0 since there are no islands until they are nucleated on the surface. As the islands grow in size the amount of perimeter where additional adsorption can occur increases and the sticking coefficient increases with it. Once the islands are large enough, however, that they begin to touch one another, the amount of free perimeter decreases and the sticking coefficient decreases as well to a value typical of the adsorbate sticking to itself when we reach a monolayer of coverage.

Experimentally, we are often interested in how the surface coverage changes for different gas pressures. When this is done at constant *T*, it yields an adsorption isotherm. We consider here three different models for adsorption isotherms.

In the first model, assume that every impinging atom on the surface will stick ($\alpha_C = 1$) and that the adsorbate atoms do not interact with one another. The surface is treated as a plane with unlimited, equivalent adsorption sites. This simple model is easy to solve within the kinetic theory of gases and is good for describing adsorption at very low pressures and in the early stages of adsorption where the assumptions we made are reasonable. The result of this calculation is that the isotherm can be expressed as

$$\sigma = k_1 P \tag{7.3}$$

where k_1 is a constant.

A second model assumes that impinging atoms that strike an adsorbed atom are reflected back but that impinging atoms that strike the surface will stick. Again we assume no interaction between the adsorbate atoms. The surface is planar and has a limited (but constant) number of adsorption sites, *N*. If we define the surface coverage at one monolayer as σ_0 then we see that this model predicts that adsorption will stop when σ reaches σ_0.

In this model, the desorption rate is proportional to the number of filled surface sites (σN) and the adsorption rate is proportional to both *P* and to the number of empty surface sites $(1 - \sigma/\sigma_0)N$. At equilibrium, the desorption rate and adsorption rates will be equal.

$$k_d \sigma N = k_a (1 - \sigma/\sigma_o) N P \tag{7.4}$$

where k_d and k_a are proportionality constants. Defining a new constant $k_2 = k_a/k_d$,

$$\frac{\sigma}{1 - \sigma/\sigma_0} = k_2 P \tag{7.5}$$

Solving for σ yields the Langmuir isotherm:

$$\sigma = \frac{\sigma_0 k_2 P}{\sigma_0 + k_2 P} \tag{7.6}$$

While we assumed no interactions between the adsorbate atoms, in this model allowing interactions results in the same form for the equation but with a different proportionality constant.

In our third model, we introduce the idea of allowing multilayer adsorption so that adatoms can adsorb on top of other adatoms. We still assume, however, that there are no interactions between the adsorbed atoms. The surface is again planar and has a constant limited number of adsorption sites. This is an extension of the Langmuir model where multiple layers are allowed, but each layer obeys the assumptions of the Langmuir model. These assumptions lead to the Brunauer-Emmett-Teller (BET) isotherm.

$$\sigma = \frac{P}{P_0 - P} \left(\frac{1}{\sigma_0 k_3} + \frac{k_3 - 1}{\sigma_0 k_3} \frac{P}{P_0} \right)^{-1} \tag{7.7}$$

where k_3 is a constant and P_0 is the saturation pressure at which an infinite number of layers builds up.

7.2 Ordered Structures

In discussing adsorption isotherms we consistently assumed that adsorbed atoms do not interact with one another. In many cases, especially for physisorbed atoms, the adsorbate-adsorbate interactions are significant and can lead to the adsorbed atoms forming two-dimensional structures on the surface. Often there will be a simple relationship between the adsorbate spacing and the substrate atom spacing. In many cases, these adsorbate structures will lead to a close-packed adsorbate structure that will have the same rotational symmetry as the substrate.

For low coverages of adsorbed gas atoms, we expect that adatoms are widely separated and randomly distributed across the surface lattice sites in a two-dimensional lattice gas. If there is no interaction between the adatoms, this would be described using a two-dimensional kinetic theory of gases and ideal gas law. The three-dimensional ideal gas law depends on P, T, and n. A two-dimensional model will depend on P, T, and σ. Since we are interested in allowing adsorbate-adsorbate interactions, we move beyond an ideal gas model, to look at two-dimensional analogs of the van der Waals equation of state used for three-dimensional interacting gases. We are interested in the case where two-dimensional crystalline solids can nucleate from this lattice gas due to the interactions between the adatoms. This is the two-dimensional equivalent of homogeneous nucleation which will be discussed in Chapter 8.

At sufficiently low coverages, where surface atoms are widely separated, or at high temperatures, where the thermal energy is much greater than the adsorbate-adsorbate interaction, we do not expect a two-dimensional crystal to form. As we increase pressure or decrease temperature the new crystalline phase should be able to form. Figure 7.3 shows regions of two-dimensional crystal and regions of gas on a surface. The number of atoms is typically varied in experiments, by adjusting the pressure of the gas and/or the exposure time. A phase diagram for a single component gas is plotted as T vs. σ as in Figure 7.4. At low coverages or high temperatures, we observe a random two-dimensional gas. At lower temperatures and higher coverages an ordered adsorbate structure will form on the surface.

We can classify the two-dimensional crystal structures by considering the relationship of the crystals to the underlying substrate structure. If the adsorbate crystal has lattice vectors (see Chapter 3) in the same direction as the surface lattice vectors and with lengths that are a simple multiple of the surface lattice vector lengths, then we call the adsorbate crystal a primitive (*p*) lattice. If the adsorbate lattice vectors are twice the surface lattice

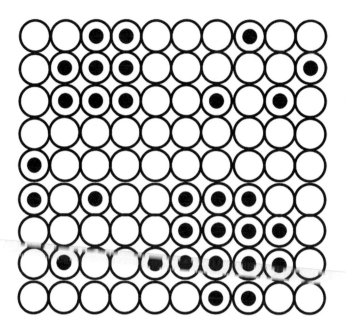

FIGURE 7.3
Two phases (lattice gas and crystal) adsorbed on a surface. White circles are substrate atoms and black circles are adsorbed atoms.

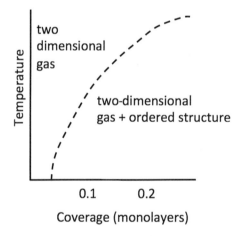

FIGURE 7.4
Phase diagram showing lattice gas and formation of a condensate.

vectors it is a $p(2\times2)$ lattice as shown in Figure 7.5. If we take a primitive lattice and add an extra adsorbate atom in the center of the unit cell, we create a centered (c) structure, with the relative length of the lattice vectors again defining the crystal. A $c(4\times2)$ lattice is shown in Figure 7.5. If the adsorbate lattice vectors are rotated from the surface lattice vectors, we add a notation "$Rx°$" to the end of the designation where x is the angle in degrees. An example of a $\sqrt{3}\times\sqrt{3}R30°$ is provided in Figure 7.5.

Figure 7.5 shows the adsorbate atoms as sitting directly on top of the surface atoms, but other positions could result in a lower energy configuration. Figure 7.6 shows three

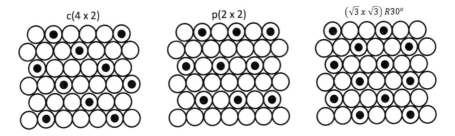

FIGURE 7.5
Three ordered structures on close-packed surfaces. White circles are substrate atoms and black circles are adsorbed atoms.

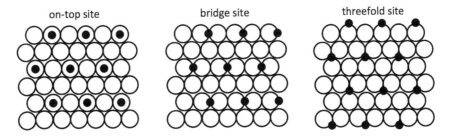

FIGURE 7.6
Three possible adatom locations that all produce a $p(2\times2)$ structure.

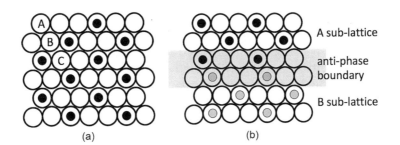

FIGURE 7.7
(a) Three equivalent sub-lattices (A, B, C) on a close-packed surface. The A-type crystal is shown. (b) Anti-phase boundary between two different sub-lattice structures.

identical $p(2\times2)$ crystalline structures sitting on three different sites on a close-packed surface. Our classification system would not distinguish between these three crystals.

As isolated adsorbate crystals begin to form on the surface, they may occupy positions on the surface lattice that are equivalent locally but will not be part of the same long-range order. An example is the three-fold sites of a close-packed surface (introduced in Section 3.5). There are three equivalent sites, labeled A, B, and C, in Figure 7.7a. Crystallites could form on any of these three equivalent sites as shown in Figure 7.7b. As these crystals grow, two-dimensional grain boundaries will form as two crystallites meet as indicated in Figure 7.7b between the A and B type crystals. If the crystal is primarily type A, then inclusions of Type B or C domains would be referred to as anti-phase domains.

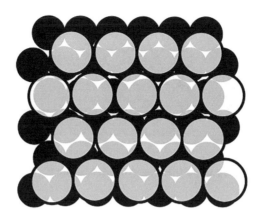

FIGURE 7.8

Incommensurate structure. The adatom structure (transparent white atoms) is not dependent on the substrate structure (black atoms).

As the surface coverage changes, different adsorbate crystal structures will be able to form on the surface. The $p(2\times2)$ in Figure 7.5 has one adsorbate atom for every four surface atoms and will thus be ideal for a surface coverage of 0.25 monolayer. The $\sqrt{3}\times\sqrt{3}R30°$ structure in Figure 7.5 would be expected at a surface coverage of 0.33 monolayer. Changing from one two-dimensional crystal structure to another involves a surface phase transition. Between coverages of 0.25 and 0.33 monolayers, for instance, we would expect a mixed phase region where both the $p(2\times2)$ and $\sqrt{3}\times\sqrt{3}R30°$ structures would coexist. For these low surface coverages, the adsorbate crystal is expected to follow the symmetry of the surface since the distance between adsorbate atoms is large compared to the lattice spacing expected for a bulk crystal of the adsorbate atoms. These structures are referred to as commensurate structures. At higher coverages, however, the interactions between the adsorbate atoms may become dominant over the adsorbate–substrate interactions and incommensurate structures may form where the adsorbate crystal structure is not directly related to the underlying surface crystal structure. An example of an incommensurate structure is provided in Figure 7.8. A simple, hard sphere model of the adsorbate structures suggests that the occurrence of incommensurate structures should be related to the relative lattice spacings (a) of the adsorbate and surface crystals. This can be described as a misfit

$$\text{misfit} = \frac{a_{\text{adsorbate}} - a_{\text{substrate}}}{a_{\text{substrate}}} \tag{7.8}$$

A large misfit will more likely lead to incommensurate structures as shown in Figure 7.8.

The transition from commensurate to incommensurate structures can be pictured in one dimension by considering that the adsorbate-adsorbate interactions are modeled by a weak spring between the adatoms. The surface is a periodic set of potential wells at each surface atom. Assume- that the adsorbate atoms would normally have a lattice spacing of a_a in the bulk and the surface atom spacing is a slightly larger value a_s. At lower coverage, the adsorbate atoms will minimize energy by being commensurate with the surface and being positioned in the potential wells and taking on the lattice spacing a_s. If we increase the number of adatoms, it is no longer possible for each adatom to sit in the potential well.

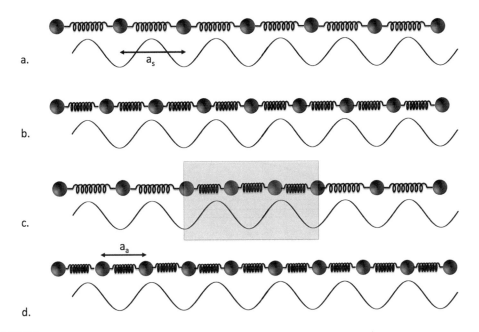

FIGURE 7.9
Transition from commensurate to incommensurate structure. (a) Commensurate structure with adatoms following substrate lattice spacing. (b) Incommensurate structure with adatoms following an intermediate lattice spacing. (c) Mixed commensurate and incommensurate (shaded) structure. (d) Incommensurate structure with adatoms following adatom lattice spacing.

One possible solution is for the adatoms to take on an incommensurate periodic structure with a lattice constant $a_a < a < a_s$. This can be shown to be a relatively high-energy configuration. Another possible solution is to have regions of commensurate structures separated by regions of incommensurate structures. As more adatoms are included in the system, the incommensurate regions will grow until eventually the entire surface is covered with an incommensurate adatom structure with lattice spacing a_a. This is demonstrated in Figure 7.9.

From this discussion, a system consisting of a single adsorbate on a surface can have quite a few different phases possible as the coverage changes. The existence of these phases can be represented on a phase diagram of T vs. σ such as Figure 7.10. Each distinct phase is typically separated from another phase by a region where both phases can coexist on the surface.

In Chapter 1 we discussed that many of the unique properties of surfaces arose from the fact that surface atoms had nothing above them. This left them in a very different bonding environment than a bulk atom. As adsorbate atoms come onto the surface, the surface atoms now have atoms above them. In effect, they are no longer "surface atoms" because they are being covered by an adsorbate layer. This changes the bonding environment. While it is still different from the bulk, it is not as different as a clean surface. Experimentally, we observe that some surface reconstructions and relaxations that are present on clean surfaces are reduced, or may disappear completely when adsorbates are added to the surface. Other times, the presence of the adsorbate may lead to a different reconstruction of the surface. These effects are also dependent on the surface coverage of the adsorbate.

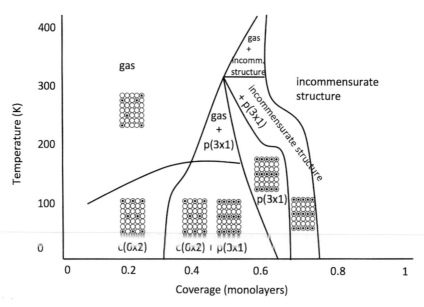

FIGURE 7.10
Phase diagram of two-dimensional structures based on Sr/W(1 1 0).

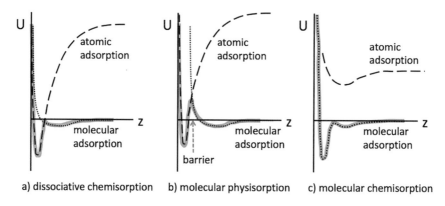

FIGURE 7.11
Potential energies of adsorption for (a) dissociative chemisorption, (b) molecular physisorption, and (c) molecular chemisorption. The potential for adsorption of separate atoms is shown as a dashed line. The potential for adsorption as a molecule is shown as a dotted line. The resultant potential is shown as a solid gray line.

7.3 Molecular Adsorption

This chapter has focused so far on atomic adsorption, but molecules can also adsorb onto surfaces. The molecule may be physisorbed or chemisorbed or it may dissociate upon adsorption into its constituent atoms. We can plot the potential energy vs distance from the substrate for molecules in a manner similar to Figure 7.1 for the physisorption and chemisorption of atoms. Figure 7.11a shows where the energy of the atoms adsorbed independently is lower than the energy of the molecule physisorbed and there is no significant

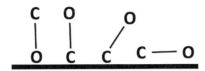

FIGURE 7.12
Several possible orientations of a diatomic molecule on a surface.

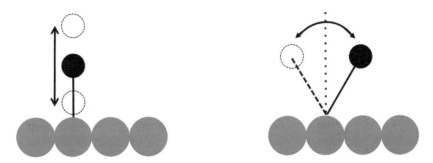

FIGURE 7.13
Vertical and horizontal vibrational modes of an atom adsorbed to a surface.

barrier between the two energy wells. This will lead to the molecule adsorbing and then dissociating to minimize energy. The resulting atoms are strongly bonded to the surface in chemisorption. Figure 7.11b shows the case where an energy barrier exists between the physisorption of the molecule and the dissociated chemisorption of the atoms. If the molecule does not have sufficient energy to overcome that barrier, the molecule will be physisorbed on the surface. Figure 7.11c examines the case where the chemisorption of the molecule is energetically more favorable than the dissociative adsorption of the individual atoms. In this case, the molecule will be strongly bonded to the surface. Which of these scenarios will occur will typically depend on the elements involved (both surface and adsorbate molecule), the crystallographic surface exposed, and the surface coverage of the adsorbate.

The adsorbed molecules are often asymmetric and, as a result, can adsorb on the surface in different orientations. An example of a simple diatomic molecule is shown in Figure 7.12.

7.4 Adsorbate Motions

As with surface atom motion discussed in Chapter 4, adsorbates can both vibrate around their equilibrium crystalline positions and move across the surface by diffusion. Typically, the vibrations of adsorbed atoms have a higher frequency than the surface atoms. The details of the vibrational modes will depend on the type of site (on-top, bridge, or three-fold) that the adatom is bonded to. The adsorbate atoms often have more degrees of freedom and larger vibration amplitudes. Typical vertical and horizontal vibrational modes are represented in Figure 7.13. The atom can vibrate in a vertical direction about the equilibrium position and can also rock along an arc with the radius defined by the equilibrium bond length. Of course, some combination of these motions is possible and the horizontal motion could be in any direction parallel to the surface.

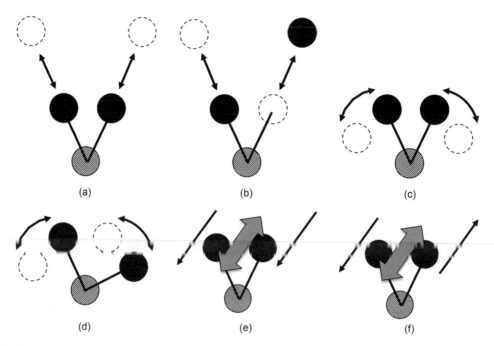

FIGURE 7.14

Vibrational modes of triatomic molecules. (a) Symmetric stretching (b) asymmetric stretching (c) scissor (d) rocking (e) wagging (f) twisting.

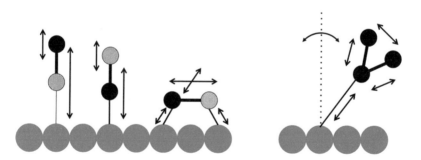

FIGURE 7.15

Vibrational modes of diatomic and triatomic molecules adsorbed to a surface.

Adsorbed molecules have even more vibrational modes since their vibrations can involve internal vibrations within the molecule as well as different types of motion around the surface site to which they are bonded. The internal vibration modes of a triatomic molecule are presented in Figure 7.14. The bottom atom is taken to be stationary and the other two atoms move. The vibration can be in the plane of the molecule such as a symmetrical or asymmetrical stretching mode, a scissor mode where the atoms move opposite to one another, and a rocking mode where they move in phase. The vibrational modes can also be out of the plane of the molecule such as a wagging mode (atoms in phase) or a twisting mode (atoms out of phase).

The internal vibrational modes can be combined with modes of the molecule bound to the surface as indicated in Figure 7.15 for both diatomic and triatomic molecules. These are similar to the stretching and rocking modes for the individual atom. The types of vibration can also depend on the type of surface site occupied (Figure 7.6). For the diatomic molecule, the amplitude and frequency of vibration may depend on which of the atoms in the molecule is bonded to the surface.

The energies (frequencies) of the vibrational modes are different. Stretching vibrations of single bonded diatomic molecules will be lower energy than double-bonded. The energies also depend on the elemental composition of the molecules (C–H vs. C–O, for example.) Bending and wagging modes are also relatively low-energy modes.

Surface diffusion of adatoms occurs by the same mechanisms discussed for surface diffusion in Chapter 4. The primary difference will be the values of the diffusion coefficient and the surface diffusion energy since these are dependent on the particular elements involved.

7.5 Characterization of Adsorbed Atoms

Low-energy electron diffraction is one of the most common techniques for studying ordered adsorbate structures. This is discussed in Chapter 12. The presence of adsorbed molecules and atoms and the surface coverage can be determined by chemical techniques such as Auger electron spectroscopy or X-ray photoelectron spectroscopy (Chapter 13). Vibrational techniques, such as electron energy loss spectroscopy or infrared spectroscopies can detect the presence of different species on the surface, the bonding sites, and the orientation of molecules on the surface. (Chapter 13).

References

Bruch, L.W., M.W. Cole, and E. Zaremba. 1997. *Physical Adsorption: Forces and Phenomena*. Mineola, NY: Dover Publications, Inc.
Gross, A. 2009. *Theoretical Surface Science: A Microscopic Perspective*. 2nd ed. Berlin: Springer Verlag.
Hudson, J. 1998. *Surface Science: An Introduction*. New York: John Wiley & Sons. Inc.
Ibach, H. 2006. *Physics of Surfaces and Interfaces*. Berlin: Springer.
Prutton, M. 1994. *Introduction to Surface Physics*. Oxford: Clarendon Press.
Zangwill, A. 1988. *Physics at Surfaces*. Cambridge: Cambridge University Press.

Problems

Problem 7.1 Kinetic Theory of gas isotherm
   a. In the simplest model of adsorption, we allow the condensation coefficient=1 and use the kinetic theory of gases. This leads to Equation 7.3, $\sigma=k_1 P$. The kinetic theory of gases gives the flux of atoms incident on the surface as

$$J = \frac{P}{\sqrt{2\rho m k_B T}}$$

where k_B is Boltzman's constant, m is the gas atom mass and T is the temperature. Using this result, determine an expression for the constant k_1 in terms of the residence time, t, the atomic mass, m, and the temperature, T. Use some typical values to calculate a typical value for k_1.

b. This isotherm can be rewritten (for low coverages) by defining the degree of coverage, θ, as

$$\theta = \sigma/\sigma_0 = N/N_s$$

where N=number of occupied adsorption sites and N_s=total number of sites. Derive (very simply) the expression for θ (P) in terms of k_1 and σ_0.

Problem 7.2 Langmuir isotherm

a. The Langmuir isotherm

$$\sigma = \frac{\sigma_0 k_2 P}{\sigma_0 + k_2 P}$$

can also be rewritten (as in problem 1) by defining the degree of coverage, θ, as

$$\theta = \sigma/\sigma_0$$

Show that the Langmuir isotherm can be described as

$$\theta = \frac{bP}{1+bP}$$

and evaluate the constant b in terms of k_2 and σ_0.

b. For low pressures ($bP \ll 1$) show that the expression in (a) reduces to the result from problem 1 and show how k_1 and k_2 are related.

Problem 7.3 Extension of Langmuir isotherm

We can extend Problem 7.2 by considering what happens if we relax the assumption that the adsorbate atoms do not have any lateral interactions. Consider a surface where each adsorption site on the surface has c nearest neighbor sites. θ (as defined in Problem 7.2) is the probability that a site will be occupied, so the probability of a neighboring site being occupied is just $c\theta$. Assume that the interaction between neighboring adsorbate atoms can be described by an energy, w, that only applies to nearest neighbors. (Atoms that are further apart do not interact). Then the average interaction energy is just $wc\theta$. This energy can be added to the adsorption energy which changes the residence time, which changes the constants k_1 and k_2, which changes the constant b (defined in Problem 7.2). The Langmuir isotherm form stays the same with only the constant changing, so now

$$\theta = \frac{b'P}{1+b'P}$$

Develop an equation relating b' to b based on these arguments.

Part II

Thin Films

8

Overview of Thin Film Growth

This chapter begins our investigation of thin films. Thin films are defined as layers of material with a thickness of less than $1\,\mu m$ (10^{-6} m, 1000 nm, 10,000 Å) grown on a substrate material. The presence of both the film and substrate, which create an interface region, makes this a much more complicated system than just looking at surfaces. The initial stages of film growth, however, are atoms adsorbed on a clean surface discussed in the previous chapter. Often the substrate and film are different materials with different structural, chemical, and thermodynamic properties. Even if the substrate and film are the same material, they may not have the same properties due to the nature of thin film growth processes. Thin film growth often results in materials that are not fully dense and are under stress. Films can have a different defect structure than bulk materials. These differences are strongly influenced by the presence of both a surface and an interface and will lead to thin films having different electrical, magnetic, optical, thermal, mechanical, and other properties when compared to the same bulk material. Additional detail on thin film growth processes is available in the literature (Brophy, Rose and Wulff 1964, Luth 2010, Machlin 1995, Ohring 2002, Venables 2000).

8.1 Introduction

Consider the process involved in making a thin film. It must involve a source of the material we intend to deposit, a way of getting that material to the substrate, and a way to get the film material to stick to the substrate. More formally, we describe thin film deposition in terms of three generic steps:

1. emission of particles from a source
2. transport of particles to a substrate
3. condensation of particles on a substrate

We will return to this simple framework when we explore different types of thin film growth methods in later chapters.

In this chapter, we focus on developing an atomic-level view of the process of forming a thin film. A simple atomic model of what happens when film particles reach a surface is illustrated in Figure 8.1. In this simple model, we treat the substrate as a perfectly flat and uniform surface with a chemical composition and a temperature. The film particles, impinging on the substrate surface, have a chemical identity, energy, and rate of impingement. Once film atoms are adsorbed on the surface they have the possibility of desorbing back into the vapor or diffusing along the surface or into the bulk. We described the adsorption, desorption, and diffusion processes in Chapters 4 and 7.

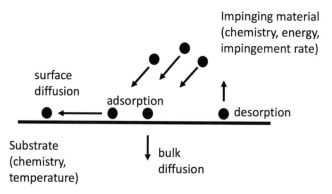

FIGURE 8.1
Simple atomic model of film deposition.

In the laboratory, the scientist or engineer has direct control of only some of the parameters in this simple model. We often choose the substrate and film material composition and control the substrate temperature as well as the energy and rate with which the film material strikes the substrate. By changing the chemistry, temperature, and deposition rate, we influence the adsorption, desorption, and diffusion, but not directly and not independently. These surface processes are critical in thin film growth and are only indirectly controlled by the knobs that we have available in the experimental laboratory.

The materials science that we described in Chapters 1–7 regarding the structure, chemistry, thermodynamics, and dynamics of surfaces and adsorbed atoms applies in these film systems. We expand on these concepts by adding the complication of heterogeneous systems where one material is forming an interface with another material. Film nucleation and growth will be explored along with the continued thickening of the film.

8.2 Homogeneous Nucleation and Growth

The nucleation and growth of a thin film can be understood as a phase change process. Information about rates of film growth is not typically found in the realm of equilibrium thermodynamics. Equilibrium phase diagrams and some simple arguments from thermodynamics, however, can provide insight into these processes. We begin by considering homogeneous nucleation of one phase in another phase and later will add the substrate and explore heterogeneous nucleation. For the homogeneous case, we discuss the formation of a solid phase inside a liquid phase, which is easier to picture and illuminates useful science.

Thermodynamic phase diagrams, developed to demonstrate the equilibrium stability of different phases, can be used to get some useful concepts about rates of phase transitions. Consider cooling a liquid into a solid through a eutectic point, T_e, as illustrated in Figure 8.2. Recall, from Chapter 5, that the eutectic point (B in Figure 8.2) is a point where the liquid and solid phases will both thermodynamically be allowed to exist.

The composition is held constant as the liquid is cooled from point A to point D passing through the eutectic point. At point A the solid is not stable so will not form and so the rate of solid formation is infinitely slow. At point B (the eutectic point) the solid and liquid

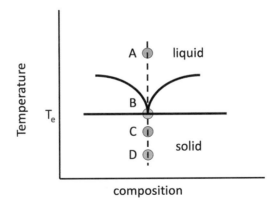

FIGURE 8.2
Phase diagram demonstrating rate of transformation to solid.

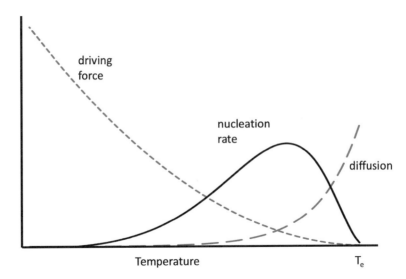

FIGURE 8.3
Nucleation rate depends on diffusion and a driving force toward solid cluster formation.

are both stable and so both can exist but there is no driving force to undergo phase transformation to the solid (rate=0). At point C, the liquid is now unstable and so there will be a tendency to form the solid phase, although the rate would be relatively slow if you are still close to point B. At point D, the liquid is still unstable and so the solid will form. The farther below the eutectic point, the greater the driving force will be to form the solid and the faster the rate will be. From this argument, we expect the rate of formation of solid in an unstable liquid to increase as the temperature decreases.

Nucleation of a new phase (creating clusters of solid atoms in the liquid), however, depends not only on this phase stability argument but also requires the solid to have enough atoms available by diffusion of atoms through the liquid into clusters. Diffusion rates, as we discussed before, typically increase with increasing temperature. In Figure 8.3, we simply sketch a rate term from the liquid instability argument that decreases with temperature and a diffusion rate term that increases with temperature. The product of

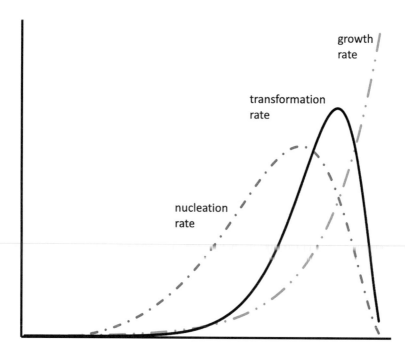

FIGURE 8.4
Overall transformation rate of individual atoms into film depends on the nucleation rate of critical clusters and the film growth rate.

these two processes yields a total nucleation rate that reaches some maximum value at a temperature below the eutectic point.

This process leads to the creation of small clusters of solid phase material in a liquid phase. The further growth of the solid phase is controlled by the diffusion of additional materials to these isolated solid clusters. The final transformation rate of turning liquid into solid will then require an additional diffusion-based term since the rate of growth of the solid clusters will again depend on diffusion. In Figure 8.4, we sketch a growth rate term that again increases with temperature. Multiplying this by the nucleation rate from Figure 8.3, we get an overall transformation rate of liquid into solid that again has a maximum value at some temperature.

We see from this simple argument that when we are growing one phase from another (as we do in thin film growth), there will be an optimum temperature that would maximize the rate of film growth. We will see in later chapters that our desire to grow films with very specific properties might lead us to operate at temperatures either above or below this optimum growth rate temperature.

Having established a simple conceptual picture, we are ready to explore the details of the nucleation of one phase inside another phase. The formation of a new phase requires consideration of both thermodynamics (will the energy of the system be minimized by forming the new phase?) and kinetics (what is the nucleation rate of the new phase?). We will initially consider homogeneous nucleation (no substrate) but will shift our terminology slightly to consider the formation of a solid phase from a vapor (rather than liquid) phase to bring us closer to the thin film deposition case.

Figure 8.5 shows, at the atomic level, a vapor of atoms (on the left) transitioning into some solid nuclei in the vapor. Thermodynamically, we seek to minimize a total energy

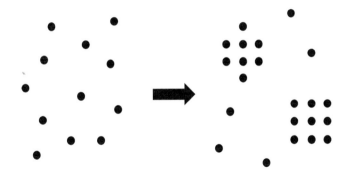

FIGURE 8.5
Vapor of atoms transitions to solid nuclei in a vapor.

that consists of two terms. One term involves the energy change involved in transforming some volume of vapor into solid phase. The other term is the energy change involved in forming a surface of the solid.

Building on Chapter 5, we define two terms that represent the change in the Gibbs free energies associated with the volume and surface creation of a spherical solid nucleus of radius, r.

$$\Delta G_{vol} = \frac{4}{3}\pi r^3 \Delta g_{vol} \tag{8.1}$$

$$\Delta G_{surf} = 4\pi r^2 g_s \tag{8.2}$$

where Δg_{vol} is the Gibbs free energy change per unit volume and g_s is the surface energy per unit area.

Equation 5.13 demonstrated that the Gibbs free energy change could be related to the temperature and concentration. For the system we consider here, the concentration can be represented by the vapor pressure and the Gibbs free energy change per unit volume can be obtained by dividing by the atomic volume, Ω,

$$\Delta g_{vol} = \frac{kT}{\Omega}\ln\frac{P_S}{P_V} \tag{8.3}$$

where P_S is the pressure above the solid and P_V is the pressure in the vapor.

Since we seek to lower the energy of the system by nucleating the solid, we want Δg_{vol} to be negative and so we need $P_V > P_S$. This produces a situation of supersaturation in the vapor that provides the driving force for the formation of the solid nuclei. The change in surface energy is always positive since we are forming surfaces from nothing. We combine these two terms to get the total Gibbs free energy upon formation of a solid, spherical nucleus or radius, r.

$$\Delta G_{total} = \frac{4}{3}\pi r^3 \Delta g_{vol} + 4\pi r^2 g_s \tag{8.4}$$

The surface, volume, and total free energy terms are sketched in Figure 8.6. Since the positive term follows r^2 and the negative term follows $-r^3$, we know that the positive term will

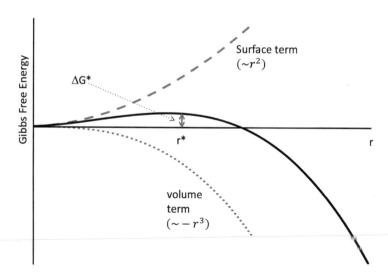

FIGURE 8.6
Gibbs free energy of a solid nucleus depends on surface and volume terms.

dominate for small r and the negative term will dominate for large r. The sum of the two terms, ΔG_{total}, reaches a maximum, positive value at $r=r^*$ and then decreases beyond that. Examining the figure starting at $r=0$, the initial formation of nuclei requires an increase in the Gibbs free energy. For $r<r^*$, the nuclei prefer to shrink to lower the free energy. For $r>r^*$, the nuclei prefer to grow to lower the free energy. Our goal is to produce nuclei with radii greater than the critical radius, r^*, so that they will continue to grow.

We can calculate r^* by taking the derivative of the expression (Equation 8.4) for ΔG_{total} and setting it equal to 0 to find $r^* = -2g_s/\Delta g_{\text{vol}}$. Since we already required Δg_{vol} to be negative for thermodynamic reasons, this yields a positive critical nucleus size. When we consider film growth in heterogeneous nucleation, we will need to add an interface energy term to the equation, but the qualitative conclusion of a critical radius for the nucleus to be stable will remain valid.

Now that we know what is needed to create a critical nucleus, we explore at what rate the critical nuclei will continue to grow. This nucleation rate depends on how quickly atoms will join the existing critical nuclei. Assuming that any atom that hits a critical nucleus will stick, we expect this rate to depend on the concentration of critical nuclei, n^*, the surface area of the critical nuclei, A^*, and the flux (rate per unit area) of atoms impinging on the critical nuclei, Φ_a. We can write the nucleation rate as: $dn/dt=n^* A^* \Phi_a$. These concepts are represented in Figure 8.7. We examine each of these three terms in greater detail.

The number of critical nuclei in the vapor will depend on the number density of possible nucleation sites and the probability of forming a critical nucleus at any site. Since each atom in the vapor is a potential nucleation site, the density of nucleation sites is just the volume density of atoms in the vapor (n_{vap}). The probability of nucleation can be written as a Boltzmann factor exponential term involving an energy barrier to forming a nucleus, ΔG^*, which is just the free energy at the critical radius, r^*, and the energy available to the atom, $k_B T$.

$$n^* = n_{\text{vap}} e^{-\Delta G^*/k_B T} \tag{8.5}$$

The surface area of the critical nucleus $A^*=4\pi(r^*)^2$.

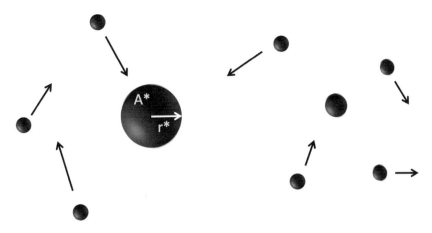

FIGURE 8.7
Homogeneous nucleation of atoms onto critical nucleus.

The flux of atoms impinging on a surface, Φ_a, can be calculated from the kinetic theory of gases (see Chapter 2).

$$\Phi_a = \frac{(P_V - P_S)}{\sqrt{2\pi m k_B T}} \qquad (2.10)$$

where m is the mass of the impinging atom or molecule. The P_S term can often be neglected since it is typically very small compared to P_V.

Combining these yields an expression for the deposition rate

$$\frac{dn}{dt} = \left[n_{\text{vap}} e^{-\Delta G^* / k_B T} \right] \left[4\pi \left(r^* \right)^2 \right] \left[\frac{(P_V - P_S)}{\sqrt{2\pi m k_B T}} \right] \qquad (8.6)$$

If we put typical values into these expressions, we find that the number density of critical nuclei, n^*, with the exponential term, dominates the expression for nucleation rate for this homogeneous case.

We assumed in this analysis that any atom that reached a critical nucleus would stick to it and result in growth of the nucleus. This is not typically true. In reality, we have competing processes with impingement on the surface competing with atoms that reach the surface but then desorb again or reflect off of the surface and do not contribute to growth. This is often described by a sticking coefficient, s, introduced in Chapter 7.

$$s = \frac{\text{number of atoms remaining}}{\text{number of atoms impinging}} = \frac{\text{mass of atoms remaining}}{\text{mass of atoms impinging}} \qquad (8.7)$$

This simple homogeneous model for nucleation already demonstrates that the process of nucleating another phase will depend on the number of nucleation sites available, the temperature, and the rate that new atoms are impinging on clusters. It also depends on various terms that will vary with the composition of the materials involved in the process. These parameters will continue to be important in examining the detailed steps in thin film formation.

8.3 Steps in Film Formation

If we picture a film atom reaching a substrate surface, there are several steps that need to be accomplished to contribute to the growth of a thin film. The first is that the extra energy that the atom may have needs to be dissipated and the atom and substrate must come into thermal equilibrium. Next that atom needs to either be bound to the surface or escape from the surface through desorption. The atom also might move along the surface by diffusion. Nucleation (this time heterogeneous) needs to occur by forming a cluster on the surface of critical radius. This cluster then needs to grow in what resembles the growth of islands of film on the substrate. Eventually, the islands will come together into a coalesced complete film that will then continue to grow by additional atoms coming down onto the film surface rather than onto the substrate surface. We examine each of these steps in turn.

8.3.1 Thermal Accommodation

The first step in creating a thin film is that the impinging atoms, which come in with some energy, E_V, must lose enough energy thermally to remain on the surface. We are concerned with three energies: the energy of the impinging atom when in the vapor, E_V, the energy of the substrate, E_S, and the energy of the atom if it is reflected from the surface, E_R. Since energy is proportional to $k_B T$, we can equivalently discuss temperatures, T_V, T_S, and T_R. These terms are indicated in Figure 8.8.

To describe this process, we define a thermal accommodation coefficient, α_T.

$$\alpha_T = \frac{E_V - E_R}{E_V - E_S} = \frac{T_V - T_R}{T_V - T_S} \tag{8.8}$$

If the atom collides elastically with the substrate, no energy is lost and $E_V = E_R$ which leads to $\alpha_T = 0$. If all excess energy is lost, then $E_R = E_S$ and $\alpha_T = 1$. If incident energy is going to be lost in colliding with the substrate, we need to consider the collision process in more detail. McCarroll and Ehrlich (1963) developed a simple one-dimensional model involving a chain of atoms connected by springs, which nicely illustrates this concept. The model is presented in Figure 8.9.

As the atom with mass m_0 comes in from the right, it collides with the surface and some of the incident energy, E_V, can be absorbed by the system of substrate masses, m, connected by springs with spring constants, k. The spring constant of the interaction of the impinging atom with the substrate surface atom will have a different spring constant, k_0. If the system rebounds with sufficient energy, the surface bond, k_0, will be broken and the atom will escape from the surface. If the system absorbs enough energy from the impinging atom, the atom will be trapped and will oscillate as it loses energy to the substrate lattice.

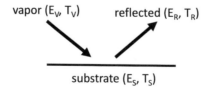

FIGURE 8.8
Energies and temperatures involved in thermal accommodation.

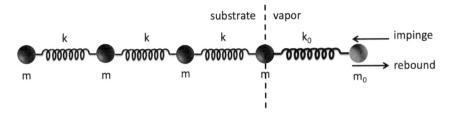

FIGURE 8.9
One-dimensional model of vapor atom impinging on substrate.

The results of this simple model suggest that the impinging atom will be trapped if it arrives with an energy less than about 25 times the energy barrier to desorption (which is related to k_0). Since these desorption energy barriers are typically 1–4 eV, the impinging atom will be trapped if it has energies less than 25–100 eV (which is equivalent to temperatures of 290,000–1,160,000 K). Since most deposition processes result in impinging energies of less than 10 eV, we can conclude that most atoms that impinge on the surface will be accommodated. This result is experimentally confirmed. In addition, we find that this process of thermal accommodation is very fast with times on the order of 10^{-14} seconds.

8.3.2 Binding and Desorption

In Chapter 7 we discussed bonds between atoms. When film atoms first interact with substrate atoms we expect two possible bonding situations:

1. Physical adsorption (physisorption) is typically a weak, Van der Waals type bond with an energy around 0.1 eV.
2. Chemical adsorption (chemisorption) creates a stronger chemical bond with energies around 1–10 eV.

An important question here is whether we can keep the impinging atom bound to the surface. It is thermally accommodated, but it can still desorb from the surface. We now have a competition between the deposition of impinging atoms and the desorption of film atoms from the substrate. At this stage, since we are talking about individual atoms rather than critical nuclei, we can consider the rate of impinging atoms, Φ_a, from the kinetic theory of gases discussed in Chapter 2.

The desorption process can be described in terms of a desorption rate, ν_d, which is the product of how frequently an atom tries to desorb and the probability of being successful. The frequency of an adsorbed atom attempting to desorb, ν_0, is the lattice vibration frequency which depends on the mass of the atoms in the lattice, their separation distance, and the strength of the bonds between these atoms. Each time the atom vibrates outward from the surface it is essentially trying to escape. The probability will again be a Boltzmann factor exponential which depends on the energy barrier to desorption, ΔG_{des}, and the temperature of the substrate, T_S, since the adsorbed atom is thermally accommodated to the substrate.

$$\nu_d = \nu_0 e^{-\Delta G_{des}/k_B T_S} \qquad (8.9)$$

We also define a mean residence time, τ_a, that an atom would stay on the surface from $\tau_a = 1/\nu_d$. As expected, higher substrate temperatures lead to shorter mean residence times on the surface. Typical residence times are on the order of 10^{-6}s. Combining the residence time and the impingement flux, we calculate the surface density (atoms/area) of adsorbed atoms, n_a.

$$n_a = \Phi_a \tau_a = \Phi_a \frac{1}{v_0} e^{\Delta G_{des}/k_B T_S} \tag{8.10}$$

This simple model suggests that if we heat the substrate we will reduce the number of film atoms on the surface. It also would predict that if we stopped depositing, atoms on the surface would all desorb until there was no film left. This clearly is not the case (or this book would never need to have been written) so we must be missing an important element. The missing element is surface diffusion, which will allow atoms to form clusters on the surface and help to stabilize the atoms against desorption. Diffusion processes are very important in understanding thin films.

8.3.3 Surface Diffusion

The analysis of the last section shows that isolated film atoms are not stable on a substrate surface. We find, however, that clusters of film atoms are stable and will allow the formation of films. To form a cluster, two atoms on the surface need to be able to diffuse across the surface and combine.

A statistical diffusion distance, X, can be determined from a random walk analysis (see for example Reif 1965 p. 486). The distance will depend on how quickly the atoms diffuse on the surface and how long they remain on the surface.

$$X \approx \sqrt{D_S \tau_a} = a_0 e^{(E_{des}-E_S)/2k_B T} \tag{8.11}$$

where D_S is the surface diffusion constant as defined in Chapter 4, τ_a is the mean residence time on the surface and a_0 is the distance between atoms on the surface. Increasing the temperature of the surface typically increases D_S but decreases τ_a.

Figure 8.10 demonstrates a case where the areas potentially covered by two atoms' motions overlap and a cluster could potentially form and a case where two atoms are too

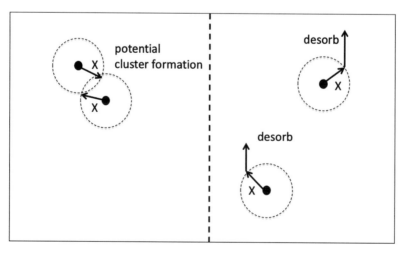

FIGURE 8.10
Surface diffusion may allow cluster formation before atoms desorb.

widely separated and the atoms will desorb before forming a cluster. Conceptually, faster surface diffusion (larger D_s) and/or higher impingement flux so that atoms arrive on the surface closer together and might have time to diffuse to form a cluster, should enhance the probability of film nucleation.

8.3.4 Heterogeneous Nucleation

The formation of clusters can be described by a heterogeneous nucleation process often referred to as the capillarity model. This macroscopic thermodynamic model is heterogeneous since it takes place on a substrate. The nuclei are assumed to be sections of spheres (spherical caps), which simplifies the mathematics.

As is common throughout science and engineering, cluster formation involves two competing processes. Promoting the formation of clusters is a condensation energy per unit volume (Δg_C) that lowers the desorption rate by providing a higher barrier to desorption. Minimization of energy, however, leads to clusters breaking up into individual atoms since clusters have a higher surface tension or surface energy per unit area (γ) than individual atoms. Figure 8.11 shows the geometry and relevant energies.

The geometry of a spherical cap can be simply described in terms of the radius, r, of the sphere and the internal angle, θ, made by the cap where it meets the substrate surface. The relevant areas and volume are:

$$\text{Surface area of the nucleus } = a_{vf}r^2 \text{ where } a_{vf} = 2\pi(1 - \cos\theta)$$

$$\text{Contact area of the nucleus and substrate} = a_{fs}r^2 \text{ where } a_{fs} = \pi\sin^2\theta$$

$$\text{Volume of the nucleus} = a_f r^3 \text{ where } a_f = \frac{\pi\left(2 - 3\cos\theta + \cos^3\theta\right)}{3}$$

The total change in the Gibbs free energy upon the formation of the cluster is

$$\Delta G = a_f r^3 \Delta g_C + a_{vf}r^2\gamma_{vf} + a_{fs}r^2\gamma_{fs} - a_{fs}r^2\gamma_{sv} \tag{8.12}$$

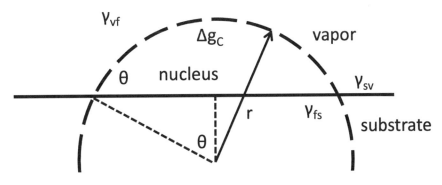

FIGURE 8.11
Energies and geometry of spherical cap of film on a substrate.

Where γ_{vf} is the surface energy per unit area of the vapor-film interface, γ_{fs} is the surface energy per unit area of the film-substrate interface, and γ_{sv} is the surface energy per unit area of the substrate-vapor interface.

The first term in the expression arises from the condensation energy associated with the formation of the cluster. Since Δg_C is negative, this term will lower the total energy of the system. The second term comes from the creation of a vapor-film interface that did not exist before. The third term, similarly, comes from the creation of a film-substrate interface beneath the nucleus. The surface energies are positive so these two terms will increase the total energy of the system. The final term is subtracted to represent the loss of vapor-substrate interface that is covered by the nucleus. Since the surface energy is positive, this subtracted term will lower the total energy of the system. Typically, all of the energy terms will depend on temperature and Δg_C will depend on deposition rate as well.

As with homogeneous nucleation earlier, we can plot ΔG against r and determine a critical nucleus size as demonstrated in Figure 8.12. Clusters with radii smaller than r^* will minimize energy by shrinking. Those with radii greater than r^* will minimize energy by growing.

Taking the derivative of this expression for $\Delta G(r)$ with respect to r and setting it equal to zero gives the critical radius for the formation of a nucleus.

$$r^* = \frac{-2\left(a_{vf}\gamma_{vf} + a_{fs}\gamma_{fs} - a_{fs}\gamma_{sv}\right)}{3a_f\Delta g_C} \tag{8.13}$$

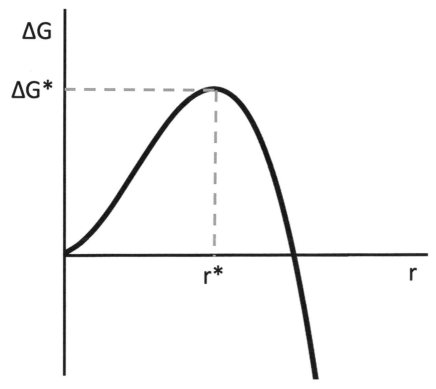

FIGURE 8.12
Clusters with $r > r^*$ are energetically stable.

The corresponding energy barrier at $r*$ can be determined by substituting $r*$ back into the expression for ΔG.

$$\Delta G^* = \frac{4\left[\left(a_{vf} + a_{fs}\right)\gamma_{vf}\right]^3}{27 a_f^2 \Delta g_C^2} \tag{8.14}$$

Substituting reasonable values into this expression results in a critical radius of only a few atoms, suggesting that stabilizing a film for continued growth is not difficult.

Although useful, this model is clearly an approximation that ignores the existence of individual atoms. A cluster a few atoms wide does not really resemble a spherical cap with uniform properties.

The continued growth of the critical nuclei will occur mainly by surface diffusion of other film atoms on the surface as indicated in Figure 8.13. It is possible for an atom to come from the vapor and strike a nucleus, but the very small area covered by nuclei in the early stages of film growth makes this unlikely until a significant fraction of the surface is covered with film.

This simple model assumes that substrates are perfectly flat. Real surfaces have steps and kinks in them as shown in Figure 8.14. A film atom located at a step can be pictured as interacting with both the substrate below it and along one side. A film atom at a kinked wall could have interactions on three sides. These extra interactions can keep the atom

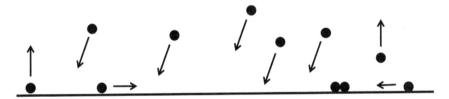

FIGURE 8.13
Early film growth is from adsorption on the substrate and diffusion.

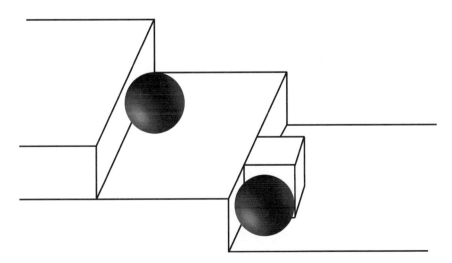

FIGURE 8.14
Steps and kinks on substrates are preferred nucleation sites.

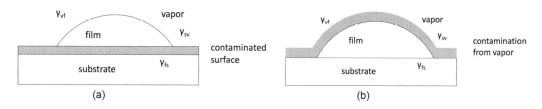

FIGURE 8.15
Contaminated surface (a) and contamination from the vapor (b) can change the energetics of film nucleation and growth.

on the surface longer increasing the probability that other atoms will join it and nucleate a cluster. As a result steps and kinks on surfaces are often preferred sites for nucleation.

The energetics described here can be altered by the presence of contamination. Contamination from poor background pressure, impure deposition sources and/or dirty substrates can change the energies associated with thin film growth and can thus significantly change the properties of the film grown. Figure 8.15 demonstrates, for instance, that a dirty substrate changes γ_{sv} and γ_{fs}. Similarly, a high background gas pressure will result in impingement of background gas atoms onto the substrate and growing film that will change γ_{sv} and γ_{vf}.

Using an analysis similar to the homogeneous nucleation discussion, we can determine how quickly nuclei (clusters of critical radius r^*) will form. The number of critical clusters on the surface per unit area, n_C, depends on the equilibrium surface coverage of clusters and their size. At equilibrium, adsorption and desorption balance yielding an equilibrium surface coverage, n_{eq}. The atoms that could potentially join a critical nucleus by impinging on the nucleus lie within an area surrounding the nucleus, A_{imp}. To get the rate of nucleation, we need to know how quickly atoms are being added to these critical clusters. We define Φ_C to be the flux at which atoms are diffusing across the surface and impinging on a nucleus.

$$\frac{dn_C}{dt} = n_{eq} A_{imp} \Phi_C \tag{8.15}$$

We examine each of these three terms in more detail.

The equilibrium surface coverage, n_{eq} depends on the surface density of adsorption sites, n_S, and the probability of a critical nucleus forming at that site. The probability is given by an exponential Boltzmann factor involving the barrier energy ΔG^* and the available energy, $k_B T$.

$$n_{eq} = n_S e^{-\Delta G^* / k_B T} \tag{8.16}$$

The circumference on the surface of a spherical cap of radius r^* is $2\pi r^* \sin\theta$. Assuming that atoms a distance, r_{imp}, beyond the cluster could reach the cluster, and assuming that $r_{imp} \ll r^* \sin\theta$ (which may not be a particularly good assumption here), the area of atoms about to impinge is

$$A_{imp} \approx \left(2\pi r^* \sin\theta\right) r_{imp}. \tag{8.17}$$

This geometry is shown in Figure 8.16. Note that r_{imp} will be comparable to the distance traveled on the surface before desorbing, X, defined in Equation 8.11.

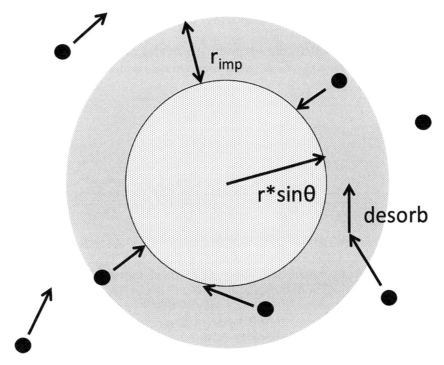

FIGURE 8.16
Geometry of atoms joining a critical nucleus by surface diffusion.

The flux of atoms reaching the cluster depends on the flux of atoms adsorbing from the vapor (Φ_a), the time they remain on the surface (τ_a) and how frequently these atoms jump across the surface (ν_S).

We defined Φ_a and τ_a earlier. The jump frequency depends on the natural vibrational frequency, ν_0, of the atoms on the surface and the probability of actually jumping on any given vibration.

$$\nu_S = \nu_0 e^{-E_S/k_B T} \tag{8.18}$$

where E_S is the surface diffusion energy barrier.

The flux of atoms impinging on the critical nucleus can now be expressed.

$$\Phi_C = \frac{P}{\sqrt{2\pi m k_B T}} e^{-(E_{des}-E_S)/k_B T} \tag{8.19}$$

Combining all these expressions:

$$\frac{dn_C}{dt} = \sqrt{2\pi} \left(r^* \sin\theta \right) r_{imp} \frac{P}{\sqrt{m k_B T}} n_S e^{-\left(E_{des}-E_S-\Delta G^*\right)/k_B T} \tag{8.20}$$

This expression gives us a good sense of what is happening on the substrate surface on the atomic scale, but most of the terms in this expression are not variables that we can control in the laboratory. By returning to the expressions for r^* and ΔG^*, however, we can make some predictions of what would happen if we change the substrate surface temperature (T)

or the deposition rate ($\dot{R}$). The deposition rate is the number of deposited atoms per unit area per second.

Since the intent of these calculations is just to get an idea of trends in behavior, we simplify our expressions for r^* and ΔG^* by assuming that the surface is inert, which is equivalent to setting $\gamma_{sv}=0$ and $\gamma_{fs}=\gamma_{vf}$.

$$r^* = \frac{2}{3}\left[\frac{-\left(a_{vf}+a_{fs}\right)\gamma_{vf}(T)}{a_f \Delta g_V\left(\dot{R},T\right)}\right] \tag{8.21}$$

$$\Delta G^* = \frac{4}{27}\left[\frac{\left(\left(a_{vf}+a_{fs}\right)\gamma_{vf}\right)^3}{a_f^2 \Delta G_V^2\left(\dot{R},T\right)}\right] \tag{8.22}$$

where the dependences on deposition rate $\dot{R}$ and temperature are shown explicitly. By taking partial derivatives of these expressions and substituting in typical values, the impact on r^* and ΔG^* when temperature or deposition rate are changed can be determined.

The partial derivatives of both terms with respect to temperature are positive and with respect to deposition rate are negative. This indicates that an increase in substrate temperature or a decrease in deposition rate will lead to a larger critical nucleus size, r^*. This may result in larger crystalline grains under those conditions. We will return to this idea later in the chapter as we explore continued film growth. Lower temperatures or higher deposition rates result in a lower nucleation barrier, ΔG^*. This will result in a higher nucleation rate of critical nuclei, which suggests that films will completely cover the surface earlier in the growth process.

While a very useful model for qualitative understanding of the processes in thin film growth, this macroscopic thermodynamic model may not be appropriate for nuclei that contain relatively small numbers of atoms since it ignores atomic structure completely. Walton and Rhodin (Walton1962) modified the capillarity model to treat clusters of atoms more like large molecules rather than as solid caps by considering the bonds between the atoms.

The critical nucleus now consists of some number of atoms, i^*. An equilibrium is established between individual atoms and clusters of atoms. We consider the reaction where i atoms of element A can form a cluster of i atoms, $iA \leftrightarrow A_i$. By analogy with the thermodynamics of chemical reactions (Equation 5.13), the free energy of formation of a critical nucleus is defined as

$$\Delta G_{i^*} = E_{i^*} + k_B T \ln\left(n_1^{i^*}\big/n_{i^*}\right) \tag{8.23}$$

where E_{i^*} is the energy to break apart a critical cluster of i^* atoms into individual atoms, n_1 is the number of single atoms per unit area on the surface, and n_{i^*} is the number of critical clusters containing i^* atoms per unit area. n_1 arises from the atoms adsorbed from the vapor that was previously developed in Equation 8.10.

Assuming $n_1 \gg n_{i^*}$, the fraction of nucleation sites occupied by critical clusters is

$$\frac{n_{i^*}}{n_S} = \left(\frac{n_1}{n_S}\right)^{i^*} e^{E_{i^*}/k_B T} \tag{8.24}$$

where n_S is the total number of adsorption sites on the surface per unit area as defined earlier.

The rate at which critical nuclei will form depends on the number of critical nuclei and the rate at which single atoms will join the critical nuclei. This second term depends on the distance that single atoms can diffuse (X), the speed at which they diffuse (X_L) and the number density of single atoms, n_1. We assume that nuclei larger than the critical nucleus will not decay.

$$\frac{dn_{i^*}}{dt} = n_{i^*} X^2 v n_1 = \dot{R} a_0^2 n_S \left(\frac{\dot{R}}{v n_s}\right)^{i^*} e^{\left[(i^*+1)E_{des} - E_S + E_{i^*}\right]/kT} \tag{8.25}$$

where we have used $n_1 = \dot{R}\tau_a$ and the definition of X from Equation 8.11.

This revised model has several advantages over the original model in that it now includes atomic scale parameters and thus has crystallographic information in it. The critical size of the nucleus, i^*, is found to depend on the substrate temperature with transitions in growth modes occurring at temperatures where the preferred value of i^* changes as the temperature increases.

8.3.5 Island Growth

Now that we have explored nucleation and the very early stages of film growth leading to island clusters of film atoms on the substrate, we consider how these film islands continue to develop. There are three possibilities: (1) the islands grow three dimensionally, (2) the islands grow two dimensionally forming a uniform film layer one atom thick before beginning the second layer, (3) a mixture of these where the film begins with growth one layer at a time, but then changes after some number of film layers to a three dimensional model for further growth. All three of these growth modes are observed experimentally.

The first growth mode of island growth is sometimes referred to as Volmer-Weber growth. The islands form three-dimensional structures prior to the coalescence of the film to completely cover the substrate. This would be expected when film atoms form stronger bonds to one another than to the substrate material. This could also occur when film atoms diffuse very slowly. The appearance of the film at several sequential times is shown in Figure 8.17.

The second option, layer-by-layer growth, is sometimes referred to as Frank–van der Merwe growth. This typically produces the highest crystalline quality and arises when film atoms bind more strongly to the substrate than to one another. It can also arise if the film atoms diffuse quickly. The appearance of the film at several sequential times is shown in Figure 8.18.

FIGURE 8.17
Island growth at three sequential times.

FIGURE 8.18
Layer by layer growth at three sequential times.

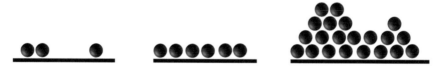

FIGURE 8.19
Mixed growth at three sequential times.

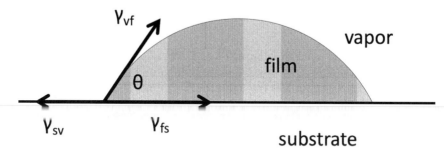

FIGURE 8.20
Surface energies and geometry of Young's equation.

The third option, mixed growth or Stranski–Krastanov growth, starts with layer-by-layer growth and changes into three-dimensional island structures on top of the complete layers. This suggests a change in the energetics of the system when the substrate is no longer exposed. The appearance of the film at several sequential times is shown in Figure 8.19.

We can approximately explore when these three conditions might be expected by considering the macroscopic thermodynamic surface energies of the substrate – vapor (γ_{sv}), film – vapor (γ_{vf}), and film – substrate (γ_{fs}) interfaces and making use of Young's equation. Figure 8.20 shows the geometry of our situation.

Young's equation relates the surface energies (surface tensions) at the point where the film, substrate, and vapor meet.

$$\gamma_{sv} = \gamma_{fs} + \gamma_{vf}\cos\theta \text{ or } \frac{\gamma_{sv} - \gamma_{fs}}{\gamma_{vf}} = \cos\theta \qquad (8.26)$$

Island growth would correspond to θ values between 0° and 90° corresponding to $\cos\theta < 1$. The surface energies would be related as $\gamma_{sv} < \gamma_{fs} + \gamma_{vf}$. So island growth might be expected when the surface energy of the substrate-vapor interface is low.

Similarly, layer-by-layer growth corresponds to a condition where $\theta = 0$ or is not defined. It would not be defined because for layer growth there really is no point where the substrate, vapor and film come together. Under these conditions, $\cos\theta \geq 1$ and the surface energies are related as $\gamma_{sv} \geq \gamma_{fs} + \gamma_{vf}$. So layer growth might correspond to a particularly high substrate-vapor surface energy.

If the surface energies are comparable in magnitude, $\gamma_{sv} \approx \gamma_{fs} + \gamma_{vf}$ then it is not clear which model would be expected and some sort of mixed mode might occur.

Typically, the scientist or engineer can only control these growth processes by changing the materials used in the film or substrate. In some applications, however, the materials may be dictated by other factors and we may have limited ability to change them.

8.3.6 Island Coalescence

In any of these growth models, the film will eventually coalesce and completely cover the substrate. In addition to atoms impinging on the sides of islands to cause them to grow, three other mechanisms (Ostwald ripening, sintering, and cluster migration) commonly contribute to film coalescence. Coalescence can be imagined as a change in the distribution of island sizes with islands joining one another so that the distribution evolves from one dominated by small islands to one dominated by large islands until the film completely covers the surface which is, effectively, a single large island.

Ostwald ripening occurs because of a difference in activity related to the amount of convex curvature of an island. The greater curvature found in small islands results in a higher activity and a greater likelihood of atoms escaping from small islands as opposed to large islands. If the islands are widely separated, the escaped atoms will desorb. Small islands will gradually disappear. If the islands are close enough together that atoms can move from one island to another then the large islands will tend to grow at the expense of the small islands as depicted in Figure 8.21. Either case results in a change in the distribution of island sizes with an increase in the relative number of larger islands.

In a sintering process, when two islands touch one another either through growth or by random movement of the islands on the surface, the surface energy of the system can be lowered by smoothing out the curvature variations of the joined islands and achieving an equilibrium shape again. Atoms will move into the region where the islands met since the island curvature and thus, the activity, will be higher than in the joined region as shown in Figure 8.22.

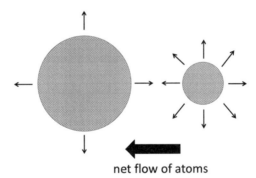

net flow of atoms

FIGURE 8.21
Film island coalescence by Ostwald ripening.

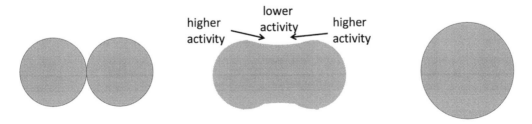

higher activity — lower activity — higher activity

FIGURE 8.22
Film island coalescence by sintering.

Cluster migration can also contribute to island formation. Small clusters (<100 Å across) tend to be particularly mobile and move randomly across the substrate surface. They are likely to encounter a large island and join it leading again to a reduction in the number of small islands while increasing the radius and height of the large island.

8.3.7 Thicker Films – Zone Models

Once the film has coalesced over the substrate, further film growth will continue to rely on adsorption and desorption of film atoms as well as their bulk and surface diffusion much as we discussed so far, except that the surface that we deposit on is now film material rather than the substrate. This will change the energies associated with the film growth.

As the film develops in three dimensions, geometric factors may also start to play a role. Depending on the geometry of the deposition chamber, depositing atoms may be approaching the surface at some angle that differs from the normal. In this case, shadowing may occur where three-dimensional features on the film surface may block deposition on other regions of the surface as shown in Figure 8.23. In Chapter 9 we will discuss the conditions under which certain deposition techniques operate in this line of sight mode.

Two key experimental parameters are available in most deposition systems: the temperature (T) of the substrate and film and the rate at which the film is growing (deposition rate typically expressed as a film thickness change per second). Deposition rate incorporates the impingement and desorption rates of film atoms. By changing these parameters we modify the atomic scale parameters of adsorption, desorption, and diffusion. Temperature tends to be the more sensitive parameter with relatively small changes in temperature producing significant changes in the relative importance of the atomic processes resulting in changes in film properties.

If we graph deposition rate vs. temperature, as in Figure 8.24, we can imagine some point on that graph representing a film with particular properties grown with a particular deposition rate and temperature. If we slightly change the deposition rate or temperature, we expect that the resulting film will be similar in properties. If we significantly change the deposition parameters, then the film properties may also change significantly. So we can divide the graph into zones that we expect to have similar film properties.

Thornton (1975) experimentally confirmed this concept and developed the detailed zone diagram shown in Figure 8.25. This work was done using a sputter deposition

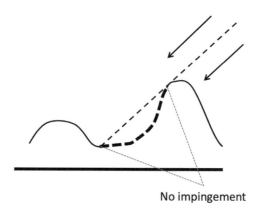

No impingement

FIGURE 8.23
Shadowing effects prevent film deposition on regions of the substrate.

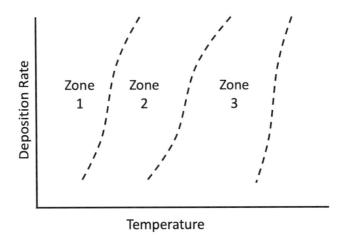

FIGURE 8.24
Film properties will be similar in zones with similar deposition rates and temperatures.

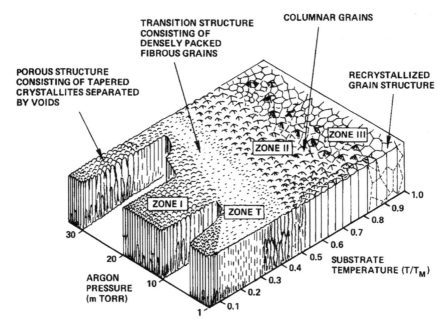

FIGURE 8.25
Film deposition zone diagram. (Reproduced from Thornton 1975 with the permission of AVS: Science & Technology of Materials, Interfaces, and Processing.)

system (see Chapter 9) and so the deposition rate is represented by the pressure of Argon in the chamber and the substrate temperature is normalized by dividing by the melting point temperature of the film material. Instead of just identifying zones, the figure represents the structure of the film in each zone with cut-away regions to show what the film interior looks like.

The processes involved in film growth in each zone (Grovenor, Hentzell and Smith 1984) are identified in Table 8.1. At low temperatures (Zone 1), diffusion is limited since the

TABLE 8.1

Characteristics of Film Deposition Zones

Zone	Temperature	Diffusion	Other Processes	Structure
1	$T<0.2$–$0.3\,T_M$	Limited		Small grains, many voids
T	$T<0.2$–$0.5\,T_M$	Surface	Renucleation during growth	Mixed small and large grains, fewer voids
2	$T<0.3$–$0.7\,T_M$	Surface	Grain boundary migration	Columnar grains
3	$T>0.5\,T_M$	Bulk (dominates + surface	Grain boundary migration; recrystallization within grains	Large grains (sometimes columnar)

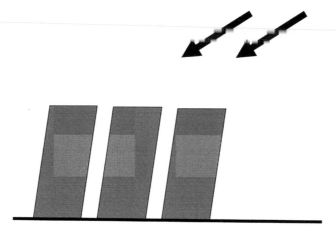

FIGURE 8.26
Tilt of columnar film growth from non-normal incidence.

thermal energy available to atoms is low compared to the barrier energy for diffusion. As a result, very little long-range structure is able to develop and small grains with many void spaces are created. As temperature increases into the transition zone (Zone T), surface diffusion (with a lower barrier energy) becomes important. Larger crystalline grains are now able to form in the film and fewer voids will occur. At even higher temperatures (Zone 2), grain boundaries are able to migrate and so larger, columnar grains are observed to form. At the highest temperatures (Zone 3), bulk diffusion (with a higher energy barrier) begins to dominate and large crystalline grains are observed.

The columnar structures, which are commonly observed, develop from conditions of limited atomic mobility. The grains are often tilted slightly toward the source of impinging film atoms as in Figure 8.26.

The presence of voids in films typically gives films a lower density than bulk materials with greater porosity observed on the macroscopic, microscopic, and atomic levels.

The importance of grain sizes is also noted from the zone models of film growth. Grain size typically increases with increasing film thickness, increasing substrate temperature, and decreasing deposition rate as indicated in Figure 8.27. Grain size can also be modified after film growth by annealing the sample at high temperatures to encourage further diffusion to occur. Grain size is found to increase with increasing annealing temperature as shown in Figure 8.27.

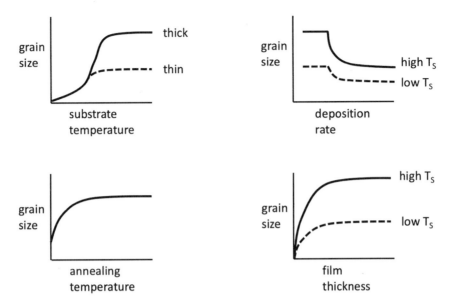

FIGURE 8.27
Dependence of film grain size on temperature, deposition rate, annealing temperature, and film thickness.

8.4 Deviations from Non-Ideal Structure

As the film grows, the surface of the film is expected, from statistical considerations, to deviate from being perfectly flat in the absence of atomic mobility. Atoms depositing randomly on a surface would result in a Poisson distribution for the average film thickness resulting in deviations from average (roughness) that would increase as the square root of the film thickness (Chopra 1969 p 185). Increased diffusion of film atoms, from higher substrate temperatures and/or lower deposition rates would allow for reduced roughness compared to this random sticking model.

A more detailed analysis of surface roughness, particularly in metals and semiconductors, is possible using the Ehrlich-Schwoebel energy barrier (Schwoebel and Shipsey1966) which leads to a kinetic roughening of the film surface as described in Chapter 4. This model applies to both the substrate surface and to the surface of the film.

In this chapter, we have only mentioned the energy carried to the substrate by the impinging film atoms. These energies, which will be discussed in more detail in Chapters 9 and 10, range from about 0.5 eV for thermal evaporation to 10–20 eV in sputter deposition and 100–1000 eV for film atoms that are accelerated in their deposition process. The interactions of these incident, energetic particles may result in sputter removal of film or substrate atoms from the surface, implantation of impinging particles below the surface of the film or substrate, increased heating of the surface as the kinetic energy of the impinging particle is converted to thermal energy, the creation of defects in the film or substrate, and shock (pressure) waves that can propagate through the film and/or substrate.

The development also neglects the atomic structure of the substrate and film. Representing them as featureless planes is certainly not accurate. In Chapter 7 we noted the importance of crystal structure and epitaxy when atoms initially deposit on a clean substrate. We expect the full range of crystalline defects (discussed in Chapter 3) to be present in both film and substrate. These defects may provide additional nucleation sites for film growth.

As the film coalesces on the substrate, the relationship of the film crystal structure to the substrate crystal structure may result in the development of stresses in the film that may be either compressive or tensile. In tensile stress, the film structure is stretched out by interaction with the substrate structure so that the interatomic spacings are greater than those observed in the bulk. The film, pushing back against this tensile stress, will put the substrate in compressive stress. A film in compressive stress has the interatomic spacings reduced from those observed in the bulk by interaction with the substrate. Again, the film opposes this compression and puts the substrate under tensile stress. The presence of these stresses in the heterogeneous system can alter the energetics described here.

8.5 Advanced Modelling

Although our focus in this book is on simple models that are often classical or macroscopic, understanding the details of thin film deposition requires much more advanced modeling typically rooted in quantum mechanics and statistical mechanics. In Chapter 6 we introduced density functional theory which can provide a more detailed understanding of the interactions within atomic systems at an appropriate atomic scale by examining the ground state electron density. This approach does not require as much prior knowledge of interaction potentials but is computationally restricted to relatively small numbers of particles. It can be combined with molecular dynamics simulations to get a detailed, but relatively small scale, understanding of systems of interest through energy minimization.

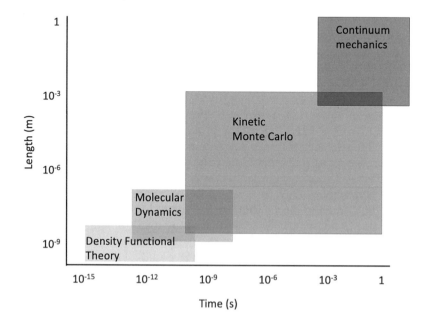

FIGURE 8.28
Comparison of length and time scales of various computational techniques. (Adapted from Kratzer 2009 with permission of the John von Neumann Institute for Computing.)

Molecular dynamic simulations on larger scales may be possible using other interaction potentials. These simulations are virtual experiments with adjustable parameters (such as temperature).

A common computational method for expanding our understanding to larger length and time scales is to use Monte Carlo techniques (Kratzer 2009) which sample possible changes in the system based on probabilities. Since thin film processes typically involve time as an important variable, the kinetic or dynamic Monte Carlo techniques are of particular interest. Figure 8.28 examines the relative time and distance scales associated with several of these computational techniques. Of course, the greater physical detail provided by the small-scale methods also comes with a higher cost in terms of computational power and time required. The kinetic Monte Carlo techniques require accurate knowledge of the rates of the various processes that may happen during film growth. As such this technique, which can be used for larger systems of atoms, can use information determined from density functional theory or other atomic-scale models to determine the probabilities (rates) of various processes.

8.6 Summary and Characterization of Thin Films

This chapter demonstrates the critical importance of atomic processes such as adsorption, desorption, and diffusion on the growth of thin films. The experimentally controllable parameters of temperature, deposition rate, deposition energy, and chemistry tend to impact all of these processes simultaneously complicating the growth of thin films. In future chapters, we will examine a range of film deposition techniques to understand how parameters under our control can be altered to attempt to grow films with the properties desired.

The sub-monolayer stages of thin film deposition can be detected and monitored by chemical techniques such as Auger electron spectroscopy or X-ray photoelectron spectroscopy (Chapter 13) as well as scanning probe microscopies (Chapter 11). Diffraction techniques such as reflection high energy electron diffraction (Chapter 12) are often used with epitaxial films. Quartz crystal monitors are commonly used to monitor film growth and are discussed in Chapter 12. Optical techniques (Chapter 14) can also be used during thin film growth.

References

Brophy, J.H., R.M. Rose, and J. Wulff. 1964. *The Structure and Properties of Materials: Volume II Thermodynamics of Structure.* New York: John Wiley & Sons, Inc.

Chopra, K.L. 1969. *Thin Film Phenomena.* New York: McGraw-Hill Book Company.

Grovenor, C.R.M., H.T.G. Hentzell, and D.A.Smith. 1984. The development of grain structure during growth of metallic films. *Acta Metall.* 32: 773–781.

Kratzer, P. 2009. Monte Carlo and kinetic Monte Carlo methods – A tutorial. In: *Multiscale Simulation Methods in Molecular Sciences,* eds. J. Grotendorst, N. Attig, S. Blügel, and D. Marx. Jülich: Institute for Advanced Simulation.

Luth, H. 2010. *Solid Surfaces, Interfaces, and Thin Films*. 5th ed. Berlin: Springer.

Machlin, E.S. 1995. *Materials Science in Microelectronics: The Relationship between Thin Film Processing and Structure*. Croton-on-Hudson, NY: Giro Press.

McCarroll, B. and G. Ehrlich. 1963. Trapping and energy transfer in atomic collisions with a crystal surface. *J. Chem. Phys.* 38: 523–532.

Ohring, M. 2002. *Materials Science of Thin Films*. 2nd ed. San Diego, CA: Academic Press.

Reif, F. 1965. *Fundamentals of Statistical and Thermal Physics*. McGraw-Hill.

Schwoebel, R.L. and E.J. Shipsey, 1966. Step motion on crystal surfaces. *J. Appl. Phys.* 37: 3682–3686.

Thornton, J.A. 1975. Influence of substrate temperature and deposition rate on structure of thick sputtered Cu coatings. *J. Vac. Sci. Technol.* 12: 830–835.

Venables, J.A. 2000. *Introduction to Surface and Thin Film Processes*. Cambridge: Cambridge University Press.

Walton, D. 1962. Nucleation of vapor deposits. *J. Chem. Phys.* 37: 2182–2188.

Problems

Problem 8.1 Cylindrical clusters

In Equation 8.4, we found the change in free energy for homogeneous nucleation of spherical clusters. Now consider a cluster shaped as a short cylinder of radius r and height h that forms on the surface of a substrate.

a. Show that, for this shape, the equivalent of Equation 8.4 is $\Delta G = \pi r^2 h \Delta g_{vol} + 2\pi r h g_s$

b. A common location for films to nucleate on a substrate is where a dislocation reaches the surface. In this case, we need to include a dislocation energy $A - B\ln(r)$ in the expression from part a. $\Delta G = \pi r^2 h \Delta g_{vol} + 2\pi r h g_s + A - B\ln(r)$ Sketch the three terms that make up this free energy change.

c. Depending on the relative magnitudes of the three terms, ΔG could have two very different shapes, one with two turning points and one with none. Sketch these two possible shapes.

d. Find the value of the critical radius, r^*.

e. Consider the quantity $\dfrac{\Delta g_{vol} B}{\pi g_s^2}$ and determine over which ranges of this quantity the ΔG solution with turning points is found and over which range the ΔG solution with no turning points is found.

Problem 8.2 Nucleation energies

Compare the energy reduction in heterogeneous and homogeneous nucleation processes.

a. In homogeneous nucleation, two spherical nuclei with surface energy per unit area, g, and radii r_1 and r_2 can combine to form a single, larger sphere. Using conservation of mass determine how much the free energy of the system is reduced by this coalescence. Do you expect the volume term to be important in this calculation?

b. In heterogeneous nucleation, the spheres are replaced with spherical caps with radii of curvature, r_1 and r_2. Determine how much the free energy of the system is reduced when these two spherical caps merge into a larger spherical cap.

Problem 8.3 Walton-Rhodin theory

The Walton-Rhodin theory predicts a series of critical temperatures at which changes in the critical nucleus size would occur.

a. Starting with Equation 8.25, show that the critical transition temperature to go from a critical size of $i^* = 1$ to $i^* = 2$ is $T_{1 \to 2} = -\dfrac{E_{des} + E_2}{k_B \ln\left(\dot{R} / v n_0\right)}$.

b. Similarly, derive expressions for the critical transition temperatures, $T_{1 \to 3}$ and $T_{2 \to 3}$.

c. Suppose when depositing at high temperatures you note that a film deposits with $i^* = 3$. If you deposit at lower temperatures, we could see a temperature range with $i^* = 2$ if $T_{2 \to 3} > T_{1 \to 3}$ or we might go directly to $i^* = 1$. What conditions can we put on the relevant E_i's if the $i^* = 2$ phase is observed?

Problem 8.4 Film contamination

In Chapter 2, we used the kinetic theory of gases to explore the rate at which molecules would impinge from a gas on a surface. Suppose that we have a chamber with a background gas, assumed to be all H_2, that is impinging on a substrate. At the same time, we deposit Fe onto the substrate with a flux of 2.5×10^{20} atoms/s m². At 500 K, calculate the ratio of Fe atoms to H_2 molecules incident on the surface for a background pressure of

a. 10^{-5} Torr

b. 10^{-9} Torr

Problem 8.5 Growth mode surface energies

Another way to consider the thin film growth modes discussed in Section 8.3.5 is to compare the total energy of each growth mode and then pick the mode with the lowest energy. This can be done by examining the layer by layer and island growth modes. If we put the same amount of material onto the surface in each case, the layer model will completely cover the surface with a thin layer. The island growth model will not cover the surface, but regions of film will be thicker than in the layer model. For simplicity, assume that the islands have the same height and cover one half of the area (A) of the substrate.

a. Using the same surface energy terms in Figure 8.20, show that the difference in energy between the two models is: $\Delta E = E_{layer} - E_{island} = \left(\gamma_{vf} + \gamma_{fs} - \gamma_{sv}\right)\dfrac{A}{2}$.

b. Use this result to derive the three conditions on the surface energies for each growth mode discussed after Equation 8.26.

9

Physical Vapor Deposition

Building on the general understanding of thin film growth from Chapter 8, we now explore in greater detail how thin films are actually produced from a vapor. In this chapter, we examine physical vapor deposition (Frey and Khan 2015, Maissel and Glang 1970, Martin 2010, Mattox 2010, O'Hanlon 1989, Ohring 2002, Vossen and Kern 1978) in which the depositing atoms enter the vapor through physical processes such as evaporation or impact with other atoms and do not generally undergo chemical reactions during the deposition process. In Chapter 10, we will explore the other major set of deposition processes, chemical vapor deposition.

Deposition processes can be divided into three stages as mentioned in Chapter 8:

1. source: getting source materials into the gaseous state
2. transport: getting the vapor of source atoms to the substrate
3. deposition: getting the source atoms from the vapor into the film

We will discuss various deposition processes by breaking them up into these three stages. This facilitates both understanding of the process and comparison with other processes.

9.1 Evaporation

Evaporation, as a film deposition process, is represented schematically in Figure 9.1. We place a substrate a distance, h, from a source (typically solid or liquid). The source is heated so that source atoms evaporate into the vapor and travel to the substrate.

9.1.1 Source

To get atoms into the gas phase through evaporation requires heating the source material until a desired vapor pressure of source material is achieved. Typically, this will be greater than about 10^{-4} Torr. In some cases, the desired vapor pressure may be achieved below the melting point of the material in which case the gas atoms enter the vapor through sublimation. In other cases, the material will melt and gas atoms will evaporate into the vapor.

If the source material is not a single element, the vapor may not have the same composition as the source material. Compounds may break apart into other molecules and/or their constituent atoms. Metal alloy sources will have the components evaporate independently from the source with the amount evaporating depending on each individual vapor pressure. As a result of the difference in vapor pressures of the components, alloy sources will change composition during heating as one of the components more quickly enters the vapor phase. These effects will result in a film that will not have the same composition as the alloy source.

DOI: 10.1201/9780429194542-11

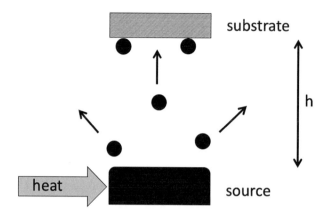

FIGURE 9.1
Schematic representation of an evaporation film deposition system.

If we approximate the gas immediately above the source as being in equilibrium, the flux of atoms leaving the source would be equal to the flux of atoms depositing back onto the source, which we determined from the kinetic theory of gases in Equation 2.12.

$$\Phi_{\text{evap}} = \frac{3.513 \times 10^{22} \, P_{\text{vap}}}{\sqrt{MT}}$$

where P_{vap} is measured in torr, T is the temperature of the gas, and the constant is chosen so that the resulting flux is in molecules/cm²s.

It is often easier to measure a mass flux rather than a particle flux, and so we convert this result to a mass flux:

$$\Gamma_{\text{evap}} = 5.834 \times 10^{-2} \sqrt{\frac{M}{T}} P_{\text{vap}} \tag{9.1}$$

where the pressure is again in torr and the choice of constant yields a mass flux in g/cm²s. If we consider air at room temperature, a vapor pressure of 10^{-2} Torr results in a mass flux of 10^{-4} g from each square cm of source per second.

9.1.2 Transport

Evaporation is typically a line-of-sight deposition process. This means that atoms entering the vapor from the source travel in straight lines, without collisions, until arriving at the substrate. In Section 2.1.3, we discussed the concept of mean free path and found that having an atom travel a long distance without collisions requires a good vacuum. The longer the path length between the source and the substrate, the better vacuum we will need. For distances in the 0.1–1 m range, typical of a vacuum chamber, we want pressures of 10^{-5} Torr or better to achieve line-of-sight deposition.

The atoms and molecules leaving the source will have kinetic energies roughly corresponding to the thermal energy of the heated source.

$$E_{\text{kinetic}} = \frac{1}{2} m v^2 \approx E_{\text{thermal}} \approx k_B T \tag{9.2}$$

This is equivalent to assuming that the temperature of the gas is the same as the temperature of the source. If the source material is heated to 800 K, the atoms will have kinetic energies around 0.1 eV corresponding to velocities of about 650 m/s.

Two quantities of great interest in film deposition are the deposition flux, Φ_{dep}, measured in atoms per unit area of substrate per unit time and the film thickness growth rate, $\dot{d}$, measured as the change in thickness per unit time. These will be related to one another by the density of the film, ρ and the mass of an evaporant atom, m_{ev}.

$$\Phi_{dep} = \frac{\rho}{m_{ev}} \dot{d} \tag{9.3}$$

The deposition flux will clearly depend on the flux of atoms evaporated from the source as given in Equation 2.12. If the substrate has an area, A_{sub}, then only a fraction, $A_{sub}/4\pi r^2$, of those evaporated atoms will be directed toward the substrate. Assuming a very small substrate located directly over the source so that we can neglect angular variations in the flux, we expect

$$\Phi_{dep} \propto \frac{P_{vap}}{\sqrt{m_{ev}T}} \frac{1}{r^2} \tag{9.4}$$

and

$$\dot{d} \propto \frac{1}{\rho}\sqrt{\frac{m_{ev}}{T}} \frac{P_{vap}}{r^2} \tag{9.5}$$

Recognizing that P_{vap} will typically increase with increasing temperature at a rate greater than $\sqrt{T}$, this simple analysis suggests that the rate of film growth will increase with increased source temperature and with decreased distance between the source and substrate.

Although the evaporant atoms may be traveling in straight lines, the direction of travel and the distribution of vapor atoms as they reach the substrate depends on the geometry of the source. We start with a mathematically simple case of a point source. This will have no preferred direction and so the evaporant will come off the source equally in all directions as in Figure 9.2. The distribution of evaporant at the source depends on the

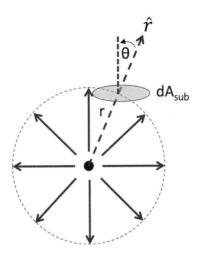

FIGURE 9.2
Deposition geometry from a point source.

orientation of the source relative to the radial direction, $\hat{r}$, indicated by the angle θ in the figure. We consider a small area element of the substrate, dA_s, which is located a distance, r, from the point source. The evaporant will interact with the projection of this area element onto the sphere of radius r, which is $dA_s\cos\theta$. In other words, if we tilt the substrate so that the evaporant direction is along the surface of the substrate, $\theta=90°$, then no material will be deposited. If we tilt the substrate so that it is perpendicular to the flux of evaporant atoms, $\theta=0°$, we will maximize the deposition.

Mathematically, if we identify dM_{sub} as the amount of mass striking the substrate and M_e as the total amount of mass evaporated in all directions from the substrate, we see that the ratio of the mass striking the substrate to the total mass is the same as the ratio of the area projection of the substrate on the sphere or radius r to the total surface area of the sphere.

$$\frac{dM_{sub}}{M_e} = \frac{dA_{sub}\cos\theta}{4\pi r^2} \qquad (9.6)$$

The uniformity of the evaporant incident on the substrate will depend on the variation of incident mass with area, and so we rearrange Equation 9.6 to yield

$$\frac{dM_{sub}}{dA_{sub}} = \frac{M_e\cos\theta}{4\pi r^2} \qquad (9.7)$$

and so we observe that the distribution of the transported evaporant at the substrate will depend on the orientation of the substrate and the distance between the source and substrate.

In many cases, the evaporant source is better described as having an extended surface rather than being a point source. This geometry is represented in Figure 9.3 where r and θ are defined as in the point source case. Since the evaporant from a surface source does not come off uniformly in all directions, we now add the angle ϕ to characterized the direction of the evaporant atoms.

This geometry is similar to a Knudsen cell as pictured in Figure 9.4a where the atoms are in a closed container with a very small opening of area dA_e. The atoms are assumed to be moving independently in random directions within the cell with a velocity v. We are

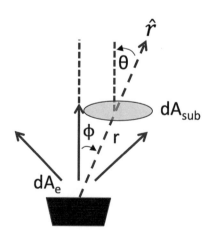

FIGURE 9.3
Deposition geometry from an extended source.

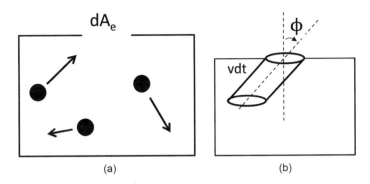

FIGURE 9.4
Knudsen cell (a) atoms in container with a small opening. (b) Cylinder of atoms traveling in the ϕ direction.

interested in the number of atoms that reach the opening and are traveling in the direction ϕ so that they might reach our substrate.

Figure 9.4b shows the cylinder of atoms that could reach the substrate. The height of the cylinder is given by vdt, which is the distance an atom of velocity v could travel in a time dt. The area of the ends of the cylinder is just the projection of dA_e in the direction ϕ that is $dA_e\cos\phi$. Thus, the volume of atoms that could escape if they are moving in the correct direction is $(vdt)(dA_e\cos\phi)$. Assuming that these atoms obey a Maxwellian velocity distribution, as described in Chapter 2, it can be shown that

$$\frac{dM_{sub}}{dA_{sub}} = \frac{M_e \cos\theta}{\pi r^2}\cos\phi \tag{9.8}$$

The distribution of evaporant now depends on the horizontal location of the substrate as well as on substrate orientation and distance.

Experimentally, we often observe

$$\frac{dM_{sub}}{dA_{sub}} \propto \cos^n\phi \tag{9.9}$$

The greater the value of the exponent n, the more narrowly directed the evaporant distribution will be. Notice that in all of these distributions, more atoms will be evaporated normally to the surface than at higher angles.

9.1.3 Deposition

In considering the final stage of deposition onto the substrate, we are often interested in questions of film thickness and purity and the variation of these across a substrate.

From the previous section, we have already seen that the mass of evaporant reaching a substrate will vary depending on r, θ, and ϕ. The film thickness, d, must also depend on these parameters. If the film has a uniform mass density, ρ, we can consider the small amount of mass, dM_{sub}, deposited on a small area element, dA_{sub}, and producing a film of thickness, d.

$$\rho = \frac{dM_{sub}}{(dA_{sub})d} \tag{9.10}$$

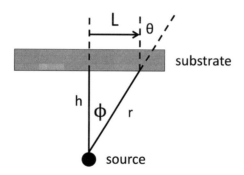

FIGURE 9.5
Dimensions and orientations in a point source deposition system.

which can be rewritten as

$$d = \frac{1}{\rho}\frac{dM_{sub}}{dA_{sub}}$$

(9.11)

If we consider a flat substrate, centered on the source and oriented perpendicular to the source, then we can introduce the distance from the source to the center of the sample, h, and the distance along the substrate, L, which is measured from the center point. These are represented in Figure 9.5. Notice that small area elements away from the center will have a distance from the source of r and an orientation of θ.

For this geometry we note that $\theta = \phi$, $\cos\theta = h/r$, and $r = (h^2 + L^2)^{1/2}$. We can now use these parameters to describe the resultant film thickness coming from a point source and a surface source.

For a point source, we use Equations 9.7 and 9.11 and the relationships defined in Figure 9.5 to show

$$d = \frac{1}{\rho}\frac{dM_{sub}}{dA_{sub}} = \frac{1}{\rho}\frac{M_e \cos\theta}{4\pi r^2} = \frac{M_e h}{4\pi\rho\left(h^2 + L^2\right)^{3/2}}$$

(9.12)

If we define the maximum thickness, d_0, as

$$d_0 = \frac{M_e}{4\pi\rho h^2}$$

(9.13)

then we can determine the film thickness variation with the distances h and L.

$$\frac{d}{d_0} = \left[1 + \left(\frac{L}{h}\right)^2\right]^{-3/2}$$

(9.14)

For a surface source, we can follow a similar procedure. The maximum thickness is four times greater

$$d_0 = \frac{M_e}{\pi\rho h^2}$$

(9.15)

and the film thickness variation for the surface source is

$$\frac{d}{d_0} = \left[1 + \left(\frac{L}{h}\right)^2\right]^{-2} \tag{9.16}$$

which indicates a slightly poorer thickness uniformity for the surface source.

These equations suggest several strategies for improving the thickness uniformity of evaporated films. The strategies will typically carry costs or constraints with them.

- We could decrease the sample size (L). Sample size, however, may be dictated by other considerations, not under our control.
- We could increase the distance to the substrate (h). This will require a larger vacuum chamber. The greater mean free path of the evaporant will require a better vacuum. More of the evaporant will miss the sample entirely and will be wasted.
- We could use multiple sources that overlap so that the overall distribution of film is more uniform. Multiple sources involve greater expense as well as issues of coordinating deposition rates from each source.
- We could move the substrate during deposition so that it is exposed to both larger and smaller fluxes of incident evaporant. This would involve complicated mechanical fixtures.
- We could use a rotating mask to reduce the evaporant near center. One could imagine a disk with a wedge cut out of it and rotated so that areas near the center of the sample see a reduced flux and those near the outer edges see a greater flux of evaporant.

A solution that exploits the geometry is to put the source and the substrate on the same spherical surface as shown in Figure 9.6. To the extent that both are small enough to be good approximations to the curved surface of the sphere, this will yield uniform film thickness. In this geometry, $\cos\theta = \cos\phi = r/2r_0$. Our expression for evaporation from a surface source simplifies to

$$\frac{dM_{sub}}{dA_{sub}} = M_e \frac{\cos\theta\cos\phi}{\pi r^2} = M_e \frac{1}{4\pi r_0^2} \tag{9.17}$$

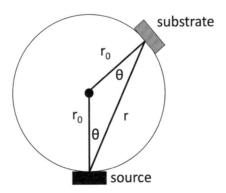

FIGURE 9.6
Planetary geometry with both source and substrate aligned with a circle.

which is a constant. This geometry has been used extensively in planetary fixtures used for commercial evaporation.

Chemical purity of evaporated films can also be a concern. Typically, the impurities arise from three possible sources: (1) impurities in the source materials, (2) impurities in the materials used to contain the source material, and (3) impurities in the vapor in the chamber.

The problem of source material contamination is typically solved by using very pure materials. For many materials, 99.99999% purity levels are commercially available. Containers for the source material can be a source of contamination especially if they diffuse into the source material during heating or outgas extensively into the vapor. Choosing a container with low rates of diffusion and low vapor pressures will typically solve this issue. Manufacturers provide tables of recommended container materials to use with most common evaporant source materials.

Residual gases in the chamber or contamination induced by outgassing of heated elements during evaporation provides us with two sources of impinging atoms onto the substrate. One source is the deposition flux of our film material given by Equation 9.3.

$$\Phi_{dep} = \frac{\rho}{m_{ev}} \dot{d} \tag{9.3}$$

The other is the impingement rate for particles from a gas from Equation 2.12.

$$\Phi_{gas} \propto \frac{P}{\sqrt{m_g T_g}}$$

The impurity concentration in the film

$$C_{impurity} = \frac{\Phi_{gas}}{\Phi_{dep}} \propto \frac{m_{ev}}{\rho \sqrt{m_g T_g}} \frac{P}{\dot{d}} \tag{9.18}$$

where the proportionality depends on the units used and the molecular structure of the gas. This suggests that a higher purity film can be grown by either reducing the pressure in the chamber or by increasing the film growth rate. Recall that P and T_g are related through the ideal gas law.

9.1.4 Evaporation Parameters and Processes

Most evaporation systems have three parameters that are under the control of the operator: substrate temperature, deposition rate, and evaporant particle energy. The substrate temperature may be directly controlled using a substrate heater. Otherwise, some heating from the deposited material may be expected. This will typically increase with the evaporation temperature of the source. The deposition rate is controlled by adjusting the temperature of the evaporant. Higher source temperatures lead to higher deposition rates. The energy of the evaporating particles also increases with increasing source temperature, although evaporation is a relatively low energy process and the effects of changes in particle energy from source temperature are often not substantial. Notice that these three parameters are not independent of one another. Changing the source temperature can impact all three of these parameters. Getting appropriate film properties typically requires experimentation with the available parameters and characterization of the resultant films.

The hardware for implementing evaporation processes depends on the type of heating used, which is typically either resistive heating or particle beam heating. Heating rates need to be controlled to prevent splattering of the source material onto the substrate. Figure 9.7 demonstrates several evaporation designs. Resistive heating can be done by designing a configuration of high melting point wire (often tungsten) and placing the material to be evaporated on the wire. A current is passed through the wire resulting in resistive heating. As the evaporant material melts it can remain on the tungsten wire from surface tension. Another option is to build a metal sheet "boat" of a high melting point metal and place the evaporant material in the boat. Passing a current through the sheet results in resistive heating. Note that both of these designs may change temperature as evaporant material melts and then evaporates. Some feedback is necessary to maintain a constant deposition rate. A modification of the sheet metal boat design that provides more of a point source is to build an enclosed box from the sheet metal with a small hole for the evaporant to escape. The box is then resistively heated by passing current through it. Another common design that depends on resistive heating is to place the evaporant in a ceramic crucible and coil a wire around the outside of the crucible. The wire is resistively heated and the heat is transferred through the crucible to the evaporant material.

An alternate means of heating the evaporant material is to use a beam of electrons as in Figure 9.7e. Electron-beam evaporation is a common technique. A beam of electrons is created by thermionic emission from a wire filament and directed onto the evaporant using magnetic fields. The evaporant is typically contained in a ceramic crucible. There is no line of sight from the filament to the evaporant and so no contamination from the filament into the evaporant material. The electron beam is rastered over the evaporant and the electron kinetic energy is transferred to the evaporant to heat it. This configuration allows higher

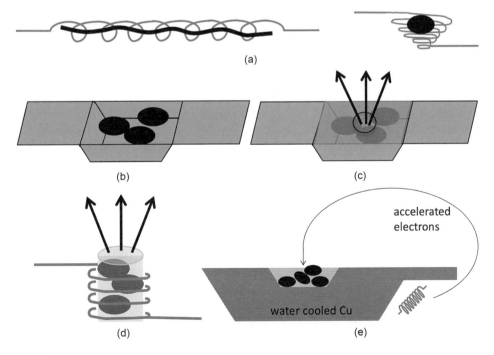

FIGURE 9.7
Evaporation hardware. (a) Resistive heated wires (b) resistive heated metal boat (c) resistive heated metal box (d) resistive heated crucible (e) electron beam evaporator.

powers and can be better for evaporating high melting point materials. Some of the evaporated atoms do pass through the electron beam and may become ionized from collisions with the electrons.

Later in this chapter, we discuss deposition using a high current of ions (arc vaporization) or an intense beam of photons (laser ablation).

9.2 Sputter Deposition

The other primary method of physical vapor deposition is sputter deposition. This is a more complicated process than evaporation. In Figure 9.8, we see that the region between the source and substrate is now an Ar plasma. We discussed the basic properties of plasmas in Chapter 2. The presence of this plasma is important in understanding all three stages of the sputter deposition process.

9.2.1 Source

The Ar^+ ions in the plasma play a key role in getting the solid source material from the source into the vapor. A high voltage on the source will attract Ar^+ items that will collide with the source with enough energy to cause atoms to be ejected from the source. As a result, the source is often referred to as the target in sputter deposition.

Many physical processes occur when Ar^+ ions collide with the target. About 1%–2% of the target atoms that are ejected come out as ions. Electrons are also emitted from the target, which help in keeping the plasma going. Some of the Ar^+ ions can be reflected back into the plasma as neutral Ar atoms. Other Ar^+ ions can be buried in the target. Light can also be emitted from the target during sputtering. Our main interest is on the ejection of neutral source atoms and so we focus on this aspect of the sputtering process.

Figure 9.9 shows a deceptively simple picture of what goes on in sputtering. The incident ion interacts with many atoms in the target through a collision cascade leading to recoils in the target atoms. Some of these recoils are in the direction of the surface and lead to the ejection of a target atom. Computer modeling of the process indicates that for incident energies <50 keV (typical of what we work with), the process can be modeled as a

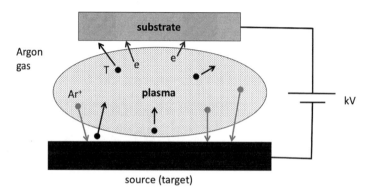

FIGURE 9.8
Schematic diagram of a sputter deposition system.

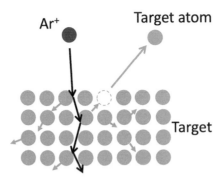

FIGURE 9.9
Simple model of sputter removal process.

momentum transfer process involving hard sphere collisions in the top 1 nm of the target. The collision cascade continues down into the target for about 10–20 nm resulting in Ar being implanted in the target.

Of the incoming energy, only about 5% is carried off by the ejected target atoms, which typically leave the surface with energies in the 5–100 eV range. The remaining 95% of the energy goes into the target and is dissipated as heat. As with evaporation, we observe a non-uniform distribution of atoms coming off the surface with more atoms ejected normal to the surface than at higher angles. They typically will follow a cosine distribution such as that described for a surface source in evaporation.

A simple parameter that can be used to characterize the sputter process is the sputter yield, S, which is the number of atoms ejected per incident atom. The sputter yield, introduced in Chapter 2, depends on the target material through the binding energy and mass of the target atoms. It also depends on the mass and incident energy of the sputter gas ions. S increases for heavier gases and for higher energies. Argon is typically used as a sputtering gas since it has a fairly high mass and is a noble gas with low chemical reactivity. The sputter yield also depends on the angle of incidence of the incident ions as shown in Figure 2.12.

Figure 2.13 provides sputter yields for normal incidence sputtering as a function of incident ion energy. There is some minimum threshold incident energy needed before any atoms are ejected from the target. The sputter yield then increases with incident ion energy up to about 10 keV and decreases slightly above this value. The difference in sputter yield between operating at 2–5 keV and operating at 10 keV is not very large. As a result, most people operate in the 2–5 keV range (where power supplies are less expensive). Typical values of S are in the 1–10 range.

Calculations of S are typically done using hard sphere collisions and require us to include the number of atoms ejected, the number of layers involved in the process, the surface density of target atoms, and the collision cross section of the incident ion with the target atom. The number of atoms ejected will depend on the transfer of momentum and energy, which will depend on the masses of the incident ion, m_{ion}, and target atoms, m_{target}, the collision angle and the binding energy of the target atom. The maximum energy transferred in a hard sphere collision can be calculated from conservation laws to be

$$E_{T_{max}} = \frac{4m_{ion}m_{target}}{\left(m_{ion} + m_{target}\right)^2} E_{ion} \qquad (9.19)$$

Bulk target composition

Surface target composition

Film composition

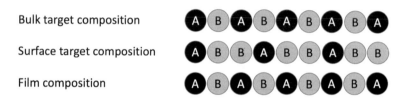

FIGURE 9.10
Compositions of target bulk and surface if $S_A > S_B$ and resulting film composition.

The number of layers involved in the process depends on the mean free path of the ion in the target material, which is typically about two atomic layers. Notice that this expression gives the maximum energy that could be transferred to the ejected target atoms. Typical energy values for the target atoms are found to be around 5% of the incident energy.

Deposition of stoichiometric alloy films is easier with sputter deposition than with evaporation since the composition of the alloy in the sputter target is approximately the same as the composition of the alloy in the sputter-deposited film. In evaporation, the rapid mixing of the components in a liquid phase allowed them to evaporate independently. In sputtering, however, the solid target components mix by a much slower diffusion process. As a result, the composition of the surface of the target reaches a steady state where the composition balances the sputter yield to produce a constant film composition.

Figure 9.10 demonstrates this process. Assume that we start with an initial alloy in the target that is 50% element A and 50% element B. Now let $S_A > S_B$ so that we will remove more A atoms as we sputter. This results in a surface that is enriched in element B. With more B on the surface, this will increase the amount of element B sputtered. The combination of enrichment of B on the surface and element A having the higher sputter yield is that the original composition of the target is sputtered to produce the film. If we let f be the atomic fraction of each element on the target surface and C be the bulk target composition, then

$$\frac{f_A S_A}{C_A} = \frac{f_B S_B}{C_B} \tag{9.20}$$

In a new target, this process needs to be initiated by sputtering a few tens of nanometers of material before depositing any films.

9.2.2 Transport

The target atoms pass through a much more complicated environment as they journey from the target to the substrate when compared with the vacuum path of evaporation. In sputter deposition, the path is through an Ar gas and plasma environment. Typically, the Ar gas pressure is in the 1–100 mTorr range. The Ar gas is only partially ionized and so there will be roughly one Ar^+ ion for every 10,000 Ar neutrals. Electrons in the plasma will be colliding with Ar neutral atoms to form more ions and additional electrons, which will help to sustain the plasma.

The target atoms entering this environment will undergo a series of collisions with Ar atoms, Ar^+ ions and electrons. The number of these collisions will depend on the density of the Ar gas (controlled by the gas pressure). The higher the Ar gas pressure introduced into the deposition chamber, the larger the number of collisions. With each collision, the target atom will change direction and will lose some energy. The result is very different from the line-of-sight deposition process in evaporation. In sputtering, the target atom loses energy

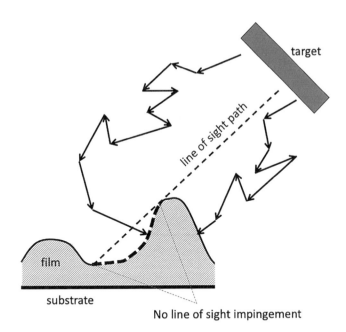

FIGURE 9.11
Random walk path of sputtered target atoms results in deposition into regions not accessible to line-of-sight deposition.

with each collision until it typically reaches the substrate with about 1–10 eV. It also may approach the surface from a much broader range of angles depending on where the last collision occurred before it reaches the substrate. As a result, sputter deposition may produce film deposition around corners. This entire process is demonstrated in Figure 9.11. We can treat the journey from target to substrate as a biased random walk diffusion through the Ar gas. While Ar, being a noble gas, will not react chemically with the target atoms, the presence of other gases in the chamber, either intentionally or as background gas, may result in chemical reactions occurring with the target atoms as they travel to the substrate.

9.2.3 Deposition

A wide variety of species will impinge on the positively biased substrate in sputter deposition. The target atoms and ions that we seek to deposit as a film will impinge with a rate dependent on both the sputtering conditions and the transport conditions. Electrons from the plasma will also be attracted to the substrate surface and will strike with high energies (but low mass). The gas of Ar, often at a pressure of 1–100 mTorr, will impinge on the surface in accordance with the kinetic theory of gases as discussed in Chapter 2. This can be a rather high flux of incident Ar and, even allowing for a lower sticking coefficient, significant amounts of Ar may be incorporated into the film.

The energetic particles striking the film as it grows may modify the growth process by transferring kinetic energy to the film. We will discuss this process in more detail in Section 9.3. These changes can be from physically altering the growing film through collisions and by heating the film surface resulting in changes in the diffusion processes on the surface of the film. Observations of 100–200 K changes in the substrate temperature are common. Figure 9.12 shows the range of processes that can be contributing to the heating

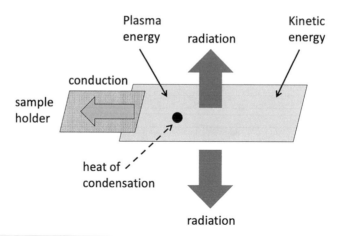

FIGURE 9.12
Processes contributing to sample heating.

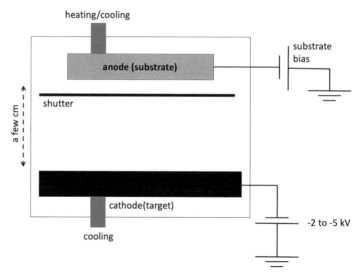

FIGURE 9.13
Schematic representation of a DC sputter deposition system.

of the substrate. In addition to heat transfer from conduction (through the sample holder) and by radiation, the kinetic energy of the film materials is included. The plasma energy represents heating by collisions from neutral atoms and electrons. The process of an atom condensing on the service will also release heat.

9.2.4 DC (Diode) Sputter Deposition

We now examine three variations on sputter deposition. We start with the simplest configuration, direct current (DC) diode sputtering. This is basically what we have been using so far in illustrating the principles of sputter deposition. We will then explore radio frequency (RF) sputter deposition and the use of magnetron sputtering in subsequent sections. Figure 9.13 is a schematic representation of a typical DC sputtering system. We explore the adjustable parameters available in DC sputtering to understand their effect on film deposition.

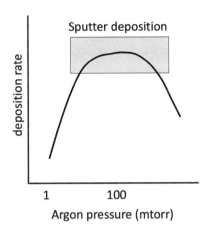

FIGURE 9.14
Deposition rate is maximized for Ar pressure around 100 mTorr.

Ar gas pressure: The Ar gas density is controlled by the pressure of Ar admitted to the chamber. The deposition rate typically reaches a maximum of around 100 mTorr as demonstrated in Figure 9.14. The maximum in the curve arises from competing effects involving the number of Ar ions available for sputtering and increased scattering from neutral Ar ions. At low pressures, an increase in pressure increases the number of Ar ions available for sputtering yielding a higher deposition rate. As the density of Ar gets higher, increased scattering of the Ar ions from neutral Ar atoms causes a decrease in the deposition rate. Notice that we could sputter at a lower pressure if we could increase the number of ions without increasing the number of neutral Ar atoms. We will return to this idea later in this chapter.

Sputter voltage: Earlier in this chapter, we saw that the sputter yield is sufficiently large for voltages in the −2 to −5 kV range.

Substrate bias voltage: In the previous section, we noted that the substrate was being bombarded by electrons and by ions from the target and the plasma. This caused physical rearrangement of the growing film and heating of the substrate. The use of a negative bias on the substrate can allow control of these processes. A high negative bias can reduce the number and energies of the impinging particles. Film properties can change significantly from changes in the substrate bias voltage. The neutral target atoms, however, will not be affected by this voltage and so the overall deposition rate will not be affected.

Substrate temperature: We have seen how important the temperature of the material is for diffusion and desorption processes. Significant heating can occur from the deposition process itself. Typically, the substrate temperature will increase with increasing sputter voltage. The substrate temperature will decrease with increasing substrate bias voltage since this reduces the energy brought into the sample from impact of charged particles. To control the substrate temperature, an ability to heat and/or cool the sample is often added to the deposition system.

Deposition rate: The deposition rate can be changed by altering the Ar gas pressure as noted above. Deposition rate also increases with increasing sputter yield, which is usually the result of increasing the sputter voltage.

Particle energy: The energy of the impinging neutral target atoms can also influence film properties. This energy typically increases with increasing sputter voltage since the target atoms come off with more energy. The energy decreases as we increase the Ar gas pressure since more collisions during the transport stage reduce the energy of the neutral Ar gas particles as they reach the substrate.

In examining these parameters, notice the interconnections between them. Changing the Ar gas pressure in order to change the deposition rate can also result in different energy of the impinging target atoms, which can change film properties and can also change the substrate temperature. Other parameters are equally interconnected. We do not have simple knobs on the equipment that will allow us to independently control atomic-level parameters such as temperature, gas density, and energy.

9.2.5 RF Sputter Deposition

RF sputter deposition can be advantageous in two circumstances in particular. The first is when using a target material that is an insulator since a positive charge will build up on such a material when using DC sputtering. The positive charge delivered by the Ar$^+$ ions will not be removed to ground on an insulating material and will build up on the surface until it repels the incident Ar$^+$ reducing the sputter yield significantly. The second circumstance involves operating at a lower Ar pressure and was noted in the previous section. If we are able to produce more Ar$^+$ ions without increasing the Ar gas pressure, we can get similar sputter yields with less total gas in the chamber. This will result in fewer collisions of the target atoms as they move to the substrate, which will affect their energy and how close to a line-of-sight process they produce.

Figure 9.15 is a schematic representation of RF sputter deposition. The primary change from DC sputtering is in the application of the high voltage to the target and substrate. Notice that the substrate has been connected to the chamber to make a very large electrode and that the two electrodes now have an alternating potential that is not symmetric about zero. The target is still primarily negatively charged to attract the Ar$^+$ ions and the substrate is still primarily the positive electrode as in DC sputtering. For a short time, however, the two electrodes reverse and the target becomes positively biased, which will attract electrons from the plasma that can neutralize the positive charge build-up on the target.

One potential problem with such an arrangement is that the film growing on the substrate could be sputtered during the short time that the substrate is negatively biased. This is minimized by creating the large electrode of the substrate and chamber and by carefully selecting the frequency of the alternating voltage.

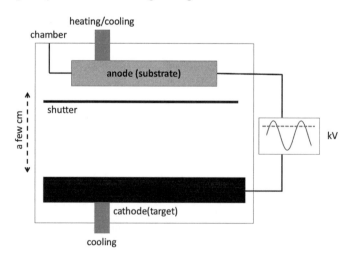

FIGURE 9.15
Schematic representation of a RF sputter deposition system.

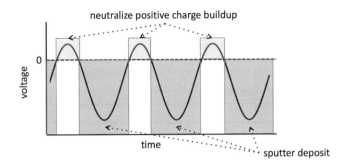

FIGURE 9.16
Asymmetric RF potential showing when charge neutralization and sputtering occur.

In selecting an appropriate frequency, we take advantage of the very different mass of the electrons and the ions in the plasma. For frequencies less than about 50 kHz, both the electrons and ions are able to move and to follow the switching of the anode and cathode. This may result in DC sputtering at both electrodes. For frequencies above about 50 kHz, however, the more massive ions are no longer able to switch directions and reach the substrate during the short time that the substrate potential is negative. The electrons, with their smaller mass, are able to switch directions and reach the target during this short time providing charge neutralization for the build-up of positive charge. Typical RF systems operate at a frequency of 13.56 MHz, which is assigned by the Federal Communications Commission. The potential of the target is shown in Figure 9.16 to demonstrate when both sputtering and charge neutralization occur.

The resulting plasma can be maintained at a lower Ar gas pressure (typically 1–15 mTorr) while still keeping up the same sputter yield. This lower pressure will result in fewer collisions on the target atoms as they move toward the substrate. This will typically mean higher incident particle energies and more line-of-sight deposition. These differences may be advantageous over DC sputtering for some applications but not for others.

9.2.6 Magnetron Sputter Deposition

Another way to increase the efficiency of Ar ionization without needing to increase Ar gas pressure is to add a magnetic field parallel to the surface of the target. Magnetron sputtering, which can be used with either DC or RF sputter deposition, allows Ar pressures to be reduced to around 0.5 mTorr. This will result in fewer collisions of target atoms with Ar atoms that will produce a more line-of-sight deposition process.

Magnetron sputtering increases the probability of electrons striking an Ar atom and causing electron impact ionization. As we saw in Chapter 2 when discussing the plasma environment, a charged particle will spiral around a magnetic field line. This increases the path length that it travels before hitting a surface, which increases the probability that it will encounter an Ar atom. By placing the magnetic field lines parallel to the target surface, the resulting ions are in an optimum location to then be attracted to the target and to cause sputtering. Magnetron sputtering can increase the flux of ions at the target by a factor of 10. The magnetic field also keeps more of the electrons close to the target surface and away from the substrate. This will reduce electron collisions with the substrate, which can reduce heating from electron impact.

A very common configuration with crossed electric and magnetic fields is indicated in Figure 9.17. Permanent magnets (typically around 200 Gauss in strength) are placed

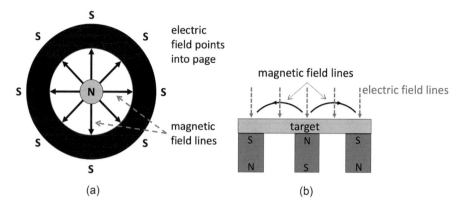

FIGURE 9.17
Crossed electric and magnetic fields for magnetron sputtering. (a) top view (b) side view.

behind the target so that the magnetic field lines are radial to the surface of the target while the electric field is perpendicular to the surface.

9.3 Modifications to Physical Vapor Deposition

Film growth by physical vapor deposition can be modified by altering the processes through the introduction of physical restructuring of the film by additional ion impacts and by chemical reactivity of the target and/or film with gas in the chamber during deposition.

9.3.1 Ion-Assisted Deposition

Earlier we mentioned that ions from the plasma could impinge on the substrate surface as the film grows. This can lead to physical rearrangement of growing film atoms and to local heating at the surface of the film. These can change the properties of the film. Earlier we regarded this as a by-product of the sputtering process. We could, however, intentionally introduce a beam of ions directed at the substrate to attempt to adjust the properties of the film in a controlled manner. This technique could be used with any other vacuum-based film deposition technique.

The ions directed at the substrate are not necessarily of the same type as in the film and are typically accelerated with relatively low energies (50–300 eV). Experimental observations suggest that film properties can be changed without the impacting ions being significantly incorporated into the film. An example of ion-assisted deposition is the use of incident ions to disrupt the formation of columnar growth. This typically requires about 20 eV of added energy per depositing atom.

9.3.2 Reactive Deposition

Another modification to evaporation or sputter deposition processes is to add a reactive gas into the chamber during deposition. Common examples of gases would be oxygen and nitrogen. These gases can react with the target, film, and/or substrate depending on what elements

TABLE 9.1

Comparison of Evaporation and Sputter Deposition

Evaporation	Sputtering
Lower energy incident atoms	Higher energy incident atoms
High vacuum path resulting in • few collisions • line-of-sight deposition • little gas incorporation in film	Low vacuum (plasma) path resulting in • more collisions • less line-of-sight deposition • more gas incorporated in film
Larger grain sizes	Smaller grain sizes
Fewer grain orientations	More grain orientations
Poorer adhesion to substrate	Better adhesion to substrate

are chosen and when the gas is introduced. The concentration of the gas can also be critical to get the desired effect. For instance, a target could be poisoned by the gas if the chemical reactions happen much faster than the sputter rate leading to a target composition that is not desired. The gas introduced into the chamber will also be impinging on the substrate in accordance with the kinetic theory of gases as discussed in Chapter 2. If the concentration of gas is too high, excess gas may be incorporated into the film. The flow of reactive gas must be carefully adjusted to get good film stoichiometry without excess gas incorporation into the film.

9.3.3 Comparison of Evaporation and Sputtering

Table 9.1 provides a brief comparison of the two primary means of physical vapor deposition based on experimental observations.

Whether to use evaporation or sputter deposition and whether to add some modification to the simple deposition process will largely depend on the particular films being grown and the desired properties of those films.

9.4 Molecular Beam Epitaxy and Epitaxial Films

We next explore a particular application of physical vapor deposition that involves depositing films at a very low deposition rate in a very good vacuum environment. Molecular Beam Epitaxy (MBE) (Tsao 1993) typically involves a carefully controlled evaporation process that is conducted in an ultra-high vacuum environment. By using a very low deposition rate, the substrate temperature does not need to be as high to allow time for atoms to diffuse to locations on the crystal lattice. The result is films with very high crystalline quality. Multiple sources are often used to allow growth of alloy or multi-layer films. The disadvantage of this technique is the high cost associated with the sources and with the ultra-high vacuum system.

The source for MBE is often an evaporation source such as a Knudsen cell as pictured in Figure 9.4. This allows for a carefully controlled, low deposition rate in an ultra-high vacuum environment. The transport process, like evaporation, will be line of sight since the pressure of gas in the deposition chamber is very low and so collisions will be rare. The low-energy evaporant atoms will arrive at the substrate and will be deposited with the same non-uniform distribution of material that we observed with evaporation.

Up to this point, we have not had to worry much about the crystalline arrangement of the film relative to the substrate. MBE allows us to produce highly crystalline films and so we now explore the issue of epitaxy, which is the growth of a film with a well-defined crystallographic relationship between the film and substrate.

If the film and substrate are the same material, this is referred to as homoepitaxy, auto-epitaxy, or isoepitaxy. If the film and substrate are different materials, then we refer to heteroepitaxy. In homoepitaxy it is not surprising to observe that the film and substrate materials have the same crystal structure. The film is able to grow with its ideal lattice in this matched structure resulting in a situation where there are no strains or defects at the interface caused by differences in structure. This situation is also sometimes observed in heteroepitaxy and is pictured in Figure 9.18.

A strained structure (pseudomorphy) occurs when the film grows with the same structure as the substrate, but this is not the natural structure of the film material. The film will then be strained. Figure 9.19 shows an example where the film material has a natural crystal structure that has a larger lattice spacing than the substrate. It is growing on the substrate with the smaller lattice constant of the substrate. The strained film is not stable and at some thickness will convert to the bulk material structure. Thin layers of these strained films can be used to create materials with unusual properties. The growth of strained layer superlattices has seen considerable interest for many years.

Another epitaxial relationship is possible when a film grows with a relaxed structure in which it keeps its natural crystalline structure that is different from the film structure. In this case, edge dislocations will be formed at the interface between the two crystalline structures as shown in Figure 9.20.

Thermodynamics governs the choice of strained vs. relaxed film growth. As usual, we seek to minimize the energy of the system, which will involve comparing the strain energy created in the strained structure growth with the dislocation energy formed in the relaxed structure growth.

In Chapter 3, we discussed the use of Miller indices to specify crystalline planes and direction. We can use these indices as well to specify the epitaxial relationship of a

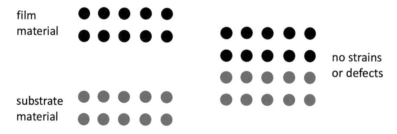

FIGURE 9.18
Film and substrate heteroepitaxy.

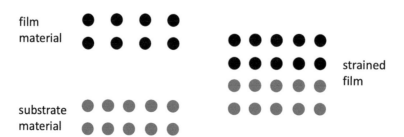

FIGURE 9.19
Strained film and substrate (pseudomorphy).

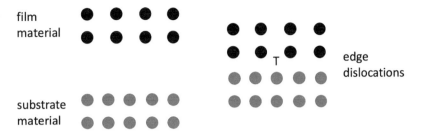

FIGURE 9.20
Relaxed film and substrate structure with edge dislocations.

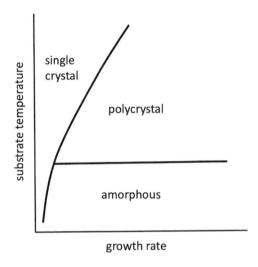

FIGURE 9.21
Generic phase diagram for films grown by MBE.

crystalline film with the underlying crystalline substrate. For example, if a NiO film is grown with a (1 0 0) orientation parallel to the surface on top of a Ni(1 1 1) surface, we can notate this as NiO(1 0 0)||Ni(1 1 1).

We can also calculate a lattice misfit, f, based on the different bulk material lattice parameters of the substrate, a_{0s}, and film, a_{0f}.

$$f = \frac{na_{0s} - ma_{0f}}{a_{of}} \tag{9.21}$$

where n and m are integers. Recall that the lattice constants are functions of temperature. If $f > 0$ then the film will be in tension. If $f < 0$ then the film is in compression.

The substrate temperature and deposition rate are critical parameters in determining epitaxial growth. A critical epitaxial temperature, T_C, can be determined above which the film will experience perfect epitaxial growth and below which will experience polycrystalline growth. Typical values for this critical temperature are 100–500°C and it will depend on the deposition rate and the materials chosen for the substrate and film. Lower deposition rates typically improve the epitaxy assuming that the vacuum is good enough.

Figure 9.21 sketches a generic phase diagram for films grown by MBE. As expected, very low growth rates or high substrate temperature allow sufficient diffusion for the single crystal film to form.

9.5 Arc Vaporization – Cathodic Arc Deposition

Another form of physical vapor deposition is arc vaporization or cathodic arc deposition (Anders 2008). In this technique, a plasma is generated and a DC voltage is applied to the target but in ways that are quite different from sputter deposition. As shown in Figure 9.22, the target is again the cathode, but an anode is placed adjacent to the cathode. A high current (current densities around 100 A/cm²), low voltage discharge is initiated by touching the electrode surfaces and then separating them. The arc heats up both the anode and cathode significantly and so the anodes need to be cooled to make sure that anode material does not come off in significant quantities and deposit on the substrate.

9.5.1 Source

The solid target of the electrically conductive material desired for the film is vaporized by the high current arc striking the target in a spot a few micrometers wide. The high power of the arc explosively evaporates the target material very quickly (10–100 ns timescales). The arc is rastered over the target using an applied magnetic field leading to a uniform evaporation of material from the target. Ions, neutral particles, and atomic clusters are generated. Droplets of target material can also be ejected, but this is typically avoided by moving the arc quickly over the target. Filters can be used to eliminate atomic clusters and droplets as well.

9.5.2 Transport

The arc produces many electrons between the cathode (target) and the substrate. The electrons are very efficient in ionizing the target atoms by electron impact ionization. As a result, almost 100% of the target atoms in the vapor are ionized. This creates a plasma region between the target and the substrate. Unlike sputter deposition, where the plasma is primarily Ar⁺ ions and electrons, this plasma is made up of target element ions and

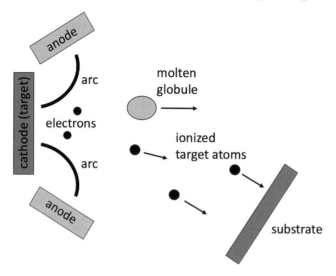

FIGURE 9.22
Schematic diagram of arc vaporization.

electrons. Since the target atoms are highly ionized, magnetic fields can be used to guide the target ions to the substrate while having any droplets that are ejected from the source miss the substrate. The substrate is typically biased negatively to the target and so the ions are accelerated toward the substrate with energies often up to 100 eV although they may lose some energy by collisions in the plasma depending on the plasma density. The transport is typically line of sight with minimal ion–ion interactions.

9.5.3 Deposition

As the ions reach the substrate, they may be at significantly higher energies than in evaporation or sputter deposition. This may lead to enhanced chemical reactions on the surface and to film densification by ion bombardment as we discussed in ion-assisted deposition. This technique can be used for depositing alloy films.

As with other deposition techniques, the atomic-level variables that we would like to control, temperature, deposition rate, and particle energy, do not correspond independently to experimental parameters under our control. The substrate and film will experience heating from the ion bombardment that depends on the energy of the target ions and the environment that they are transported through. The energy delivered by the incident ions can be controlled by applying a bias to the substrate. The resulting temperature can be adjusted using substrate heating and cooling capabilities.

The deposition rate will be altered by changes in a substrate bias since more ions can be collected at the substrate with a higher negative bias. The deposition rate can also be changed by adjusting the arc current. The arc current will affect the energy of the depositing particles as well, which can be modified by changing the substrate bias.

The high energy of the condensing target ions may lead to very dense, adherent films that could be under high compressive stress. The target ions have sufficient energy that they may be implanted into the growing film rather than being added to the top of the film.

Reactive deposition is common in cathodic arc deposition. Reactive gases, such as oxygen or nitrogen, can be activated by exposure to the plasma and then react with the target material on the substrate to produce the desired film composition.

9.6 Pulsed Laser Deposition – Laser Ablation

Another physical vapor deposition technique (Krebs 2003, Schneider and Lippert 2010) relies on a high-power, pulsed laser beam to melt, evaporate and ionize target material. The resultant plasma expands rapidly away from the target in a plume of ablated material, as shown in Figure 9.23.

9.6.1 Source

The laser beam is often from an ultraviolet excimer laser or a longer wavelength Nd:YAG laser and enters the vacuum chamber by being focused through a window onto the solid target. Excimer lasers have a power around 100 W/cm^2 and a pulse rate of around 10–20 Hz. Each pulse is very short, typically on the order of 1 ns. The laser is often rastered over the target. The target may be rotated to improve uniformity of wear. The energy from the laser rapidly heats a region of the target causing the material to be rapidly vaporized and

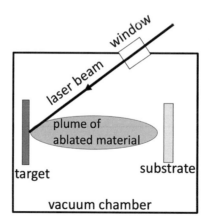

FIGURE 9.23
FIGURE 9.23
Schematic representation of pulsed laser deposition.

expand away from the target in a partially ionized plasma plume. The plume consists of neutral target atoms, target ions, and electrons that are traveling with an average velocity of about 100 cm/s.

The first step in this process is the interaction of the laser beam with the target material, which is dependent on the wavelength of the laser. Typically, shorter wavelengths (ultraviolet) have greater ablation. The ablation process happens quickly compared to thermal diffusion and so a small region on the target is melted and vaporized. The small size of the ablation region also allows for a small overall target size (<1 cm^2), which may be an advantage when experimenting with complex target compositions. The high temperature in the ablation region (up to 5000 K) and the very high heating rates (10^{12} K/s) evaporate all components of the target efficiently independent of their binding energies and vapor pressure. This allows for stoichiometric deposition of the target composition in the film.

9.6.2 Transport

Note that the plasma in pulsed laser deposition is very different from that used in sputter deposition. The sputter deposition plasma is primarily an Ar plasma through which the target material must travel. Here the plasma consists of target material that is expanding away from the target toward the substrate. As a result, the transport region in pulsed laser deposition is quite different. The chamber may be under high vacuum or may be backfilled with some pressure of gas allowing for reactive deposition. A non-reactive gas can also be used in the chamber to modify the kinetic energy of the expanding plume.

9.6.3 Deposition

As the plume reaches the substrate, some target material will condense onto the substrate forming a film. The deposition rate will depend on the target material (and how it interacts with the incident laser wavelength), the laser power, the laser pulse rate, and the distance from the target to the sample. The shape of the plume tends to produce a highly non-uniform thickness on the substrate. Rastering the laser beam on the target and moving the substrate during deposition can improve the film thickness uniformity. Globules of target material can come from the target in the ablation process and can be deposited in the film. The high energy of the plume can interact with the growing film to change the film structure. Different elements will also have different angular and velocity distributions within the plume.

The advantages of pulsed laser deposition are good control of the deposition rate by varying the laser pulse rate and power, the ability to transfer target material to the substrate with good reproducibility of the stoichiometry of the target in the film, the ability to deposit almost any material, and the cleanliness of the laser as a source (no hot filaments outgassing in the chamber).

References

Anders, A. 2008. *Cathodic Arc Plasma Deposition: From Fractal Spots to Energetic Condensation*. New York: Springer Inc.

Frey, H. and H. R. Khan, eds. 2015. *Handbook of Thin-Film Technology*. Berlin: Springer Verlag.

Krebs, H-U., et al., 2003. Pulsed laser deposition (PLD) – a versatile thin film technique. In *Advance in Solid State Physics, Vol 43*, ed. B. Kramer, 505–518. Berlin: Springer Verlag.

Maissel, L.I. and R. Glang, eds. 1970. *Handbook of Thin Film Technology*. New York: McGraw-Hill Book Company.

Martin, P.M. ed. 2010. *Handbook of Deposition Technologies for Films and Coatings: Science, Applications and Technology*. Amsterdam: Elsevier Inc.

Mattox, D.M. 2010. *Handbook of Physical Vapor Deposition (PVD) Processing*. 2nd ed. Burlington, MA: Elsevier: William Andrew Applied Science Publishers.

O'Hanlon, J.F. 1989. *A User's Guide to Vacuum Technology*. 2nd ed. New York: John Wiley & Sons.

Ohring, M. 2002. *Materials Science of Thin Films*. 2nd ed. San Diego, CA: Academic Press.

Schneider, C.W. and T. Lippert. 2010. Laser ablation and thin film deposition. In *Springer Series in Material Science Vol. 139: Laser Processing of Materials*, ed. P. Schaaf, 89–112. Berlin: Springer Verlag.

Tsao, J.Y. 1993. *Materials Fundamentals of Molecular Beam Epitaxy*. Boston, MA: Academic Press, Inc.

Vossen, J.L. and W. Kern, eds. 1978. *Thin Film Processes*. San Diego, CA: Academic Press, Inc.

Problems

Problem 9.1 Consider the differences between films grown from a point source and a surface source.

    a. Derive Equation 9.14 from Equations 9.12 and 9.13.

    b. Similarly derive Equation 9.16 from Equations 9.12 and 9.15.

    c. What values of L/h would provide 5% uniformity across the sample for each type of source?

Problem 9.2 The flux of Aluminum being deposited onto a substrate is 1.5×10^{20} atoms/s/m² (which corresponds to a growth rate of 2.5×10^{-9} m/s). In addition, a background gas of N_2 strikes the surface. At a temperature of 300 K, calculate the ratio of Al atoms to N_2 molecules incident on the surface for:

   i. $P=10^{-3}$ Pa (low vacuum, decorative coatings)

  ii. $P=10^{-5}$ Pa

 iii. $P=10^{-7}$ Pa (ultra-high vacuum, molecular beam epitaxy)

Problem 9.3 Sputter-deposited alloy films of Ti (12 wt%)-W (88 wt%) are used as diffusion barriers.

a. Using the W and Ti vapor pressure curves below, discuss how easy it might be to grow these alloy films using evaporation rather than sputter deposition. Recall that one of the problems with growing alloys by evaporation is that the composition of the alloy source is very different from the composition of the alloy in the film (You may find the discussion in Section 5.1 helpful and you could assume that the activities of these two materials are the same). For ideal solutions, we can assume P_A(alloy)$=X_A P_A$(pure) where the P_A values are the vapor pressure for element A in the alloy and in the pure form and X_A is the mole fraction of A in the alloy. This can be combined with Equation 2.10 to examine the behavior in ideal solution alloys. Using the vapor pressure curves below determine the composition of the Ti-W alloy evaporation source assuming that we seek a vapor pressure of 10^{-4} mbar for reasonable evaporation rates. (You could confirm from a W-Ti phase diagram that the source is molten at the high temperatures required for W to reach this vapor pressure.) Is evaporation a feasible method for growing these films?

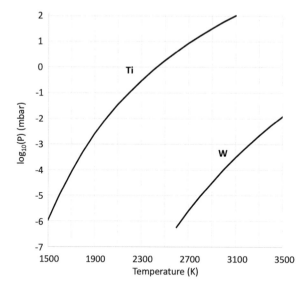

b. If we want to grow the film using sputter deposition with 0.5 keV Ar ions, we start with a target with the desired composition of the film (12 wt% Ti). If we sputter with Ar ions and the sputter yields are $S_{Ti}=0.51$ and $S_W=0.57$, what composition (in atomic %) will the sputtering target surface reach in steady state?

Problem 9.4 Fe thin films are grown on a single-crystal Al(1 0 0) substrate surface. The films grow with no dislocations (if they are thin enough) which suggests that the atoms find a way to line up nicely across the Al/Fe interface. Fe is bcc with a lattice parameter of 2.867 Å and Aluminum is fcc with a lattice constant of 4.050 Å. Describe the crystalline orientation of the Fe on the Al(1 0 0) surface. (Either draw a picture or describe which Fe surface is parallel to the Al(1 0 0) surface and what direction in the Fe is parallel to what direction in the Al. (This problem is rigged to make it very easy. Look at the ratio of the lattice constants and see if it is a familiar number.)

Problem 9.5 GaAs – AlGaAs multilayers by MBE

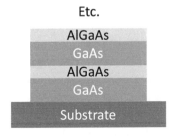

Consider the growth of alternating layers of GaAs and $Al_xGa_{1-x}As$ using MBE. This can be accomplished by depositing Ga and As continuously during the entire deposition. The Al beam can be turned on and off depending on which material is desired. Consider a deposition which lasts for 2.0 hours with the Al beam alternating on for 0.5 minutes and then off for 1 minute during the entire deposition. The relative deposition rates of the Ga and Al will determine the composition (x) of the AlGaAs layers. At the end of the deposition, the total thickness of GaAs films deposited was found to be 2.4 μm and the total thickness of AlGaAs films was 1.8 μm.

a. What are the growth rates (in nm/minute) of GaAs and of $Al_xGa_{1-x}As$?
b. What is x, the atom fraction of Al, in $Al_xGa_{1-x}As$? (This can be determined easily from the growth rates.)
c. What is the thickness of one layer of GaAs?
d. What is the thickness of one layer of $Al_xGa_{1-x}As$?
e. How many layers of each type of film were deposited?

Problem 9.6 Reactive evaporation of high temperature superconductors

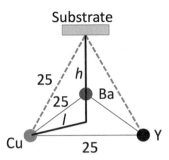

Consider the deposition of an alloy consisting of three elements such as the Cu, Y, Ba system shown in the figure. Three independent point sources can be placed at the corners of an equilateral triangle with a small substrate centered above the middle of the triangle. The lengths of the sides of the triangle are 25 cm. Assume that any evaporant material that hits the substrate sticks to it and that the system is configured to achieve the maximum deposition rate.

a. We seek to grow a YBa_2Cu_3 alloy. The Y source is heated to 1750 K and produces a vapor pressure of 10^{-3} Torr. Using the vapor pressure curves determine what temperature the Cu source must be heated to in order to give the proper film stoichiometry.

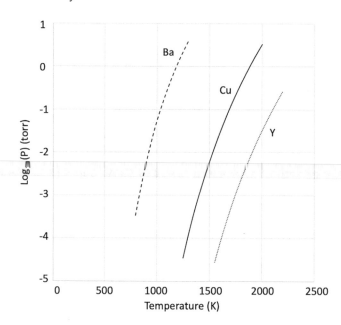

b. Instead of estimating the temperatures from vapor pressure graphs, we can use the equations for the vapor pressure curves and solve numerically (by iteration for example) to determine the temperature from part (a). Many vapor pressures can be fit using a formula $\log(P) = A + B/T + C\log_{10}(T) + D/T^3$ where T is in Kelvin and P is in atmospheres. For our materials, in the temperature ranges of interest:

Element	A	B	C	D
Ba	12.405	−9690	−2.289	0
Cu	9.123	−17748	−0.7317	0
Y	9.735	−22306	−0.8705	0

Determine the temperature for the Cu source and compare to your results from part (a).

c. Adding room temperature O_2 gas into the chamber during deposition allows for the creation of a $YBa_2Cu_3O_7$ high temperature superconductor material. What minimum O_2 partial pressure would be required assuming that all of the oxygen reacts?

Problem 9.8 Using a single evaporation surface source, located directly below the substrate, we seek to grow a film on a 20 cm wide substrate such that the thickness at the edge of the film is 90% of the thickness at the center of the film. How far below the substrate should the source be positioned to achieve the desired film thickness uniformity?

Problem 9.9 Molecular beam epitaxy is being used to grow GaAs films. Separate evaporation sources of Ga and As are available and are located 10 cm directly under the substrate. Each evaporation source has a surface area of 3 cm². The Ga source is heated to 1300 K and the As source is heated to 650 K.

a. Using the graph below, what are the vapor pressures of each source?

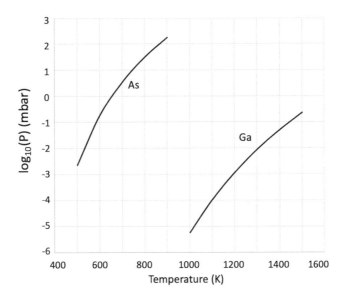

b. Calculate the atomic fluxes from each source. Which source will limit the deposition of GaAs?

c. Assuming that the sources can be treated as surface sources, determine the deposition rate (nm per second) of GaAs. Assume a very small substrate and start with Equations 9.8 and 9.11.

Problem 9.10 A Cu film was deposited using physical vapor deposition at a rate of 10 nm/second at a temperature of 400 K. Characterization of the film after deposition indicated that the film contained 0.1% oxygen. What was the partial pressure of O_2 in the deposition chamber during deposition?

Problem 9.11 Sputter deposition can be either line of sight or not depending on whether gas phase collisions occur. In sputter deposition of Fe at 1 keV we find that the Fe initially comes off the surface of the target with an energy of 9 eV and then undergoes two collisions with Ar atoms while traveling from the target to the substrate in a particular deposition system. What is the energy of the Fe atoms when they reach the substrate? You can assume elastic collisions and that the Ar atoms are stationary. The standard formula for collisions between atoms of type A and B (at rest) under these conditions is $\dfrac{E_B}{E_A} = \dfrac{4M_A M_B}{(M_A + M_B)^2} \cos^2\theta$ where θ is the angle of the collision. In our case you can simply average over the angular term.

10

Chemical Vapor Deposition

The other broad area of thin film deposition that we explore is chemical vapor deposition (CVD) (Chapman and Anderson 1974, Frey and Khan 2015, Jones and Hitchman 2009, Maissel and Glang 1970, Martin 2010, Ohring 2002, Vossen and Kern 1978, Waits 1998). As the name implies, these techniques will make much greater use of chemical reactions at different stages in the deposition process to produce thin films. CVD still relies on the same three basic steps of thin film deposition of getting source material into a vapor phase, transporting it to the substrate, and depositing a film on the substrate.

10.1 Overview and Chemical Reactions

A schematic representation of a CVD process is presented in Figure 10.1. Specific CVD systems may not need all of the components represented in the figure. Note that chemical reactions are important first in the source zone where source material enters the vapor through a reaction process. Chemical reactions are important again in the reaction zone where the source gas reacts on the substrate surface to deposit a film. The pressure of the gases in CVD can vary over a wide range from around 1 Torr up to slightly above atmospheric pressure. The source zone can be eliminated completely if an appropriate gas is available to introduce directly into the reaction zone. Later in this chapter, we will explore the various methods of activating the gas to enhance reaction rates and film growth.

Given the importance of chemistry in CVD, we begin by briefly reviewing some classifications of chemical reactions and exploring how they are used in the deposition of thin films. The most common types of chemical reactions will be introduced and thin film deposition examples will be provided.

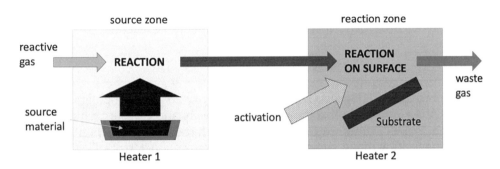

FIGURE 10.1
Schematic diagram of generic chemical vapor deposition process.

DOI: 10.1201/9780429194542-12

Thermal decomposition (pyrolysis) is one of the most direct methods. Above some critical temperature, some gas phase molecules will break up to leave a solid on the surface and a gas phase waste product. The chemical reaction for elements A and X is

$$AX(g) \rightarrow A(s) + X(g)$$

where A is the element to be deposited as a film and X is a reactive element that will eventually be disposed of as a waste gas. A common example is the deposition of Si from silane gas at 650°C. In this case the reaction is

$$SiH_4(g) \rightarrow Si(s) + 2H_2(g)$$

Other examples of thermal decomposition reactions allow the deposition of many metals (Al, Ti, Fe, …), semiconductors (Si, Ge, …) as well as compounds (SiO$_2$, Al$_2$O$_3$, BN …).

Oxidation reactions, most commonly using oxygen gas as the oxidant, are also common. The chemical reaction is

$$AX(g) + O_2(g) \rightarrow AO(s) + [O]X(g)$$

The product side O in the gaseous waste product is in square brackets to indicate that it is not chemically required to be part of this compound but is often observed. An example is a relatively low-temperature method of depositing an SiO$_2$ film from silane at 450°C.

$$SiH_4(g) + O_2(g) \rightarrow SiO_2(s) + 2H_2(g)$$

Oxidation reactions are commonly used for depositing oxide compounds (Al$_2$O$_3$, TiO$_2$, ZnO, …).

Reduction reactions are frequently used with the most common ones involving H$_2$ as the reducing gas. The general chemical reaction is

$$AX(g) + H_2(g) \leftrightarrow A(s) + HX(g)$$

Notice that these reactions are reversible, which could allow for the removal of film materials as well. One difference between thermal decomposition and reduction reactions is that the reduction reactions often can be run at lower temperatures. If you seek to limit diffusion processes, this could be an advantage to reduction reactions. An example of a reduction reaction is the deposition of W at 300°C through the reaction

$$WF_6(g) + 3H_2(g) \leftrightarrow W(s) + 6HF(g)$$

Notice the presence of HF gas as a product of this reaction. This highlights one disadvantage of CVD processes in that they often produce chemical waste products that must be dealt with. Some of these waste products can be hazardous. Reduction reactions can be used for depositing many metals (Al, Ti, Fe …), semiconductors (Si, Ge, …), and compounds (SiO$_2$, TiB$_2$, …).

Chemical compounds can also be formed on the surface using reactions with ammonia or water vapor. The chemical reactions are

$$AX(g) + NH_3(g) \rightarrow AN(s) + HX(g)$$

$$AX(g) + H_2O(g) \rightarrow AO(s) + HX(g)$$

An example of this type of reaction would be the deposition of the wear resistant film, BN, at 1100°C through the reaction

$$BF_3(g) + NH_3(g) \rightarrow BN(s) + 3HF(g)$$

These reactions are used to deposit a variety of compounds (TiN, SiC, Al_2O_3, SiO_2, …).

Disproportionation reactions involve elements that can exist in multiple valence states allowing the formation of different compounds from the same two elements. A simple example of the reaction is

$$2AX(g) \leftrightarrow A(s) + AX_2(g)$$

An example is the deposition of a Ge film through compounds GeI_4 and GeI_2 as demonstrated in Figure 10.2. Notice that the Ge source reacts with the GeI_4 gas to produce a compound with a higher Ge/I ratio in GeI_2. After transport to the substrate, this molecule undergoes thermal decomposition on the substrate leaving a Ge film with GeI_4 as the gaseous waste product. Since GeI_4 is the compound needed to initiate the process as well, it can be recycled and reintroduced into the source side to transport further Ge to the substrate. These disproportionation reactions are used to deposit a variety of materials (Al, Si, C, …).

More complex reactions are possible as well. Figure 10.3 shows a process for depositing a GaAs film. Notice that the Ga source is a solid, but the As source is $AsCl_3$ in the gas phase. These react to form a mixed gas of As_2 and GaCl that is transported to the substrate. At the substrate additional gases, As_4 and H_2, are introduced that lead to chemical reactions that leave a solid GaAs film and have gas phase HCl and As_2 as the byproducts.

Chemical reactions typically express equilibrium thermodynamics. Although flowing gas systems such as those used in CVD are not truly in equilibrium, thermodynamics can still be useful in establishing which reactions are possible and under what conditions. In Section 5.1, we introduced Ellingham plots as a way of comparing chemical reactions. An Ellingham plot for selected metal-Cl compounds is presented in Figure 10.4.

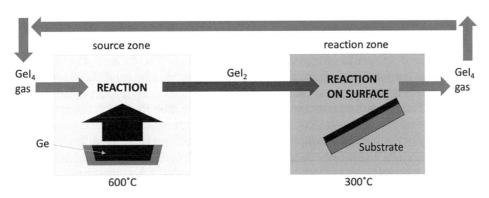

FIGURE 10.2
CVD of Ge by disproportionation reaction.

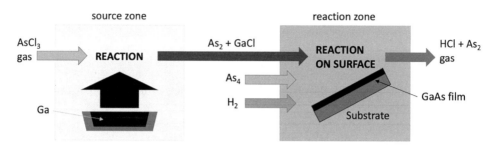

FIGURE 10.3
CVD of GaAs with multiple sources.

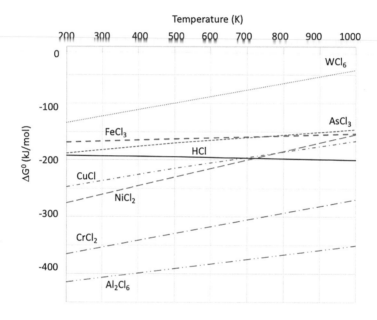

FIGURE 10.4
Schematic Ellingham plot of metal chlorides.

In examining this figure we note that WCl_6, $FeCl_3$, and $AsCl_3$ are all reduced by H to metals and HCl over the entire temperature range of the figure and so would be good candidates for CVD to produce metal films. Ni and Cu chlorides will be reduced at higher temperatures and so would be possible CVD candidates. The Cr and Al chlorides presented here are very stable at all temperatures and so will not be candidates for CVD.

If we wanted to deposit Cr metal films from $CrCl_2$ the thermodynamics is close at higher temperatures but not quite favorable. We could try to adjust the partial pressures of the components to force a reaction to occur. From thermodynamics, we are trying to change the free energy of the metal-Cl compound so that a deposition is allowed. Typically we can deposit a metal from a chloride compound if $\Delta G^0(MCl) - \Delta G^0(HCl) < 42$ kJ/mol. We know that $\Delta G = \Delta G^0 + kT \ln(P_{CrCl_2}/P_{Cl_2})$ and so we can change ΔG_{CrCl_2} by changing P_{CrCl_2}/P_{Cl_2}. To meet the 42 kJ criterion, we need $P_{CrCl_2}/P_{Cl_2} = 1000$ at 1400 K. Manipulation of the chemical reactions such as this may allow for a CVD process to be developed even when the thermodynamics would appear to prohibit it.

10.2 Source

We examine the details of the CVD process by again breaking it down into source, transport, and deposition stages. With CVD, we will also need to consider what to do with the waste byproducts of the reactions.

Several different types of sources are possible in CVD. The easiest is the case where the source material can be obtained already in the gas phase. In this case, the source zone of Figure 10.1 is not needed. Volatile liquids, which can produce the desired vapor either directly or through gas-phase chemical reactions, are also possible sources. Solids, which can sublime to produce the desired gas-phase elements for reactions in the gas phase, are also common sources. Depending on the complexity of the CVD process, some combination of these types of sources may be needed.

In considering source materials we seek sources that are sufficiently volatile that they can readily produce the desired vapors but also materials that are stable at room temperature so that we can control the reactions and not need to constantly cool the source. Solid and liquid sources should have high enough vapor pressures so that we can get good growth rates. Materials of low toxicity are preferable. The chemicals should dissociate and/or react on the substrate to give the desired film with sufficient purity.

10.3 Transport

In the transport of gas to the substrate, we seek to optimize the flow of gas so that we get efficient deposition. We need to deliver the gas quickly enough so that the reactions at the surface can proceed at an optimum rate, but we do not want to deliver the gas so quickly that some is wasted. We would also prefer to deliver the gas in such a manner that it reaches the substrate uniformly and deposits uniform films.

Gas flow can be divided into two regimes. At lower gas densities the distance between gas atoms is sufficiently large that the atoms largely do not interact with one another. In this molecular flow, regime deposition will be governed by diffusion in the gas phase. The kinetic theory of gases predicts that the diffusion coefficient will depend on the temperature and pressure of the gas as

$$D \sim T^{3/2} / P \tag{10.1}$$

In this case, we can reduce the gas pressure to obtain a higher deposition rate.

Many CVD systems operate at higher gas densities in the viscous flow regime. At relatively low flow rates this results in laminar flow, which is the desired flow. At higher flow rates turbulent flow will set in.

We explore laminar flow by considering the simple case of the flow of a fluid (gas or liquid) past a flat plate as represented in Figure 10.5. Initially, a spatially uniform flow of gas approaches the plate with a horizontal velocity of v_0. When it reaches the horizontal plate, a new boundary condition is introduced at the surface of the plate. The flow profile can be determined by dividing gas into thin elements and applying viscous forces to each element. Gas atoms at the plate surface will not be moving due to the viscous forces at the plate. The atoms just above them, will be moving slightly and so on until we get up to the

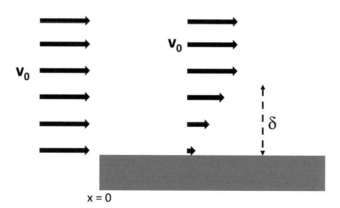

FIGURE 10.5
Gas velocity profile for laminar flow past a flat plate.

parts of the flow that are not affected by the presence of the plate. The result is that gas is not being delivered to the plate at a speed, v_0. Gas will only reach the surface of the plate through a diffusion process vertically in the gas. We can define a stagnant layer near the surface of the plate through which gas would need to diffuse to reach the surface.

This simple model suggests that mass transport to the surface of a substrate will depend on the reactant concentration in the gas (which we can control by adjusting the pressure of the gas) and by the diffusivity and stagnant layer thickness (which we can control by adjusting the gas velocity, temperature distribution, CVD chamber geometry, and gas properties such as viscosity).

Let us explore this in more detail using a simple model proposed by Grove (1967). If we consider a simple reaction such as $AB(g) \rightarrow A(s) + B(g)$ then Figure 10.6 provides a model for the CVD film growth of element A.

In Figure 10.6, Φ_{AB} is the flux of AB to the surface and Φ_{film} is the flux consumed in making the film. C_G is the concentration of AB in the gas flowing into the reaction zone and C_S is the concentration of AB at the surface of the substrate. The flux of AB delivered to the surface is just proportional to the difference in concentrations in the gas and at the surface. If we define a gas diffusion rate constant, h_G, then

$$\Phi_{AB} = h_G (C_G - C_S). \tag{10.2}$$

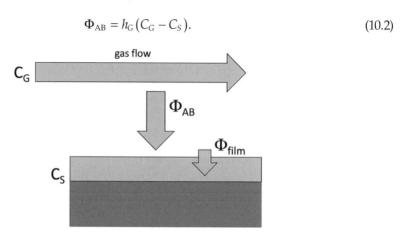

FIGURE 10.6
Simple model of mass transport in CVD film growth.

The flux consumed in the film should be proportional to the concentration available at the surface. Defining a surface rate constant, k_S, gives us

$$\Phi_{\text{film}} = k_S(C_S - 0). \tag{10.3}$$

Once we reach a steady state situation, $\Phi_{\text{AB}} = \Phi_{\text{film}} = \Phi$, and we find

$$\Phi = \frac{C_G}{\dfrac{1}{k_S} + \dfrac{1}{h_G}} \tag{10.4}$$

The surface rate constant, k_S, contains information about the surface reactivity. The gas diffusion rate constant, h_G, contains information about the ability for the reactants to reach the surface through gas phase diffusion. Since the growth rate of the film is proportional to Φ, we see that the transport and deposition stages in CVD are both possible limiting processes in film growth.

10.4 Deposition

Equation 10.4 suggests two rate-limiting cases that will control the deposition of a thin film. If h_G is very small, then the growth of the film will be limited by the diffusion of gas to the substrate. We refer to this as the mass transfer (or diffusion) limited case. Experimentally, h_G is observed to have minimal temperature dependence. This mass transfer limit is commonly observed at higher temperatures. Film growth rates will be controlled by gas flow and gas pressure.

The other rate-limiting case is surface reaction limited, which will happen when k_S is small. The growth will now be controlled by processes on the surface. A great deal of physics is buried into this rate constant. As we saw in Chapter 8, processes such as adsorption, decomposition, surface diffusion, chemical reaction, and desorption of products will all be occurring on the surface and will go into the parameter k_S. Since these processes are all highly temperature dependent, it is no surprise that k_S is highly temperature dependent and increases with temperature. As a result, this will tend to be the limiting process at lower temperatures. The growth rate will be controlled by temperature. This is often the preferred limit in which to operate.

The film growth rate can often be represented by an Arrhenius expression

$$\frac{dz}{dt} = A_0 e^{-E_{\text{reaction}}/k_B T} \tag{10.5}$$

where E_{reaction} is the limiting energy barrier to film growth. Sometimes this barrier can be lowered by activating the process by introducing energy into the system through a plasma, laser, or hot wire as discussed later in this chapter. By lowering the limiting energy barrier, films can be produced at acceptable growth rates while operating at lower temperatures.

Figure 10.7 is an Arrhenius plot of the natural logarithm of the growth rate vs $1/T$. The slope is proportional to the reaction energy barrier. The lower solid line represents the regular CVD process without any form of activation. If we want a higher deposition rate, we need to increase the temperature from T_A to T_B. An activated process, represented by the upper solid line, will have a shallower slope (lower energy barrier). A higher growth rate can be achieved in the activated process at the same temperature. The same growth rate as the regular CVD process could also be achieved at a lower temperature (T_A vs. T_C in Figure 10.7).

We can explore the mass transfer limited case a little further if we assume that the transport to the surface depends only on the diffusion across a stagnant layer. This is equivalent to setting $k_S = \infty$ in Equation 10.4. We can then approximate the gas flow as in Figure 10.8 where we assume that the gas near the surface has a zero velocity throughout a stagnant layer of thickness, δ, and a uniform velocity above that.

With these approximations, the gas diffusion rate constant is simply the diffusion constant for the gas phase compound AB divided by the distance traveled, δ. Equation 10.2 becomes

$$\Phi_{AB} = h_G \left(C_G - C_S \right) = \frac{D_{AB}}{\delta} \left(C_G - C_S \right) \tag{10.6}$$

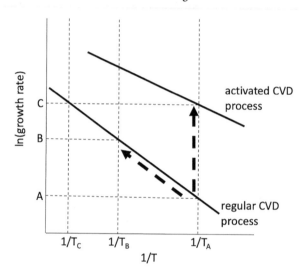

FIGURE 10.7
Achieving higher growth rates by thermal processes or by activated processes.

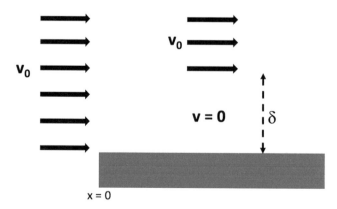

FIGURE 10.8
Gas velocity profile for stagnant layer model of mass transfer in laminar flow.

It can be shown that the stagnant layer thickness depends on the gas properties and the distance along the surface, x, as

$$\delta(x) = \sqrt{\frac{\eta x}{\rho v}} = \sqrt{x}\sqrt{\frac{d}{R_e}} \qquad (10.7)$$

where η is the gas viscosity, ρ is the gas density and v is the gas velocity. A characteristic dimension of the chamber is represented by d and the Reynolds number, $R_e = \rho v d/\eta$. This indicates that the thickness of the stagnant layer will get larger as we move along the substrate resulting in non-uniform film growth.

We can explore the variations along the flow direction in more detail using the model in Figure 10.9 where we have introduced a height, b, to the chamber in the y direction and introduced a gas into the reaction zone with an initial concentration, C_i. We assume that the chamber is large enough in the z-direction (perpendicular to the page) that we can neglect any variations in that direction. We assume that temperature is uniform throughout the chamber and that the gas velocity in the x direction is constant.

We consider the movement of gas molecules into and out of a volume and compare this to the time rate of change of the concentration (as in Chapter 4) and the resultant differential equation is

$$\frac{\partial C(x,t)}{\partial t} = D\left(\frac{\partial^2 C(x,y)}{\partial x^2} + \frac{\partial^2 C(x,y)}{\partial y^2}\right) - \frac{\partial C(x,y)}{\partial x}\frac{dx}{dt} \qquad (10.8)$$

The first term on the right side of this equation represents the diffusion and the second term represents the gas flow.

We next apply boundary conditions, independent of time, in order to solve this equation. We typically assume that all gas reacts at the substrate ($y = 0$) such that $C(x, 0) = 0$. We assume that the initial concentration (at $x = 0$) is constant so $C(0, y) = 0$. We also assume that the gas can not flow out of the top of the chamber so $dC(x, y)/dy = 0$ at $y = b$. Further considering only the steady state solutions $\left(\dfrac{\partial C}{\partial t} = 0\right)$ leads to a rather complex equation that is not particularly illustrative of the underlying concepts. If we make one additional

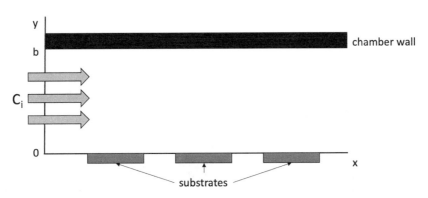

FIGURE 10.9
Model for exploring lateral variations along flow direction.

simplification and assume that the velocity of the gas is large and/or that the chamber height, b, is large such that $v_{ave}b \gg D$ then the solution simplifies to

$$C(x,y) = \frac{4C_i}{\pi} \sin\left(\frac{\pi y}{2b}\right) e^{-\pi^2 Dx / 4v_{ave}b^2} \tag{10.9}$$

We can examine the behavior of this solution at several limits to confirm that it is reasonable. It is proportional to C_i, which is expected. At $y = b$, the sine function becomes $\sin(\pi/2) = 1$ and so we see that the concentration varies in the y direction with a maximum value at $y = b$ and a concentration of 0 at $y = 0$. This is consistent with our previous models. We see that the concentration decreases exponentially with x. In Chapter 4, we saw that the flux of atoms diffusing in the y direction is proportional to $\dfrac{\partial C}{\partial y}$. The growth rate of the film should be proportional to the flux evaluated at $y = 0$. This suggests that the films will be non-uniform in the x direction with the same exponential dependence.

Experimentally several things can be done to attempt to improve film uniformity. The substrate can be tilted into the flow to change the boundary conditions. The temperature can be increased continuously along x to try to increase the reactivity. We can also process one substrate at a time in which case the gas flow could be normal to the sample in a "showerhead" configuration.

In general, we find that the deposition of a thin film will depend on the chamber geometry and substrate orientation, on the process conditions such as initial gas concentration, pressure, and velocity, on the temperature and the temperature distribution, and on the choice of materials used for the film and substrate. The choice of materials may be somewhat more complicated since chemical reactions are important in CVD. Clearly, we need a substrate and film whose melting points are greater than the temperature needed to have the desired chemical reaction occur on the surface. The substrate needs to be one on which adsorption of the gas phase molecules readily occurs and one where the desired surface reactions will occur. For example, W can be deposited from WF_6 on a Si substrate but not on a SiO_2 substrate.

CVD deposition allows us to deposit some materials that are hard to deposit by evaporation. It allows for very high growth rates and good reproducibility. It is possible to grow films epitaxially using CVD. Many CVD reactions, however, take place at high temperatures, which may lead to the interdiffusion of multiple film layers or the film and substrate. The processes are sometimes rather complex and can involve toxic and corrosive gases.

10.5 Modifications of CVD

The fundamental processes described for CVD suggest some modifications that could be made that would enhance our film growth process.

10.5.1 Low-Pressure CVD

Reducing the gas pressures by a factor of 10^3–10^6 to a range of 10^{-3} to 1 Torr can put us into the molecular flow regime rather than the viscous flow that we discussed in detail above. These lower pressures are often compensated by increasing the initial concentration of the gas. As we saw earlier in this chapter, a reduction in pressure, in the molecular flow regime, results in greater diffusion of gas to the substrate. This often results in the surface reaction becoming the rate-limiting step in Low-Pressure CVD (LPCVD). Working at even lower pressures is often described as ultra-high vacuum CVD (UHVCVD).

LPCVD can result in better film uniformity, better conformal coverage over steps on the substrate, and fewer defects in the film. The deposition rates tend to be lower than normal CVD reactions. Many LPCVD deposition reactions are also operated at higher temperatures to achieve a reasonable deposition rate. An example of an LPCVD reaction would be depositing Si_3N_4 on Si at a temperature of 800°C and a gas pressure of 400 mTorr. A mixture of dichlorosilane and ammonia gases is used and a typical deposition rate would be 6 nm/minute.

10.5.2 Plasma Enhanced CVD

We can increase the surface reactivity by introducing energy into the system and/or by causing molecules to dissociate in the gas phase before reaching the surface. One method for doing this is to introduce a plasma in the vicinity of the substrate (Jansen 1997). The plasma will break up the gas molecules removing dissociation of the molecules from the steps that need to occur on the substrate. The plasma can also deliver energy to the surface to enhance processes such as nucleation, diffusion, or reaction kinetics. This may allow for lower temperatures and/or lower gas pressures in a Plasma Enhanced CVD (PECVD) system.

We discussed the nature of plasmas in Chapter 2. The electrons in the plasma will ionize atoms in the gas to keep the plasma going. The electrons also will "activate" that gas by causing the dissociation of molecules. Typically about 1% of the gas molecules will be activated in PECVD.

In Chapter 9, we relied on a plasma to create the conditions for sputter deposition. In PECVD, the pressures are higher than in sputter deposition. As a result, the ions suffer more collisions in the gas phase. Since they lose energy with each collision, they will not typically have enough energy when they reach the substrate to cause any significant sputtering. The energy of these ions depends on the gas pressure and the high voltage applied to the substrate (cathode).

Rather than a DC plasma, as described above, an RF plasma can have advantages, especially when depositing insulating films. Just as in the sputter deposition of insulators, charge can build up on the insulating surfaces. We can reverse the polarity of the high voltage to allow the electrons to impact the substrate and neutralize the charge. If we operate at low frequencies (<1 MHz) the ions will be able to reach the cathode and strike it. If we operate at high frequencies (>1 MHz) the massive ions will reverse direction before they are able to reach the substrate. By adjusting the zero of the RF oscillation, the anode and cathode may be symmetric, or they may be asymmetric allowing greater ion bombardment on one electrode.

The RF-PECVD process has more parameters available for us to adjust than regular CVD. The substrate temperature can be controlled by a substrate heating/cooling system. There is very little heating resulting from the RF-PECVD process itself. Gas flow rates can be increased, which will typically increase the deposition rate and film uniformity but may lead to wasting some of the gas. Changing the gas pressure impacts several variables. It changes the energy of the ions that reach the electrodes and can change the deposition rate. An increase in the gas pressure (and the density of gas molecules) may lead to chemical reactions occurring in the gas phase. Pressure effects are typically also dependent on the gas concentration. Changing the RF power will affect the number of electrons available for activation of the gas and the energy of those electrons. Increased power may lead to chemical reactions in the gas phase and may also increase the deposition rate. The frequency of the RF oscillation may be changed to change the plasma characteristics, which will also change the ion bombardment characteristics. Some systems will use a dual frequency system to attempt to control these two processes independently.

10.5.3 Laser Enhanced CVD

In the previous section, energy was introduced into the system through a plasma. An alternate source of energy is to use a laser (often Nd:YAG or Ar ion) to activate the CVD process. Laser Enhanced CVD (LECVD) is also referred to as Photo-assisted CVD (PCVD). The goal is to use a laser to increase the rate of surface reactions. This enhancement is achieved through two processes. The laser heats the substrate to increase the reaction rate through a pyrolytic process. Secondly, the laser (especially if it has a wavelength in the higher energy ultraviolet part of the spectrum) may be able to dissociate molecules in the gas phase to activate the gas and enhance chemical reactivity. These two processes were represented in Figure 10.7 where the pyrolytic process is the lower line and the activated process is the upper line. The choice of laser wavelength is important in determining which processes will occur in the activation.

The ability to focus the laser on the substrate surface allows for thermal decomposition CVD in a selected area without needing to heat the entire substrate. Uniformity of film thickness, however, may not be achieved with a stationary, focused beam. The laser can be moved over parts of the surface where the film will be deposited (or the substrate can be moved under the laser). Scanning the beam over a linear path on the sample, for example, can enhance the deposition of a metal strip on the substrate by heating the path to higher temperatures where chemical reactions can occur on the surface. The laser wavelength should be chosen based on the optical absorption characteristics of the substrate and growing film. Thermal diffusion in the film also needs to be accounted for since this can dissipate heat quickly. Sometimes rapid pulsing of the laser can achieve better film deposition.

Photolysis, the breaking apart of molecules in the gas phase to create more reactive species, can also be accomplished using LECVD. The efficiency of this process depends on the ability of the molecule to absorb light, which is strongly dependent on the wavelength of the incident light. Many molecules have relatively broad absorption bands in the ultraviolet part of the spectrum. An example of a simple reaction is Cd deposition by a photo-assisted process dissociation process.

$$(CH_3)_2 Cd + light \rightarrow CH_3 + CH_3Cd \rightarrow 2CH_3 + Cd$$

The organometallic precursor is dissociated by light. The resultant metallic molecule is able to be dissociated thermally at relatively low temperatures leaving Cd to diffuse to the surface and deposit.

10.5.4 Hot Wire CVD

Another way to activate the gas molecules in CVD is by decomposition using a hot wire filament (Schropp 2009) in the vicinity of the substrate (1–10 cm). The filament is kept at a high temperature (1500–2300 K) that breaks apart molecules into more reactive species by pyrolytic and catalytic processes when the molecules pass close to it. The gas inlet is positioned so that gas flows past the hot filament. Gas pressures can be lower than normal CVD (<1 mTorr). Since no plasma is involved in this process, there are no complications from energetic plasma ions bombarding the surface. The reactive species are not ionized as they can be in PECVD and so no electrical bias is required in the chamber. High deposition rates can be achieved.

The transport of the reactive species to the substrate will depend on the pressure in the chamber. Typical pressures are in the range that extends from the molecular flow (low pressure) region where gas phase collisions are rare into the viscous flow (high pressure) region where these collisions are common. In molecular flow, the reactive species move without collisions through the space between the filament and the substrate. In the viscous flow region, the motion is a diffusion process through the gas.

One disadvantage of this technique is the hot filament being close to the substrate restricts the possible range of substrate temperatures since the substrate is radiatively heated by the filament. Substrate temperatures below about 200°C will require substrate cooling. The filaments are typically tungsten or tantalum and so reactions of CVD gases with the hot filaments, which could reduce the lifetime of the filament, need to be considered.

10.5.5 Metalorganic CVD

Another modification to CVD is based on the use of special source materials that are highly volatile at relatively low substrate temperatures. Metalorganic CVD (MOCVD), which is also referred to as Organometallic Vapor Phase Epitaxy (OMVPE), uses very high purity organometallic source gases to allow for growing films by CVD at lower temperatures. Common examples of such gases are tri-methyl Gallium, $(CH_3)_3Ga$, or bis(cyclopentadienyl) iron, $(C_5H_5)_2Fe$. High purity organometallic compounds that are stable at room temperature can be produced for many elements. In choosing appropriate compounds factors to consider include decomposition temperature, vapor pressure, possibility of incorporating contaminants into the film, possibility of gas phase reactions, and toxicity. Many of these compounds are liquids or solids with very high vapor pressures (0.5–100 Torr) at room temperature. They are transported in the vapor phase to the substrate by a carrier gas (often H_2).

This technique can grow high-quality films at lower substrate temperatures (400–1200°C) than standard CVD although the gases used in MOCVD are often hazardous. Gas pressures are typically in the 10 Torr – atmosphere range.

MOCVD applications are common in the growth of compound semiconductors and light-emitting diodes due to the ability to produce epitaxial layers and very thin films. Multilayer structures are grown by changing the source gases. Film thickness can be precisely controlled. For example, the chemical reaction for producing a compound semiconductor might look like

$$O_n M(g) + O'_n E(g) \rightarrow ME(s) + nOO'(g)$$

where O and O' are organic molecules (or sometimes H), M is a metal (for instance from Group III on the periodic table) and E is an element that will react with the metal to form the compound semiconductor (for instance from Group V on the periodic table). A specific example to form GaAs is

$$(CH_3)_3 Ga + AsH_3 \rightarrow GaAs + CH_4.$$

For epitaxial films, MOCVD often competes with MBE (described in Chapter 9). MOCVD typically has a faster growth rate and operates at a higher temperature than MBE. The presence of a higher gas pressure in MOCVD limits the availability of *in situ* characterization techniques that can be used in the ultra-high vacuum environment of MBE.

10.6 Atomic Layer Deposition

Atomic layer deposition (ALD) is a layer-by-layer growth method that can be self-limiting and can produce highly conformal films even for high aspect ratio structures with excellent film thickness uniformity (Johnson, Hultqvist, and Bent 2014, Kääriäinen, Cameron, Kääriäinen, and Sherman 2013, Leskelä and Ritala 2002). The technique offers an ability to tune film composition as well. ALD films are built up from sequential gas phase chemical process steps that are self-limiting allowing for excellent control of very thin films (nanometer range). The films are typically very high quality and pinhole free. The basic concept of ALD is represented in Figure 10.10. A precursor gas (1–100 mTorr) is pulsed into the chamber leading to the adsorption of gas molecules on the surface. The chamber is then purged by the introduction of an inert gas (often Ar or N_2). A second precursor gas is then pulsed into the chamber and these gas molecules react with the adsorbed gas atoms to create a molecular species on the surface. The chamber is again purged and the cycle is repeated as necessary to achieve the desired film thickness. If a different film layer is desired, the gases are changed for those cycles. This cyclic exposure to alternating gases is a distinction of ALD compared to CVD.

As in CVD, the source materials can be gas, liquid or solid as long as they can be put into the vapor phase. An ALD system may have a source zone such as that pictured for CVD in Figure 10.1. The precursor gas should be able to chemisorb on the surface or react chemically with another precursor molecule already on the surface. The speed of these reactions will control deposition rate and so rapid reactions are often desirable. The precursor gas, or the byproducts of the reactions, should not react with the substrate or etch the substrate or film. Precursor gases such as H_2S, H_2O, or NH_3 are common. Metal halide and metal alkyl precursors are also commonly used. Not all materials have chemical precursors suitable for ALD film growth. Many oxides, nitrides, and sulfides have been

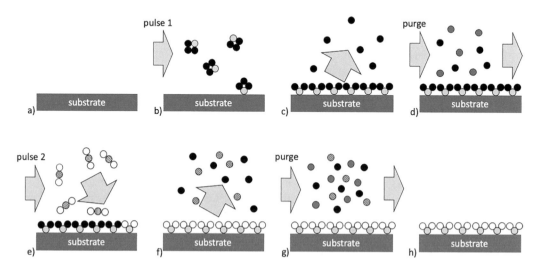

FIGURE 10.10

One cycle of ALD process. (a) Bare substrate. (b) Pulse 1 introduces a precursor gas that chemisorbs to surface. (c) Surface chemical reaction leads to monolayer of material on surface and byproducts in gas. (d) Purge with inert gas removes byproducts. (e) Pulse 2 introduces a second precursor gas that reacts with the chemisorbed monolayer. (f) Desired film monolayer is formed on surface with byproducts in gas. (g) Purge with inert gas removes byproducts. (h) One monolayer of desired film on surface.

grown by ALD as well as some pure elements. For example, TiN can be grown from precursor gases of $TiCl_4$ and NH_3.

The gases are pulsed into the reaction chamber with the substrate for an established period of time (0.1–3 seconds), which is long enough for a monolayer to form and the surface reactions to stop. As such the transport of the gases is somewhat less critical since it is only the total exposure that is important and not the uniformity in time and position. As long as the surface is exposed to enough of the precursor gas, the self-limiting nature of the reactions will insure film uniformity and quality. There are no shadowing effects in ALD and so the films develop in a conformal manner over the entire surface.

Deposition in ALD occurs in a controlled layer-by-layer manner due to the self-limiting nature of the chemical reactions. This typically occurs within a temperature window. If the temperature is too low, the chemical reactions will happen very slowly. If the temperature is too high, the compounds may thermally decompose and/or may desorb quickly from the surface. Typically the temperature range of ALD deposition processes is lower than other CVD processes, which is an advantage of ALD. The deposition rate (100–300 nm/hour) is typically lower than CVD making ALD an optimum method for very thin films. The thickness of the film structure is very well controlled since each cycle will add a single monolayer of film material. The desired thickness is achieved by simply controlling the number of cycles.

The chemical composition of a single layer film can be controlled by alternating two different ALD cycles. The first ALD cycle deposits material A. This is followed by a cycle of material B. This sequence is then repeated. In the end, the film is annealed to blend the monolayers and create the desired composition. The composition could be varied, for instance, by doing two cycles of B for each cycle of A in order to generate a more B-rich film. The exact composition of the film may not be as simple as just taking the ratio of the cycles since the chemical reactions are complex, but the general procedure can be worked out experimentally.

References

Chapman, B.N. and J.C. Anderson, eds. 1974. *Science and Technology of Surface Coating*. New York: Academic Press.

Frey, H. and H.R. Khan, eds. 2015. *Handbook of Thin-Film Technology*. Berlin: Springer Verlag.

Grove, A.S. 1967. *Physics and Technology of Semiconductor Devices*. New York: Wiley.

Jansen, F. 1997. *Plasma-Enhanced Chemical Vapor Deposition*. New York: American Vacuum Society.

Johnson, R.W., A. Hultqvist, and S. F. Bent. 2014. A brief review of atomic layer deposition: From fundamentals to applications. *Mater. Today*. 17: 236–246.

Jones, A.C. and M.L. Hitchman, eds. 2009. *Chemical Vapour Deposition: Precursors, Processes and Applications*. Cambridge: Royal Society of Chemistry.

Kääriäinen, T., D. Cameron, M.-L. Kääriäinen, and A. Sherman. 2013. *Atomic Layer Deposition: Principles, Characteristics, and Nanotechnology Applications*. 2nd ed. Salem, MA: Scrivener Publishing LLC.

Leskelä, M. and M. Ritala. 2002. Atomic layer deposition (ALD): From precursors to thin film structures. *Thin Solid Films*. 409: 138–146.

Maissel, L.I. and R. Glang, eds. 1970. *Handbook of Thin Film Technology*. New York: McGraw-Hill Book Company.

Martin, P.M., ed. 2010. *Handbook of Deposition Technologies for Films and Coatings: Science, Applications and Technology*. Amsterdam: Elsevier Inc.

Ohring, M. 2002. *Materials Science of Thin Films*. 2nd ed. San Diego, CA: Academic Press.
Schropp, R.E.I. 2009. Hot wire chemical vapor deposition: Recent progress, present state of the art and competitive opportunities. *ECS Transac*. 25: 3–14.
Vossen, J.L. and W. Kern, eds. 1978. *Thin Film Processes*. San Diego, CA: Academic Press, Inc.
Waits, R.K. 1998. *Thin Film Deposition and Patterning*. New York: American Vacuum Society.

Problems

Problem 10.1 Silane (SiH_4) is a common gas used in CVD processes.
  a. Calculate the mass of a silane molecule.
  b. Calculate how many molecules are in 5×10^{-2} kg of silane.
  c. Assuming a reaction chamber with a volume of 4.0 L and a temperature of 25°C, what would be the pressure from this many molecules of silane?

Problem 10.2 CVD equations
  a. Consider producing SiO_2 films by CVD reactions of a gas containing $SiCl_4$, NO and H_2 by reduction and oxidation reactions. Write out a balanced chemical equation for the total reaction. (Note that HCl is a common product in CVD reactions involving Cl.)

  b. Consider producing VC films by CVD reactions of a gas containing VCl_4, $C_6H_5CH_3$ and H_2 by reduction and compound formation reactions. Write out a balanced chemical equation for the total reaction. (Hint: C_6H_6 is a stable benzene ring and is one of the products.)

Problem 10.3 Stagnant Layer thickness
  A Reynolds number (Re) up to 1100 is characteristic of laminar flow. Consider a CVD reactor tube with a diameter of 20 cm. For Reynolds numbers of 100 and 1000, calculate the thickness of the stagnant layer at distances of 10 and 100 cm into the CVD reactor tube.

Part III

Characterization of Surfaces and Thin Films

11

Characterization: Overview and Imaging Techniques

11.1 Overview of Characterization

Now that we have explored the creation of thin films on substrate surface, we conclude the book with several chapters on how to characterize surfaces and thin films. Accurate characterization and interpretation are key to building an understanding of our systems. We are interested in a wide range of properties and several length scales. Depending on our applications we may wish to characterize materials on a macroscopic ($\geq 10^{-3}$m), microscopic ($\sim 10^{-6}$m), or atomic ($\sim 10^{-10}$m) scale. We are interested in what the sample looks like, the structure of the sample (atomic locations, film thickness, and density …), the composition (what elements are present) in what quantities, the chemistry (how are the elements bound together), the optical properties, electrical properties, magnetic properties, mechanical properties, thermal properties and other aspects of the surface or thin film. We may also wish to measure the fundamental properties of materials to help in modeling the behavior of the materials.

As we change scale or change what properties we seek to measure, we will typically need to employ new techniques. Each technique, however, rarely depends on just one property, and so we will find that our measurements are based on combinations of properties that will need to be separated. Making accurate measurements can be challenging and the interpretation of those measurements to learn what we seek to know can be even more challenging.

We can probe our samples in a variety of ways. Electromagnetic radiation (infrared, visible, ultraviolet, X-ray) is a very common probe. From the world of mass, we use charged particles such as electrons and ions (including nuclei and protons) as well as electrically neutral particles (neutrons and neutral atoms). We use various types of forces to "touch" the sample. The goal of characterization is to develop techniques using these probes that explore the properties that we seek to understand. Sometimes the same probe is both incident on the sample and detected. Other techniques may send in one type of probe which leads to a different type of probe leaving the sample and being detected.

The ingenuity of scientists and engineers over the years in doing this is truly remarkable. As an example, Table 11.1 lists some techniques that are based on electrons. The table indicates what particle is directed into the sample and what is detected and identifies the resultant technique. This table is not even complete for all of the electron-based techniques but gives a sense of the variety of methods that can be developed from the available probes.

Many characterization techniques can be broadly classified into two categories. One type counts the magnitude of a property such as the intensity, force, number of particles, polarization of light, or other properties. The other type, spectroscopic techniques, typically look at the variation of some property as a function of energy, wavelength, frequency, mass, or some other variable. The spectroscopic techniques, as we shall see, require good knowledge of how the measured parameters vary for the source and for the detector.

DOI: 10.1201/9780429194542-14

TABLE 11.1

Several Common Electron-Based Techniques

IN		DETECT	Technique
High energy electrons (30 keV)		Backscattered and secondary electrons	Scanning electron microscopy
Moderate energy electrons (5 keV)		Secondary electrons	Auger electron spectroscopy
X-rays (1 keV)		Secondary electrons	X-ray photoelectron spectroscopy
Low energy electrons (100 eV)		Diffracted electrons	Low-energy electron diffraction
Moderate energy electrons (5 keV)		Diffracted electrons	Reflection high-energy electron diffraction

These are often entered into the characterization through a set of sensitivity factors. We will explore these in more detail when we examine specific techniques. In addition, many of these measurements have a sufficiently high spatial resolution that some sort of mapping across the surface of the sample is possible to look for variations.

Experimental measurements also typically involve trade-offs. We want to make measurements that are accurate, fast, have a high signal-to-background ratio, and have high resolution. These are frequently not all possible to achieve simultaneously. It then requires some skill to achieve the best balance of these factors to best answer the questions you are exploring. Another limitation may be cost. The real world does not work with unlimited budgets and so we may need to consider alternative ways of making a measurement if our ideal method exceeds our budget. In the remainder of this chapter and in subsequent chapters, we will seek to give an idea of all of these considerations while exploring many of the most common techniques for surface and thin film characterization. We introduce a wide range of techniques with the goal of raising awareness about some of the most common methods of characterizing materials, the range of capabilities, and the limitations. A deeper understanding of any technique will require exploration of the references provided.

TABLE 11.2

Spatial Resolution of Several Imaging Techniques

Technique	Limits	Resolution
Eye	Retina	700,000 Å = 0.07 mm
Optical microscope	Diffraction of light	3000 Å
Scanning electron microscope	Diffraction of electrons	30 Å
Transmission electron microscope	Diffraction of electrons	1 Å
Near-field scanning probe microscopies	"Aperture" size	0.1–100 Å

11.2 Imaging Techniques

We begin our examination of surface and thin film characterization by exploring how to "see" our samples. Table 11.2 notes a series of techniques, the limiting factor in the length scales that they can resolve and what that resolution limit is. The eye is quite acceptable for macroscopic investigation of materials. Figure 11.1 depicts the typical range of analysis region sizes for various imaging techniques. Moving to the microscopic scale the optical microscope and electron microscope provide good resolution. At atomic dimensions, however, techniques such as transmission electron microscopy or one of the various near-field scanning probe microscopies (such as scanning tunneling microscopy or atomic force microscopy) are needed to "see" the atomic structure of the sample.

11.2.1 Optical Microscopes

The basic structure of a compound optical microscope is represented in Figure 11.2. Light reflected from the sample (or transmitted through the sample) passes through two lenses (the objective and the eyepiece) magnifying the image of the sample for the eye to observe. A digital camera is often mounted on the microscope in place of the eye.

The ability to resolve objects horizontally depends on the wavelength of the electromagnetic radiation used in the microscope. Two definitions of when two objects are resolved are shown in Figure 11.3. Mathematically these are very similar.

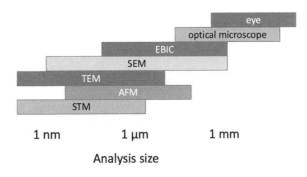

FIGURE 11.1

Typical analysis size for common imaging techniques. Abbreviations are defined throughout the chapter.

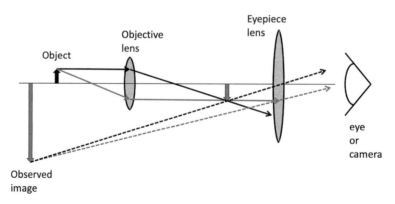

FIGURE 11.2
Schematic description of a compound optical microscope.

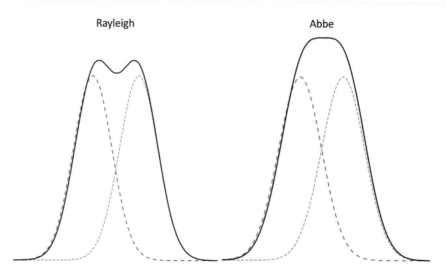

FIGURE 11.3
Rayleigh and Abbe limits of resolution. The dashed lines represent the two independent object images and the solid curve is the resulting combined image.

$$\text{Rayleigh limit: } d = 0.61\frac{\lambda}{\text{NA}} \tag{11.1}$$

$$\text{Abbe limit: } d = 0.50\frac{\lambda}{\text{NA}} \tag{11.2}$$

where d is the distance between the resolved objects, λ is the wavelength of the light, and NA is the numerical aperture of the lens, which depends on the index of refraction of the medium (often air) between the objective lens and the sample and the shape of the lens. For example, if we consider visible light of 500 nm wavelength and a NA value of 1.4, the Rayleigh limit for resolving two objects is 218 nm and the Abbe limit is 179 nm. As a rough estimate based on these equations, the resolution of a microscope is approximately $\lambda/2$.

TABLE 11.3

Resolution and Depth of Field of an Optical Microscope at Various Magnifications

Magnification	Horizontal Resolution (µm)	Depth of Field (µm)
100×	5	25
250×	2	3.8
500×	1	0.9
1000×	0.5	0.08

A major limitation of the compound optical microscope is the limited depth of field (ability to simultaneously focus over a large distance range in the vertical direction) that results from high magnification (high resolution in the horizontal direction). If we seek to resolve objects that are very close together horizontally, we end up with a very limited depth of field that restricts our ability to keep parts of the sample at different heights in focus simultaneously. Table 11.3 provides some typical values for the magnification, horizontal resolution, and depth of field. The ultimate resolution limit for the optical microscope is provided by the diffraction of light and is on the order of 3000 Å (0.3 µm) for visible light.

Advances in computer technology allow the depth of field limitations of the optical microscope to be overcome. With a digital camera replacing the eye, the microscope can be focused at different heights, and then the images can be combined to form a composite image that has a much greater apparent depth of field than any one image would have.

Confocal microscopes also overcome this limitation by using a point light source (often a laser with a pinhole aperture) rather than broadly illuminating the sample and by looking at the signal from the sample through a pinhole aperture that eliminates out-of-focus light. This geometry gives better depth of field although with a trade-off of lower intensity and the need to map the sample. Confocal microscopy has seen particular applications in fluorescent biological systems but is also used in microelectronic applications to determine surface contours and profiles resulting from thin film deposition or to examine the roughness of surfaces. The structure of a confocal microscope is shown in Figure 11.4. If the detected signal is from fluorescence of a different color than the incident beam,

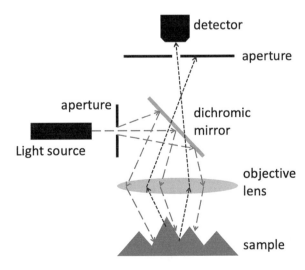

FIGURE 11.4
Schematic diagram of a confocal microscope.

a dichromic mirror can be used to reflect the incident beam but allow the fluorescent signal to pass through. Magnifications of 24,000X can be achieved.

By using a laser source and scanning the laser over the sample additional information about surface roughness and topography can be obtained similar to that from a stylus profilometer (discussed in Chapter 12). With a laser beam radius of less than 0.5 μm, the lateral resolution is good (0.1–0.2 μm) and vertical differences can be determined down to about 0.5 nm. Step height measurements can be made allowing the measurement of thin film thickness if a region of the bare substrate is available. The instrument can be operated either to perform a line scan or to create a topographic map of a region of the sample. Further information on optical microscopy and confocal microscopy is available in the references (Brundle, Evans and Wilson 1992, Mertz 2010, Murphy and Davidson 2013).

11.2.2 Scanning Electron Microscope

The diffraction limit of light, which limits the optical microscope, can be overcome by using electrons as our probe rather than light. From quantum mechanics, electrons have a wave nature as well as a particle nature and so electron microscopes will be limited by the diffraction of electrons. If we estimate the diffraction limit as $\lambda/2$ and use the deBroglie wavelength of the electron $\lambda = h/p$ along with the kinetic energy of the electron being related to the momentum by $p = \sqrt{2m_e E}$, then

$$d = \frac{\lambda}{2} = \frac{h}{2p} = \frac{h}{\sqrt{2m_e E}} \tag{11.3}$$

where h is Planck's constant. For a typical electron energy of 10,000 eV, the diffraction limit would be approximately 0.01 nm. This limit is rarely achieved except in specialized research electron microscopes. A resolution of about 1 nm is a more typical lower limit. Magnifications ranging from 10X to 500,000X are possible.

A schematic diagram of a typical Scanning Electron Microscope (SEM) is provided in Figure 11.5. A beam of electrons is shaped and then scanned over the sample in two dimensions. Some incident electrons are scattered and secondary electrons (which are electrons originally in atoms in the sample) are emitted. In addition, electromagnetic radiation (X-rays through infrared wavelengths) may be emitted depending on the materials. Various detectors in the system can detect the electrons or electromagnetic radiation leaving the sample. The intensity of these signals is correlated to the location of the beam being rastered over the sample to produce a map of intensity vs. location that is displayed as an image on the screen. The electron microscope is typically used for generating topographic images of a surface and for microstructural analysis of parts of a film or surface.

Samples for the SEM typically need to be no smaller than 0.1 mm in size and the maximum sample size will depend on the chamber design of the particular SEM. The samples must be conductive or coated with a thin conductive layer (usually Au or C) and must be compatible with a vacuum environment. Insulating samples may build up a negative charge from the electron beam, which can deflect the beam resulting in a distorted image and lower resolution.

This is the first of many electron-based techniques that we will be discussing. We will explore the interactions of electrons in materials (Ertl and Küppers 1974, Kanaya and Okayama 1972) here and refer back to this discussion when needed in future chapters. Electrons do not travel very far, on average, through solid materials. Figure 11.6 shows

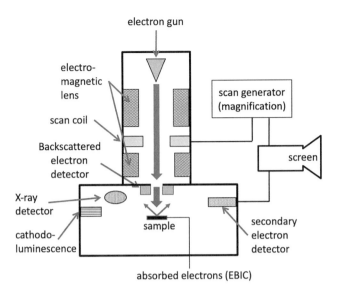

FIGURE 11.5
Schematic diagram of a scanning electron microscope.

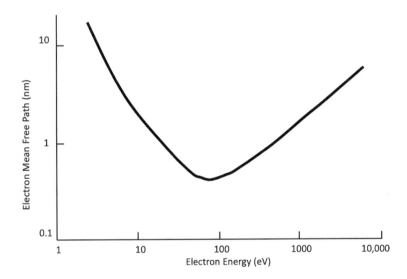

FIGURE 11.6
Universal curve for electron mean free path at various energies.

typical values for electrons of various energies incident normally on a material and how far they travel into the material before undergoing a collision. This is their mean free path as discussed for gases in Chapter 2. Experimental data from a wide range of metals can be fit with a universal curve which suggests that electrons behave similarly in a wide range of materials. The universality of the curve arises because the main energy loss mechanism for electrons in solids is excitation of valence band electrons in the solid. Since the valence band electron density is very similar for most materials, the mean free path is not very sensitive to the material in which the electron is traveling. Both at low and high energies the electrons are able to travel larger distances before experiencing a collision. The universal curve has a minimum for electron energies of around 70 eV where the electrons only travel through about 1 atomic layer of the material. This short travel distance before scattering suggests that electrons may be a useful tool for surface and thin film characterization.

Electrons that are incident on a material may remain in the material (and be pulled away to an electrical ground if the sample is grounded and conducting), they may scatter back out of the material where they could be detected, and they may interact with atoms in the materials causing the emission of secondary electrons, which were bound in the material, where the secondary electrons could be detected. If the sample is thin enough, electrons may also be emitted from the backside. We will defer the discussion of these transmitted electrons to the next section. The resulting assortment of detectable electrons is indicated in Figure 11.7. The incident electron beam typically has energies in the 1000–10,000 eV range. The secondary electrons that provide us with useful information typically have lower energies in the 50–1000 eV range. As a result, the incident electrons are able to penetrate relatively deeply into the material. The backscattered electrons can also come from more deeply in the material. Secondary electrons, however, have lower energies and shorter mean free paths. They will only escape from the material if they are created near the surface. As a result, they tend to come from the top 100 Å of the material. Secondary electrons can be created in two ways. The true secondary electrons (sometimes called Type I secondary electrons) are those created by the incident beam electrons as they enter the material. Type II secondary electrons are those created by backscattered electrons that produce secondary electrons on the way out of the material. The significance of these two types is that the Type II secondary electrons will be created over a much wider area than the Type I electrons. Even if the incident beam is highly focused, the detectable electrons that emerge from the sample will come from a larger region than the incident beam cross section due to the scattering nature of electron interactions within the material.

Several important differences exist between the backscattered electrons and secondary electrons. The backscattered electrons will tend to emerge at angles that are relatively close to the origin of the incident beam. The secondary electrons emerge at a much wider range of angles. This allows some ability to distinguish between the two types of electrons by where the electron detector is placed. Backscattered electron detectors are often a ring

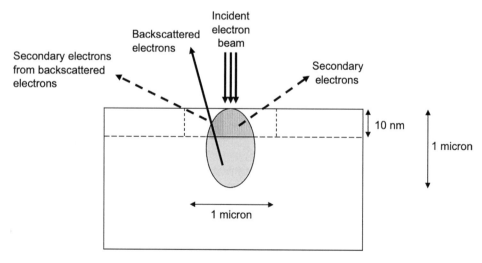

FIGURE 11.7
Detectable electrons emitted from a material with an incident electron beam.

around the incident beam whereas the secondary electron detector can be placed much further from the incident beam. The quantity of backscattered electrons emitted from the surface increases with a higher atomic number of the material. For secondary electrons the dependence on the atomic number of the material is minimal. This is demonstrated in Figure 11.8. This means that the intensity of the detected electron signal will provide some chemical information for backscattered electrons but not for secondary electrons.

The different properties of backscattered and secondary electrons suggest that SEM images formed by secondary electrons will be more sensitive to surface topographic features since they originate near the surface but the images will contain minimal chemical information. Backscattered electron images, on the other hand, will provide some chemical information but will not be as sensitive to surface topography. Figure 11.9 shows two images of the same region of a Zr-Al alloy film. The top image is from backscattered electrons and clearly shows chemical segregation with the higher atomic number Zr showing up as brighter regions on the image. The lower image is generated from secondary electrons and provides more topographic detail but no chemical distinctions.

When examining SEM images, it is important to remember that the image is a map of the emitted electron intensity from various points on the sample. The light and dark regions do not represent light and shadow but indications of where more and fewer electrons are being emitted. The images look so much like photographs formed by light, that it is possible to be misled. This tendency is reinforced by the location of the detector. Since the secondary electron detector is typically located to one side of the sample, there may be locations on the sample where electrons are emitted and are unable to reach the detector. For instance, if there is a raised feature on the sample, the top of the feature or a side sloped toward the detector will emit electrons that are easily detected. Electrons emitted from the side of the raised feature away from the detector, however, will not be able to reach the detector because they are blocked by the raised feature itself. This side of the feature will then appear to be dark. The result is an image that looks as though a "light source" is located at the detector and sample features that are on line of sight with the detector will appear to be bright while those where the line of sight is blocked will appear to be in shadow.

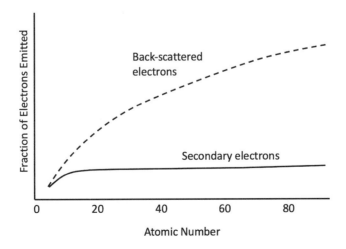

FIGURE 11.8
The fraction of backscattered electrons emitted is strongly dependent on the atomic number of the material unlike the fraction of secondary electrons.

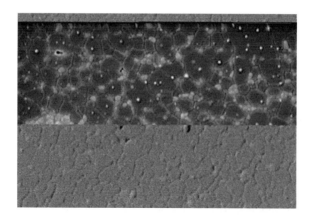

FIGURE 11.9

SEM images of Zr-Al film. Top image is from backscattered electrons. Bottom image is from secondary electrons. (From Wikipedia 2022.)

When working with non-conducting samples, problems may arise from charging of the sample by the incident beam of negatively charged electrons. If the number of emitted electrons (backscattered and secondary) is equal to the number of incident electrons then no charging would occur. Figure 11.10 shows a typical curve for the ratio of the number of secondary electrons exiting the sample to the electrons incident on the sample from the primary electron beam. At low incident beam energies, very few secondary electrons are generated and the sample will become charged negatively. At higher incident energies (above about 50–100 eV) the generation of secondary electrons becomes favorable and more electrons leave the sample than arrive. This results in the sample becoming positively charged. At higher energies (above about 1000–3000 eV depending on incident angle and material), the primary beam penetrates deeper into the sample and the secondary electrons are produced at greater depths, which reduces the probability of their escaping. The sample again is negatively charged. An SEM typically operates at high enough energy that the number of emitted electrons is less than the number of incident electrons and so negative charge will build up on a non-conducting surface. The most common way to deal

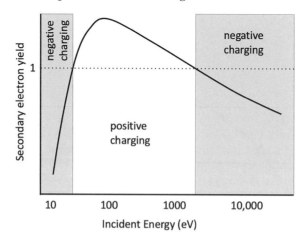

FIGURE 11.10

Secondary electron yield vs incident electron beam energy showing representative regions of negative and positive surface charging.

with this is to coat the samples with a very thin layer of gold or carbon to make the surface conductive. Clearly, this is modifying the sample both structurally and chemically, but if the sample features are primarily on a scale (0.1 µm) that is large compared to the thickness of the conducting film, then useful data can still be obtained.

Several additional types of detectors may be found on SEMs. Energy Dispersive Analysis of X-rays (EDAX, EDX) is a common addition that can provide chemical information about the surface using the analysis of X-rays that are emitted when the incident electron beam strikes the sample. We will defer discussion of these capabilities until Chapter 13 when we consider chemical characterization techniques. A cathodoluminescence detector examines the light emitted from the sample when it is struck by a beam of electrons. A classic example of such a material is a phosphor that glows in the visible wavelengths when struck by an electron beam. Many semiconducting materials exhibit cathodoluminescence. The technique can be used to explore the local density of states, defect structures, surface plasmon resonances, and other effects with the spatial resolution of the scanning electron microscope. Another type of detector measures the Electron Beam Induced Current (EBIC) that goes to ground from the sample. As the incident electron beam strikes different locations on the sample, the current to ground will vary depending on the properties of the material at that location. This technique is useful for examining semiconductors including issues of defect structure, properties of minority carriers, and is often useful in failure analysis. An Electron Backscatter Diffraction detector can provide information about crystallography and crystal orientation, and texture of regions of the sample. The principles of diffraction will be discussed further in Chapter 12. The diffraction pattern is detected on a phosphor screen placed on one side of the SEM chamber. The sample is tilted at a high angle toward the screen and backscattered electrons are diffracted toward the screen resulting in a diffraction pattern on the screen.

Some SEMs are integrated with a Focused Ion Beam (FIB) system allowing for a small region on the sample to be cut away using the FIB and then imaged using the SEM. This allows the acquisition of cross-sectional SEM images of film-substrate or film-film interfaces. Details on SEM theory and operation are available in the references (Brundle, Evans and Wilson 1992, Egerton 2016, Goldstein et al. 2003, Hawkes and Spence 2007).

11.2.3 Transmission Electron Microscope

In the previous section, we dealt primarily with electrons that were emitted from a sample on the same side as the incident beam of electrons. The Transmission Electron Microscope (TEM) requires very thin samples and explores the properties of electrons that are transmitted through the sample. This technique provides useful information on the microstructure of a sample and is commonly used for the analysis of interfaces. The crystalline structure of a very small region can be determined and images with magnifications of 15,000,000 can be obtained giving this technique atomic scale resolution (1–2 Å). The incident beam is typically at a higher incident energy (100–200 keV) than in SEM in order to have the beam of electrons traverse the thinned sample without collisions.

A significant challenge with this technique is to thin samples down to a thickness of about 0.1 µm (1000 Å) without damaging the sample. This is typically a very time-consuming process. Any technique that works at such high magnification also raises the issue of whether the regions examined on the sample are truly representative of the sample as a whole.

A very simple schematic diagram of a TEM is shown in Figure 11.11. The well-focused, high-energy beam of electrons is directed toward the thinned part of the sample. An aperture behind the sample is used to operate the instrument in either diffraction or imaging

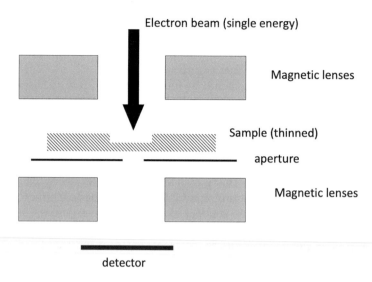

FIGURE 11.11
Schematic diagram of a transmission electron microscope.

modes. With the aperture removed, the instrument operates in diffraction mode and the electrons diffracted from crystalline planes oriented such that Bragg's Law (described in Chapter 12) is satisfied will be detected as in Figure 11.12. The resulting diffraction pattern provides information about the crystallinity of a small region of the sample.

With the aperture inserted, the instrument operates in imaging mode where the transmitted intensity is studied as a function of position. When the central diffracted beam is used for this measurement, the instrument is operating in a "bright field" mode. This uses the diffracted beam that is in the same direction as the incident beam as indicated in Figure 11.13a. Alternately, we could select a single, non-central, diffracted beam and monitor the intensity of that beam instead. Then the instrument is operated in a "dark field" mode. The dark field image can be achieved by moving the aperture over to allow the diffracted beam to be transmitted (Figure 11.13b) or by tilting the incident beam so that the desired diffracted beam is transmitted along the optical axis and the aperture can remain in the center (Figure 11.13c). This can be useful for examining defects in materials.

Figure 11.14 shows an example of TEM images from a TiN film on Si(1 0 0). The crystalline structures of these materials result in a 22% misfit producing dislocations at the interface that can be observed in the image. In the upper left corner, the diffraction pattern is also presented.

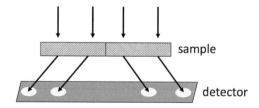

FIGURE 11.12
Diffraction from different grains in a transmission electron microscope leading to a diffraction pattern at the detector.

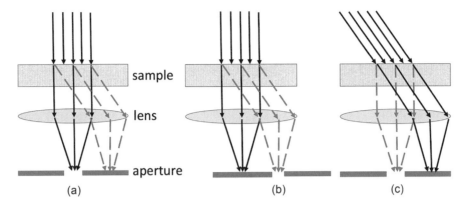

FIGURE 11.13
(a) Bright field image geometry. (b) Dark field image geometry (displaced aperture). (c) Dark field image geometry (centered aperture).

FIGURE 11.14
TEM image of a TiN film on Si(1 0 0). Location of edge dislocation is shown. Inset in upper limit is corresponding diffraction pattern. (Reprinted from: Narayan and Larson 2003 with the permission of AIP Publishing.)

As in the case of the SEM, additional detectors can be added to the TEM to make other measurements. EDAX (see Chapter 13) can be added to allow chemical identification and mapping of the sample. Electron Energy Loss Spectroscopy (EELS) can also be performed to learn more about vibrational properties. This technique will be discussed in more detail in Chapter 13. Further details about TEM operation and theory are available in the references (Brundle, Evans and Wilson 1992, Egerton 2016, Hawkes and Spence 2007, Sardela 2014, Williams and Carter 1996).

11.2.4 Low Energy Electron Microscope

A more surface-sensitive imaging technique can make use of the electron mean free path in solids (Figure 11.5). If a normal incidence beam of high energy (10–20 keV) electrons is slowed down as it approaches the sample to energies of 1–200 eV, the electrons will interact with only the surface region. By varying the incident energy, this interaction can be with only the top layer of surface atoms or can include a few layers. A technique that accomplishes this is Low Energy Electron Microscopy (LEEM). One operating mode of LEEM is similar to an SEM. Elastic backscattering of electrons is used to image the surface. Low energy electrons have a particularly large cross section for backscattering resulting in the ability to collect data quickly and monitor dynamic processes such as surface reconstruction, film growth, and motion of steps. The cross section for reflection of electrons depends on the band structure of the material and so good contrast can be obtained between different types of materials by selecting an appropriate incident electron energy. Spatial resolutions of 10 nm are possible. This allows tracking dynamical surface processes with good time and spatial resolution. Scanning probe microscopies, to be discussed in the next section, have higher spatial resolution, but are slower and have a smaller total field of view when compared to LEEM.

A major difference of LEEM with SEM is the ability to use the diffraction of the primary electrons, which requires a surface with a high degree of crystallinity. Low energy electron diffraction will be discussed in more detail in Chapter 12. Here, we focus on the imaging capabilities of LEEM. Once the incident electrons are diffracted from the surface structure, they are accelerated again and imaged on a phosphor screen or electronic detector. The instrument can be run in several different diffraction modes. By looking at the bright field image (discussed for TEM as well), the central diffracted beam is used to create an image of the surface. The intensity of the beam will vary depending on phase differences in the diffracted electron waves arising from the surface structure. This is often used to image steps, terraces, and islands on surfaces. If dark field imaging is used, a particular diffraction spot is selected and the intensity of that spot is mapped across the surface. This is particularly useful for looking at the coverage of sub-monolayer films on a surface. If the chosen diffraction spot is from the film, the intensity will be observed where the film covers the surface but not for the open surface regions.

We return to a variation of LEEM in Chapter 14 where we examine spin-polarized electron techniques, which can be used for magnetic characterization. Another electron microscopy, PhotoElectron Emission Microscopy (PEEM), uses UV or X-ray light as a source and examines the photoelectrons emitted from the sample. (Photoemission will be discussed in more detail in Chapter 13 where we consider X-ray photoelectron spectroscopy.) Differences in photoelectron emission between different materials provide one form of contrast for imaging the surface. This contrast will depend on the incident light energy. Topographic features can also provide contrast from distortions of the electric field at the surface that depends on the curvature of the features. The technique is surface-sensitive since the photoelectrons will only escape from the material if they are generated near the surface. The lateral resolution of PEEM can be about 10 nm. Additional information about LEEM and PEEM is available in the references (Bauer 1994, Hawkes 2007).

11.2.5 Scanning Probe Microscopes

Most of the characterization techniques operate in the "far field" limit, which makes use of science as it is typically taught at the undergraduate and much of graduate levels. In this limit, distances are large compared to the wavelength of the probe. For example, in an optical or electron microscope, the distances between the lenses or detector and the sample are very large compared to the wavelength of light or electrons. In the last several decades, the development of a series of scanning probe microscopies has exploited the world of "near field" physics where the probe is a very small "aperture" placed very close to the sample as represented in Figure 11.15. The characteristic distances here are small compared

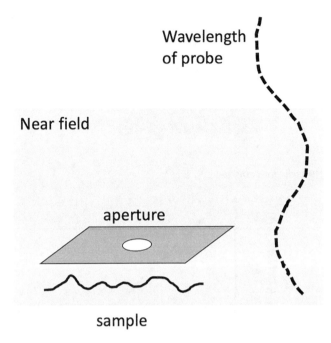

FIGURE 11.15
Schematic of scanning probe microscope illustrating the near-field limit.

to the wavelength of the probe and this allows us to avoid the diffraction limits found in far-field techniques. We will focus our attention on the scanning tunneling microscope (STM) and atomic force microscope (AFM), which are the two most common of these scanning probe microscopes, but an amazing variety of similar instruments (magnetic force microscope, scanning thermal microscope, scanning near-field optical microscope...) have been developed that exploit near-field properties. Each one uses a different probe for the "aperture" but each one operates very close to the sample surface.

The STM provides atomic resolution imaging of surfaces. It can be used to provide elemental identification of atoms on surfaces and can map out the surface density of states. It is possible to use the STM to manipulate atoms on a surface in order to rearrange the atomic order in some cases. Samples need to be conductive and relatively flat for reasons that will become apparent as we describe how the technique works.

The basic concept of the STM is to take an atomically sharp metallic tip and position it a few Å from the surface to be studied. Etched tungsten wire and cut Pt-Ir wire are common tips and are available commercially. Various procedures have been published for preparing tips in your own laboratory. A potential (mV–V) is then applied between the tip and the sample that produces a current between the tip and the sample as electrons tunnel across the gap between the tip and sample as suggested in Figure 11.16. This current is very sensitive to the size of the gap between the tip and the sample as shown in Figure 11.17. The tip is rastered across the sample to produce a map of the surface.

The technology involved in the STM, and other scanning probe microscopies, is a remarkable achievement. Bringing a tip within a few Å of the surface and then scanning horizontally with Å resolution is accomplished using piezoelectric devices. Piezoelectric materials contract and expand in a very well-controlled manner when voltages are applied to them. A typical piezoelectric scanner arrangement is shown in Figure 11.18. Vibration can also be an issue when trying to stably position a tip a few Å from a surface, but anti-vibration

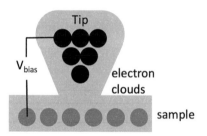

FIGURE 11.16
Schematic of scanning tunneling microscope.

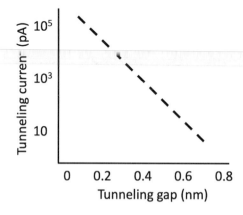

FIGURE 11.17
In an STM, the tunneling current depends very strongly on the gap between the tip and sample.

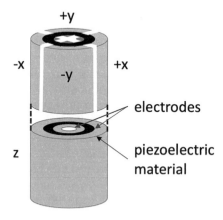

FIGURE 11.18
Piezoelectric scanner allowing precise three-dimensional motion.

technology has long been common in optics. Thermal drift in measurements may need to be controlled since changes in temperature during an experiment can cause thermal expansion or contraction of more than a few Å.

The STM can be operated in two modes represented in Figure 11.19. In constant current mode, the tunneling current is kept constant while the sample is rastered. This is accomplished by moving the tip in and out to maintain a constant gap. This is often a useful

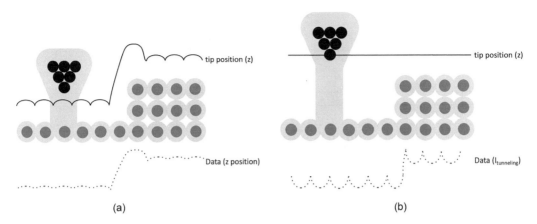

FIGURE 11.19
(a) STM in constant current mode. (b) STM in constant height mode.

initial measurement because the tip is protected from running into features on the sample by the attempt to keep a constant distance between the sample and tip. The quality of the data depends on the ability to move the tip and to measure that motion. This mode tends to have poorer resolution than the alternate mode.

The constant height mode keeps the tip position constant and measures the tunneling current changes between the sample and the tip as the distance between the tip and sample features changes. There is some danger in this mode of crashing the tip into features on the surface whose height is greater than the selected tip height. The data, however, is of higher resolution since it depends on the ability to measure the current variations that are very sensitive to changes in the surface height. This mode is often good for studying small regions on the sample that are known to be flat.

An STM image of Pb deposited on a Cu(1 1 0) surface (Nagl et al. 1995) is shown in Figure 11.20. The Pb atoms have undergone a surface reconstruction to form the rows of atoms shown in the figure.

The STM can also be used to explore the local surface density of states by positioning the tip over a single atom and varying the potential between the tip and the sample. The electrons in this location on the surface have various energies and the number of electrons allowed per energy level of the sample is the local density of states. By varying the bias potential between the tip and the sample, the local density of states can be determined for any location as a function of energy. The tip can then be moved to a new location and the local density of states at the new position can also be determined. The tunneling current (I) is measured as a function of the applied voltage (V). This curve is then differentiated to give a dI/dV curve that gives the density of states. Additional information about STM is available in several books (Bhushan 2004, Brundle, Evans and Wilson 1992, Chen 1993, Hawkes and Spence 2007, Ng 2004, Vickerman and Gilmore 2009) and their references.

The Atomic Force Microscope (AFM) is the other most commonly used scanning probe microscope. The resolution of the AFM is typically not quite as good as the STM but it can still provide very high-resolution images of surfaces on the nanometer scale. Samples should still be relatively flat but any material can be studied.

The basic concept is still to bring a sharpened tip close to the surface. The tip, often made of a doped Si or Si_3N_4, will have a radius of curvature in the 2–60 nm range and so will not be as sharp as the STM case. This tip is attached to a cantilever that will flex as the interatomic forces between the tip and the sample cause a deflection of the tip. The motion

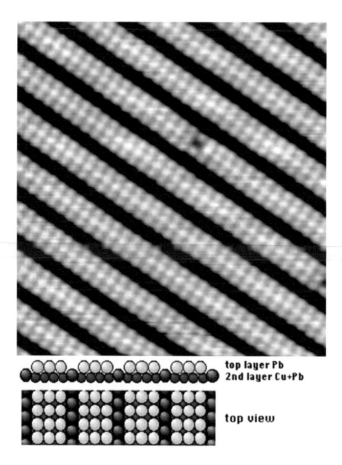

FIGURE 11.20
STM image of Pb on Cu(1 1 0) showing surface reconstruction. (From Institut für Angewandte Physik.)

of the cantilever can then be detected through optical or electronic means. This concept is presented in Figure 11.21.

Figure 11.22 shows a typical AFM in more detail. In many instruments, the tip remains stationary while the sample is moved beneath it using piezoelectric devices. Other

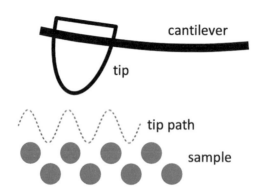

FIGURE 11.21
Basic concept of an AFM (tip is reduced in scale relative to the atoms to fit on the diagram).

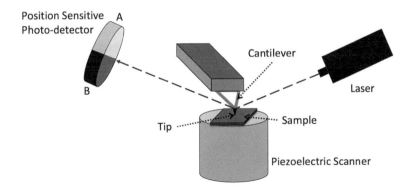

FIGURE 11.22
Schematic representation of a typical AFM.

instruments leave the sample stationary and scan the tip and detection components. The deflection of the cantilever is detected by a position-sensitive photo-detector, which determines how much of the reflected laser light strikes detector A or B. If the photo-detector is divided into four sections rather than two, then torques of the tip due to frictional forces could also be detected.

The interatomic forces were discussed in Chapter 3 and can be attractive or repulsive depending on the separation distance. As the tip approaches the sample it initially feels an attractive force and will be deflected toward the sample. This force becomes stronger as the tip gets closer to the sample causing a greater deflection toward the sample. If the tip gets very close to the sample, the force becomes repulsive as shown in Figure 11.23.

The AFM can also be operated in a tapping mode. This can produce higher resolution images with less force applied to the sample making it especially useful on samples that could easily be damaged. The cantilever is now oscillated near its resonant frequency (typically 50,000–500,000 Hz). The tip "contacts" the surface each time it vibrates toward the surface. Energy loss from the contact with the surface results in a measurable frequency reduction in the vibration of the cantilever. By operating in a constant oscillation frequency mode, the tip can be moved up and down and the surface topography can be mapped out. A recent development is peak force tapping where the tip is vibrated at a

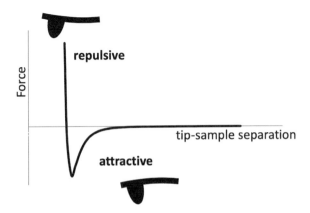

FIGURE 11.23
Interatomic forces in the atomic force microscope.

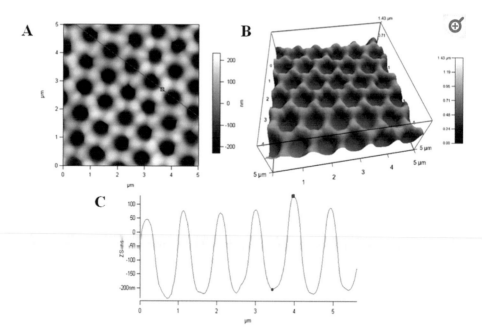

FIGURE 11.24

Plasmonic nanostructure built from 1000 nm latex particles and imaged by AFM. (a) Two-dimensional AFM image. (b) Three-dimensional representation of image. (c) Line scan diagonally down a row from a. (Reprinted by permission of Springer Nature from Kahraman et al. 2013.)

constant, lower frequency (around 2000 Hz). Each time the tip approaches the surface and then retracts, the force curve is measured and the peak force is determined at each point in a map of the sample.

An AFM image can be created by plotting the relative height of each point of the entire area being examined. Alternately, a single line scan can be examined through a particular surface feature to examine the properties of that feature more closely. The AFM is frequently used to measure surface roughness on the nanometer scale. The background noise in the instrument can be suppressed to the point of having sub-nanometer vertical resolution.

Figure 11.24 shows an AFM image of 1000 nm silver-coated latex particles. Image (a) shows the two-dimensional AFM image. Image (b) provides a three-dimensional representation. Image (c) is a line scan diagonally down a row from image (a). Additional details about AFM are available in various books (Bhushan 2004, Brundle, Evans and Wilson 1992, Eaton and West 2010, Hawkes and Spence 2007, Ng 2004, Vickerman and Gilmore 2009).

References

Bauer, E. 1994. Low energy electron microscopy. *Rep. Prog. Phys.* 57: 895–938.

Bhushan, B., ed. 2004. *Springer Handbook of Nanotechnology.* Berlin: Springer-Verlag. (see Part B: Scanning Probe Microscopy).

Brundle, C.R., C.A. Evans Jr., and S. Wilson eds. 1992. *Encyclopedia of Materials Characterization: Surfaces, Interfaces, Thin Films.* Boston, MA: Butterworth-Heinemann.

Chen, C.J. 1993. *Introduction to Scanning Tunneling Microscopy.* New York: Oxford University Press.

Eaton, P. and P. West. 2010. *Atomic Force Microscopy*. Oxford: Oxford University Press.

Egerton, R.F. 2016. *Physical Principles of Electron Microscopy: An Introduction to TEM, SEM, and AEM*. 2nd ed. Switzerland: Springer International Publishing.

Ertl, G. and J. Küppers. 1974. *Low Energy Electrons and Surface Chemistry*. Weinheim: Verlag Chemie.

Feldman, L.C. and J.W. Mayer. 1986. *Fundamentals of Surface and Thin Film Analysis*. New York: North-Holland.

Goldstein, J.I., D. Newbury, D. Joy, C. Lyman, P. Echlin, E. Lifshin, L. Sawyer, and J. Michael. 2003. *Scanning Electron Microscopy and X-Ray Microanalysis*. 3rd ed. New York: Springer.

Hawkes, P. and J.C.H. Spence, eds. 2007. *Science of Microscopy*. New York: Springer-Verlag.

Institut für Angewandte Physik. *Growth of Lead on Copper*. https://www.iap.tuwien.ac.at/www/surface/stm_gallery/pb_on_cu (accessed 26 July 2016).

Kahraman, M., P. Daggumati, O. Kurtulus, E. Seker, and S. Wachsmann-Hogiu. 2013. Fabrication and characterization of flexible and tunable plasmonic nanostructures. *Sci. Rep.* 3: 3396. http://www.ncbi.nlm.nih.gov/pmc/articles/PMC3844966/ (accessed 26 July 2016).

Kanaya, K. and S. Okayama. 1972. Penetration and energy-loss theory of electrons in solid targets. *J. Phys. D: Appl. Phys.* 5: 43–58.

Mertz, J. 2010. *Introduction to Optical Microscopy*. Greenwood Village, CO: Roberts and Company Publishers.

Murphy, D.B. and M.W. Davidson. 2013. *Fundamentals of Light Microscopy and Electronic Imaging*. Hoboken, NJ: John Wiley and Sons, Inc.

Nagl, C., M. Pinczolits, M. Schmid, P. Varga, and I.K. Robinson. 1995. p(nx1) superstructures of Pb on Cu(1 1 0). *Phys. Rev. B* 52: 16796–16802.

Narayan, J. and B.C. Larson. 2003. Domain epitaxy: A unified paradigm for thin film growth. *J. Appl. Phys.* 93: 278–285.

Ng, K.-W. 2004. Scanning probe microscopes. In: *Introduction to Nanoscale Science and Technology*. New York: Springer.

Sardela, M. ed. 2014. *Practical Materials Characterization*. New York: Springer Science+Business Media.

Vickerman, J.C. and I.S. Gilmore, eds. 2009. *Surface Analysis – The Principal Techniques*. Chichester: John Wiley and Sons, Ltd.

Wikipedia Contributors. *Scanning Electron Microscope. Wikipedia, The Free Encyclopedia*. https://en.wikipedia.org/w/index.php?title=Scanning_electron_microscope&oldid=1080962895 (accessed 3 May 2022).

Williams, D.B. and C.B. Carter. 1996. *Transmission Electron Microscopy: A Textbook for Materials Science*. New York: Plenum Press.

Problems

Problem 11.1 The NA of an optical microscope can be expressed as $NA = nD/2r$ where n is the index of refraction of the medium between the objective lens and the sample (often air with $n = 1$), D is the diameter of the objective lens and r is the distance from the lens to the sample. Consider a microscope operating with light of wavelength 500 nm, an objective lens with a diameter of 5 cm and a specimen which is 10 cm away from the lens.

a. What is the resolution of this microscope using the Rayleigh criterion?

b. How could we modify the microscope to improve the resolution?

Problem 11.2 In a scanning tunneling microscope, the relationship between tunneling current and tip to sample distance shown in Figure 11.17 can be expressed approximately as $I = I_0 e^{-2ks}$ where I is the tunneling current, k is a constant which depends on work function and applied voltage, and s is the tunneling gap between the tip and sample. For a typical k value of $1\,Å^{-1}$, by what factor will the current change if the separation distance changes by $1\,Å$?

Problem 11.3 For a piezoelectric scanner tube, we can determine the amount of length change (ΔL) of a given applied voltage (V) from $\Delta L = d_{31} L \dfrac{V}{t}$ where d_{31} is a property of the piezoelectric material with a value typically around -10^{-10} m/V, and t is the distance between the electrodes (typically the diameter of the piezoelectric tube). For a tube with length 0.5 cm and a diameter of 0.5 cm, estimate the voltage needed to change the length of the scanner by 0.1 nm. Is this a practical value to work with?

Problem 11.4 For both scanning and transmission electron microscopes, we are interested in how far the incident electron might penetrate into the sample. This will depend on the incident electron beam energy (E_0) and properties of the sample material such as density (ρ), atomic weight (A) and atomic number (Z). An equation (Kanaya and Okayama 1972) provides appropriate values for the energy range of 10–1000 keV: $r(\mu m) = \dfrac{2.76 \times 10^{-2}\, A\, E_0^{5/3}}{\rho\, Z^{8/9}}$ where E_0 is in keV, A is in g/mole, and ρ is in g/cm³. Calculate r for 10, 50 and 100 keV for the following materials:

a. Silicon

b. Silicon dioxide

c. Iron

Problem 11.5 Electron diffraction in a crystal follows Bragg's Law (see Section 12.1), $n\lambda = 2d\sin\theta$. We can use this to explore the resolution of transmission electron microscopy.

a. The wavelength of the electrons can be found from $\lambda(nm) = \dfrac{0.0388}{\sqrt{V}}$ where V is in kV. Find the wavelength of electrons accelerated with a 100 kV potential.

b. The wavelengths are typically quite small and so we can use a small angle approximation $\sin\theta \approx \theta$ and find that $\theta \sim \dfrac{\lambda}{d}$. For atomic planes separated by $d = 0.10$ nm, estimate the angle of diffraction for the 100 kV electrons from part a. (Convert your answer from radians to degrees.)

c. The parameter d is approximately the diameter of an atom. Would you expect TEM to have atomic resolution?

12

Characterization: Structural Techniques

The division between the imaging techniques of Chapter 11 and the structural techniques of this chapter is rather arbitrary. Typically, those techniques which produce a two-dimensional map of the topography of the surface are considered as imaging techniques. Structural techniques provide greater information about the direction normal to the sample surface. Measurements of film thickness and crystalline structure are included in this chapter. We will see in future chapters that techniques that are discussed with chemical or optical characterization can also provide structural information. Each characterization technique is influenced by many properties of the sample resulting in both opportunity and complexity. Many of these techniques are discussed in books on materials characterization (Brundle, Evans, and Wilson 1992, Feldman and Mayer 1986, Sardela 2014, Vickerman and Gilmore 2009) and on thin films (Ohring 2002).

We begin this chapter by looking at diffraction techniques and then examine real space scattering and, finally, some common techniques for the measurement of film thickness.

12.1 X-Ray Diffraction

We observe diffraction phenomena from crystalline materials when the spacing between atomic planes is comparable to the wavelength of our probe. X-rays are high-energy electromagnetic radiation with a wavelength on the order of 1 Å. This makes them ideal probes for atomic-scale structure. A complication for the study of surfaces and thin films is the ability of X-rays to penetrate deeply into materials (typically 1–100 μm). X-Ray Diffraction (XRD) can be used to determine the crystal structure and interatomic spacings in materials. Phases can be identified and defects in the crystal structure can be measured. The samples must be crystalline or polycrystalline.

Diffraction arises from the scattering of X-ray radiation from inner shell electrons surrounding atoms in atomic planes of a crystalline material. This scattering depends on the wavelength of the incident radiation, the spacing between the planes, and the angle of incidence of the radiation. The geometry is shown in Figure 12.1. The scattering from the top layer and from the underlying layers can interfere constructively resulting in a strong signal for certain incident (and reflected) angles. Since the beam scattered from the lower plane of atoms has gone an extra distance of $2(d\sin\theta)$, the condition for this constructive interference is that $2d\sin\theta = n\lambda$ where n is an integer and λ is the wavelength of the incident radiation. This relationship is known as Bragg's Law.

$$n\lambda = 2d\sin\theta \qquad (12.1)$$

In XRD, the integer n is typically represented by the choice of (h, k, l) values to determine the lattice spacing d_{hkl}. For example, in a cubic system, Equation 3.2 demonstrates

DOI: 10.1201/9780429194542-15

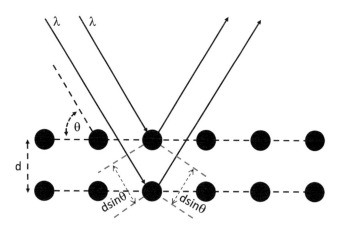

FIGURE 12.1
Geometry of diffraction for the derivation of Bragg's Law.

that $d_{100} = a_0$ corresponding to the $n = 1$ case in Bragg's Law. The $n = 2$ case corresponds to $d_{200} = \frac{1}{2} a_0$. For XRD we typically write

$$\lambda = 2d_{hkl} \sin\theta \qquad (12.2)$$

In a typical experiment, the wavelength of the X-rays is constant (often Cu Kα radiation with $\lambda = 1.54$ Å). The incident and reflected angle and/or sample position are varied to explore diffraction from different planes within the sample. Figure 12.2 shows diffraction for the planes separated by a distance d_1 but rotating the sample until the planes separated by d_2 are parallel to the surface and adjusting the angle θ could also lead to constructive interference.

Since the wavelength is known, the experiment involves measuring the angles (θ) at which we see diffraction and calculating the lattice spacings from Equation 12.2. The appropriate values of ($h\,k\,l$) can be determined from Equation 3.2. The resulting plot of intensity vs. angle can be quite complicated since many sets of planes may contribute diffraction peaks in a polycrystalline thin film/substrate system. The relative intensities of the peaks, their widths, and slight shifts in their positions can all provide us with useful information about our material.

A schematic representation of an XRD system is presented in Figure 12.3. The standard (Bragg-Brentano) geometry of modern XRD systems starts with an extended beam of X-rays about 1.2 cm long and 0.4 mm wide. This divergent beam of X-rays illuminates

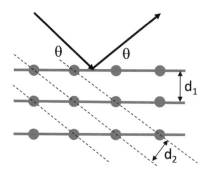

FIGURE 12.2
Examples of different planes for diffraction in a crystal.

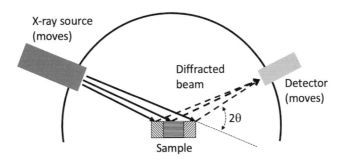

FIGURE 12.3
Schematic diagram of a standard X-ray diffraction system.

the surface of the sample with the extent of its divergence limited by slits. The diffracted beams pass through additional exit slits before converging at the detector. The sample typically remains stationary while both the incident and exit arms are swept together through the θ angles.

The 2θ values for the peaks are given by Braggs Law (Equation 12.1), but this does not provide any information about the peak intensities. In general, the peak intensities for pure materials decrease with increasing angle, and so it is not possible to just look at peak heights and see relative quantities of crystallites oriented in different directions relative to the surface. In addition to this general trend, different crystal structures have structure factors that change the relative amplitudes. Indeed, many peaks are not observed at all because of the destructive interference of those peaks. For simple cubic crystals, h, k, and l can take on any integer values for constructive interference. For fcc crystals, the h, k, and l values must all be odd or all even to observe a peak. For bcc crystals, the sum $h + k + l$ must be an even integer. Table 12.1 shows the observable peaks for several cubic crystal structures. Tables of intensities for randomly oriented powder diffraction peaks are available for comparison with acquired data to see how the data deviates from a randomly oriented structure. Polycrystalline thin films, for instance, will often show a preferred orientation that can be determined by comparing the data to the random orientation peaks. Examples of different types of materials and the planes producing diffraction are shown in Figure 12.4.

TABLE 12.1

Allowed Diffraction Peaks for Different Cubic Crystals

$h^2 + k^2 + l^2$	$h\ k\ l$	Simple Cubic	fcc	bcc
1	1 0 0	Allowed	–	–
2	1 1 0	Allowed	–	Allowed
3	1 1 1	Allowed	Allowed	–
4	2 0 0	Allowed	Allowed	Allowed
5	2 1 0	Allowed	–	–
6	2 1 1	Allowed	–	Allowed
7	None	–	–	–
8	2 2 0	Allowed	Allowed	Allowed
9	2 2 1 or 3 0 0	Allowed	–	–
10	3 1 0	Allowed	–	Allowed
11	3 1 1	Allowed	Allowed	–
12	2 2 2	Allowed	Allowed	Allowed

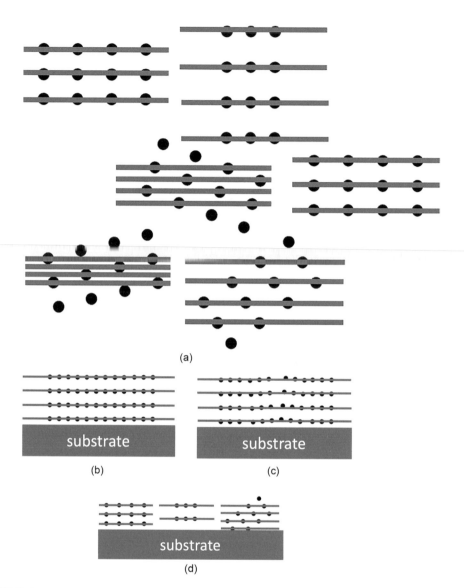

FIGURE 12.4
Planes that will diffract in (a) powders (all planes randomly represented), (b) single crystal film (only one plane),
(c) single crystal film with grains slightly misaligned (only one plane but broadened peak), (d) polycrystalline
film (multiple peaks – some planes may be preferred).

Figure 12.5 shows an example of an oriented 75 nm Cu film sputter deposited simultaneously on two different substrates with a thin (2–3 nm) Cr adhesion layer. When grown on a Si substrate with only the thin native oxide layer, the grains in the polycrystalline film are randomly oriented (the XRD pattern is very similar to a powder diffraction pattern). When grown on a 500 nm amorphous SiO_2 layer on Si, the Cu film shows a distinctly different XRD pattern with the $(2\,0\,0)$ and $(2\,2\,0)$ peaks enhanced relative to the $(1\,1\,1)$ peak. This data was collected using an in-plane diffraction arm, and so peaks from the Si substrate are not observed.

Various other experiments can be performed with XRD. Rocking curve measurements select a particular peak and fix the detector on that position. The sample is then rocked slightly around this angle in order to examine the width of the diffracted peak and determine

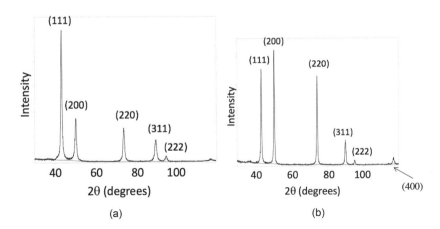

FIGURE 12.5

(a) Cu film on Si shows a random grain orientation compared to (b) Cu film on SiO_2 showing a preferred (2 0 0) and (2 2 0) orientation (Christensen 2015 unpublished).

defect concentrations and grain size. The wider the peak, for instance, the smaller the grain size as indicated in Figure 12.6. For sufficiently thin films, the grain size is often limited by the film thickness, and so rocking curves of very thin (<100nm) films are often quite wide.

When dealing with thin film samples, the substantial penetration of X-rays into the sample mean that the majority of the signal will come from the substrate rather than the thin film. If the X-rays are incident at a grazing angle, as shown in Figure 12.7, more of the X-rays will be in the thin film rather than in the substrate. For highly crystalline substrates, it is also unlikely that the conditions for diffraction from the substrate will be met with such a shallow incident angle. In this geometry, the divergent X-ray source discussed above is reflected from a mirror so that a parallel beam of X-rays is incident on the sample. The incident angle is fixed and the exit arm is swept through the θ angles. This geometry may not show diffraction from all planes if a sample is highly crystalline. Many thin films, however, are polycrystalline with crystalline grains in many different orientations so that all planes will be diffracted by some grain. Some XRD systems are also able to measure diffraction in the plane of the sample from lattice planes oriented perpendicular to the surface. This also maximizes sensitivity to thin films.

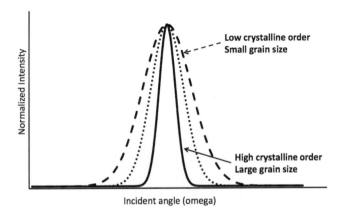

FIGURE 12.6

Rocking curves showing variation in crystalline order and grain size.

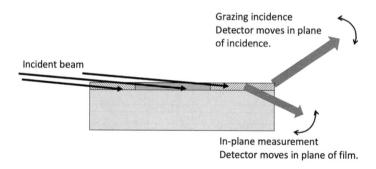

FIGURE 12.7
Thin film configurations for X-ray diffraction.

Polycrystalline thin films may not have their crystalline grains randomly oriented. Often there is a preferred direction to the film, which is referred to as texture. This can be explored using a pole figure. The geometry for a pole figure measurement is shown in Figure 12.8. The sample is tilted away from the surface normal by an angle χ and the sample is rotated about that axis. The arms of the X-ray diffractometer are set to an angle, θ, corresponding to the diffraction of a known lattice plane. If the crystallites are randomly oriented, this should produce a circular intensity profile. If there is a preferred orientation, the intensity will vary around the circle. If signals can be seen from both the substrate and film, the relative orientations of the film and substrate can be determined as demonstrated in Figure 12.9 for Ag films grown on a sapphire (Al_2O_3) substrate (Bock, Christensen, Rivers, Doucette and Lad 2004). The fcc Ag grows with the (1 1 0) plane parallel to the surface, and so Ag(1 1 1) type planes would be observed tilted 35.3° from the surface. The presence of two sets of {1 1 1} peaks separated by 109.5° suggests that the Ag is growing with two equivalent orientations relative to the sapphire surface. Sapphire peaks are also observed tilted 43° from the sample surface allowing the exact orientation of the Ag film on the sapphire surface to be determined.

Diffraction can be conveniently represented in a geometric form using the Ewald sphere. If we consider the wave vector of the incident X-rays, $\vec{k}$, and the wave vector of the diffracted X-rays, $\vec{k}'$, then we know, for elastic scattering, that they have the same magnitude, $k = k' = 2\pi/\lambda$. These two wave vectors thus define a circle in reciprocal space as indicated in Figure 12.10. This circle is one cross-section of the Ewald sphere.

In Figure 12.11, we add in some reciprocal lattice points, as defined in Chapter 3. The Bragg diffraction condition is equivalent to saying that we get constructive interference for any reciprocal lattice point that lies on the surface of the Ewald sphere.

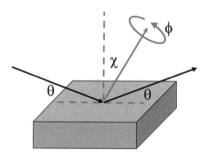

FIGURE 12.8
Geometry of a pole figure measurement.

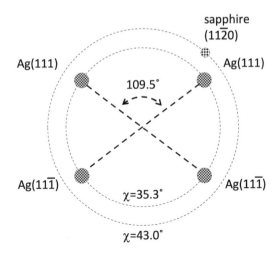

FIGURE 12.9
Pole figure of Ag film deposited on sapphire ($2\theta = 38.2°$) showing Ag(1 1 1) planes oriented in two equivalent configurations relative to the sapphire surface.

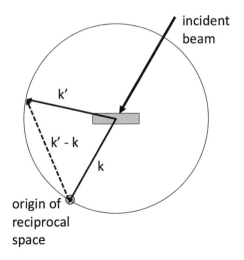

FIGURE 12.10
Cross-section of Ewald sphere of radius k.

Counting from the origin of reciprocal space, the reciprocal lattice point on the left side of the Ewald sphere surface would correspond to $h = -1$ and $k = 2$ with $l = 0$. We thus expect diffraction from the $(\overline{1}20)$ planes.

As we change the angle θ, we change the direction of $\vec{k}$ and $\vec{k}'$ allowing us to sweep through all of the possible conditions for diffraction to occur. The Ewald sphere can be a very useful way to visualize diffraction. For example, an incident beam with poor energy resolution would mean that the length of k and k' are not well defined. This would be the equivalent of saying that the Ewald sphere has some thickness to its surface. Additional information about XRD is available in the references (Bowen and Tanner 2005, Brundle, Evans, and Wilson 1992, Sardela 2014, Waseda, Matsubara and Shinoda 2011).

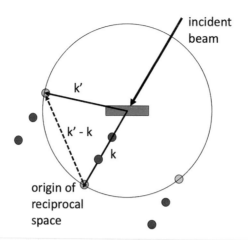

FIGURE 12.11
Ewald sphere with reciprocal lattice points. Diffraction occurs for reciprocal lattice points on the surface of the Ewald sphere.

12.2 Low Energy Electron Diffraction

The penetration depth of X-rays presented challenges in using XRD for surface and thin film characterization. We have already seen in Chapter 11 that electrons have much shorter penetration depths in materials. Electrons can be treated quantum mechanically as having a wavelength given by the deBroglie relation

$$\lambda = h/p \tag{12.3}$$

where h is Planck's constant ($h = 6.63 \times 10^{-34}$ J s) and p is the momentum of the electron. We find that electrons with an energy of 100 eV have a wavelength of 1.23 Å that again will lead to diffraction from typical crystalline lattice spacings.

Low Energy Electron Diffraction (LEED) takes advantage of these properties to study crystalline surfaces, adsorbed atoms, and very thin films. LEED can be used to determine the crystalline quality and orientation of a surface and can monitor the very early stages of thin film growth. Samples must be crystalline and the surfaces must be very clean since the electrons will only penetrate a couple of atomic layers into the sample.

If we consider an electron gun with a potential energy, V, conservation of energy tells us that

$$V = \tfrac{1}{2}mv^2 \tag{12.4}$$

Substituting this into the de Broglie relation (Equation 12.3) yields

$$\lambda = \frac{h}{\sqrt{2mV}} \tag{12.5}$$

As we change the voltage on the electron gun, we change the velocity and momentum of the electron which changes the wavelength of the electron. Typical values for the wavelength and mean free path of the electrons at different energies are shown in Table 12.2.

TABLE 12.2

Electron Energies and Associated Mean Free Path and Wavelength

Energy (eV)	Mean Free Path (Å)	Wavelength (Å)
100	4	1.23
1000	5	0.55
10,000	24	0.12

Surface sensitivity calls for a small mean free path and diffraction is best when wavelengths are on the order of 1–5 Å, so LEED is typically performed with electron energies below about 1000 eV.

The basic design of a LEED system is indicated in Figure 12.12. A beam of electrons is incident normally on the sample surface with energies in the 50–1000 eV range. These electrons are diffracted from the crystalline sample and leave the sample with constructive interference at specific angles. The result is an array of bright spots on a hemispherical fluorescent screen that intercepts the diffracted electrons. The pattern is indicative of the crystallography of the top couple of atomic levels.

Figure 12.13 show a LEED pattern from a hexagonal Be(0 0 0 1) surface after exposure to 500 Langmuirs of O_2. The outer spots are from the Be substrate and the inner spots are from the BeO layer that is growing on the substrate with a slightly larger lattice spacing.

The analysis of a LEED diffraction pattern can be understood by starting with exploring the diffraction from a one-dimensional row of scattering atoms. As in the previous section, we let $\vec{k}$ be the wave vector of the incident electrons and $\vec{k}'$ be the wave vector for the diffracted electrons. The incident beam is normal to the surface and the angle between the two wave vectors is θ. Let $\vec{d}$ be the lattice vector. This geometry is represented in Figure 12.14.

For elastic scattering, energy is conserved and $k = k' = 2\pi/\lambda$. As in the discussion of Bragg's Law in the previous section, the condition for constructive interference is that the phase difference between the incident and diffracted beam is an integral multiple of 2π.

$$\left(\vec{k}' - \vec{k}\right) \cdot \vec{d} = 2\pi n \tag{12.6}$$

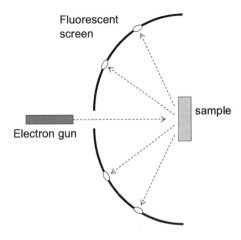

FIGURE 12.12
Schematic representation of a typical LEED system.

FIGURE 12.13
LEED pattern from BeO film (inner spots) growing on Be(0 0 0 1) surface. Beam energy = 80 eV. (From Christensen 1985.)

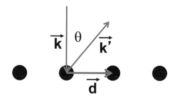

FIGURE 12.14
Geometry of diffraction from a row of atoms.

where n is an integer.

In LEED, the incident beam is normal to the surface, and so $\vec{k} \cdot \vec{d} = 0$ and the diffraction condition reduces to

$$\vec{k}' \cdot \vec{d} = 2\pi n \tag{12.7}$$

using the definition of a vector dot product and $k' = 2\pi/\lambda$, this can be rewritten as

$$\frac{2\pi}{\lambda} d \sin\theta = 2\pi n \tag{12.8}$$

that simplifies to

$$d \sin\theta = n\lambda \tag{12.9}$$

which is the standard formula for scattering from a row of point scatterers. Rewriting this expression as

$$\sin\theta = \frac{n\lambda}{d} \qquad (12.10)$$

allows us to see that the image on a LEED screen is actually an image of reciprocal space. The spacing of the spots on the screen is proportional to $1/d$ so that larger values of θ correspond to smaller lattice spacings on the real lattice.

It is often useful to rewrite this expression using the de Broglie relation (Equation 12.3) as

$$\sin\theta = \frac{n}{d}\sqrt{\frac{150}{V}} \qquad (12.11)$$

where V is the energy of the electron beam in eV and d is measured in Å. As the energy of the electron beam is increased, the LEED pattern will move to smaller angles.

We can explore two-dimensional diffraction further using the Ewald sphere formulation introduced with XRD in the previous section. Figure 12.15 displays a rectangular two-dimensional lattice. No additional planes of atoms are observable in the direction perpendicular to the page. This is the equivalent of saying that the lattice spacing in that direction is infinite.

The reciprocal space representation of this lattice, as described in Chapter 3, will have lattice spacings of $2\pi/a$, $2\pi/b$, and $2\pi/\infty = 0$. Notice that a larger separation in real space results in a smaller separation in reciprocal space as we observed in Section 3.2. With a reciprocal lattice spacing of 0 in the direction perpendicular to the surface, we see that the lattice points have no spacing between them. They form a line (typically referred to as a rod) that will extend to infinity into and out of the page, as shown in Figure 12.16.

If we transfer these ideas to the Ewald sphere representation introduced in the previous section, then the reciprocal lattice points are actually rods that will intersect the Ewald sphere. The incident beam direction, $\vec{k}$, will be parallel to the rods. If the surface structure is perfect, the rods would be infinitely thin and a lattice rod would intersect the sphere at a point. Real surfaces, however, have steps and terraces that will broaden the rod diameter. This results in an elongation of the diffracted spots that can be large enough to look like streaks. This is demonstrated in Figure 12.17.

The streaking carries information about the average terrace size of the surface. If the average terrace size is given by Wa_0 where a_0 is the spacing between atoms and W is a number, then the breadth of the reciprocal lattice rod will be

$$\delta k = 2\pi/Wa_0 \qquad (12.12)$$

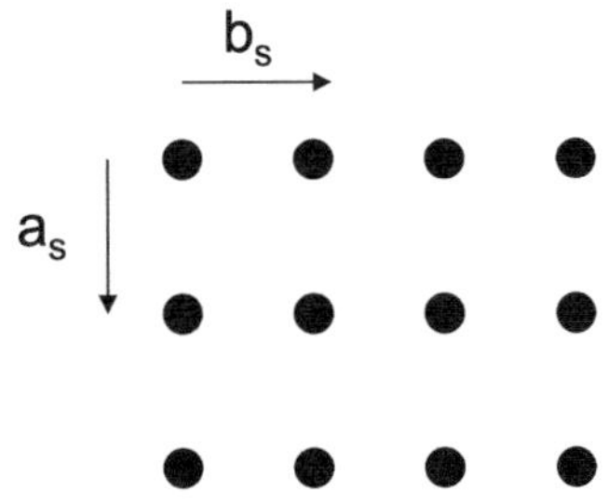

FIGURE 12.15
Rectangular two-dimensional lattice.

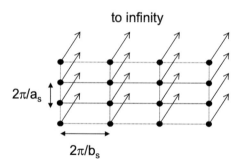

FIGURE 12.16
Reciprocal space representation of two-dimensional rectangular lattice. (Arrows to −∞, out of the page, are omitted to reduce clutter.)

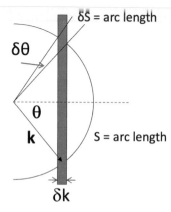

FIGURE 12.17
Broadened rod intersecting the Ewald sphere.

We can examine the arc lengths, δS and S, associated with the angles $\delta\theta$ and θ.

$$\delta S = \delta\theta|k| = \delta k/\sin\theta \tag{12.13}$$

$$S = \theta|k| \tag{12.14}$$

The ratio of the arc lengths is

$$\delta S/S = \delta\theta/\theta = 2\pi/Wa_0(\sin\theta)\theta \tag{12.15}$$

Since the angles are small, we can use the small angle approximation, $\sin\theta = \theta$, to determine the average terrace size.

$$Wa_0 = (D/L)(2\pi/\theta^2) \tag{12.16}$$

In Chapter 11, low-energy electron microscopy was introduced. This technique can also perform LEED experiments in a traditional sense or can be used to create an image of the sample using the intensity of a particular LEED spot as the source of contrast. The LEEM experimental configuration also allows for LEED experiments to be done on a micron dimension region.

The analysis so far assumes that LEED is a purely two-dimensional phenomenon. Some vertical lattice spacing information can be collected from LEED using Bragg's Law (Equation 12.1). To distinguish this from the lateral spacing discussion, let the distance between the vertical planes be z.

$$2z \sin \theta = n\lambda \tag{12.17}$$

If we only consider the reflection coming straight back out of the sample at normal incidence ($\theta = 90°$), then this simplifies to $2z = n\lambda$. Applying the de Broglie relation (Equation 12.3), this can be rewritten as

$$2z = n \frac{h}{\sqrt{2mE_B}} \tag{12.18}$$

where E_B is the electron energy in eV at a maximum in the intensity of the LEED spot, z is measured in Å, and n is an integer that indexes the intensity maximum. This can be solved for E_B.

$$E_B = \left(\frac{h^2}{8m} \frac{1}{z^2} \right) n^2 \tag{12.19}$$

By measuring the intensity of the normally reflected beam as we vary the energy of the incident beam, we can plot E_B vs. n^2 and determine the lattice spacing perpendicular to the surface from the slope.

LEED can also be used to examine the thermal vibrations of atoms at the surface. The vibrations reduce the ordered structure of the surface leading to a reduction in the intensity of the diffracted beams. This is accompanied by an increase in the diffuse background intensity. Examining a LEED diffraction spot, the intensity, I, can be modeled as a function of temperature,

$$I = I_0 e^{-2M} \tag{12.20}$$

where the exponential e^{-2M} is called the Debye-Waller factor and

$$2M = \frac{12h^2}{m_{atom}k_B} \left(\frac{\cos\theta_{incident}}{\lambda_{electron}} \right)^2 \frac{T}{\theta_D^2} \tag{12.21}$$

where m_{atom} is the mass of the vibrating surface atom and θ_D is the Debye temperature. The Debye temperature can be related to the amplitude of vibration of the surface atom.

$$x^2 \propto \frac{T}{\theta_D^2} \tag{12.22}$$

By plotting $\ln(I)$ vs. T, the slope of the straight line graph will yield the Debye temperature, which can then be related to vibration amplitude. Typically the Debye temperature of the surface is less than in the bulk, and so the vibrational amplitude at the surface will be greater as discussed in Section 4.1. The change in Debye temperature with depth into the surface can be probed by increasing the incident electron energy. The electrons will penetrate deeper into the surface and the resulting Debye temperature will be more representative of the lower layer.

If the vibrations get to be large enough, the surface may melt, as discussed in Section 4.1, leading to a loss of atomic order at the surface and thus a loss of diffraction. The intensity of a LEED spot will experience a sudden decrease in intensity at the temperature of the surface melting that will typically be at a temperature below the bulk melting point. Reducing the temperature should lead to a reversible intensity plot with diffraction returning below the surface melting temperature. Further details about LEED are available in various books (Brundle, Evans, and Wilson 1992, Ertl and Küppers 1974, Feldman and Mayer 1986, Vickerman and Gilmore 2009) and their references.

12.3 Reflection High Energy Electron Diffraction

Reflection High Energy Electron Diffraction (RHEED) differs from LEED in both geometry and electron energy as shown in Figure 12.18. The incident beam of electrons is directed at the sample at grazing incidence and the electrons have typical energies in the 10–30 keV range. The electrons are diffracted by the sample and a diffraction pattern is observed on the opposite side of the sample.

A particular advantage of RHEED is that it can be operated during thin film deposition since there is nothing obstructing the sample. 10 keV electrons still have an inelastic mean free path of only about 24 Å. Combined with the grazing incidence geometry, the electrons typically are only interacting with the top 1 or 2 atomic layers, and so RHEED is a very surface-sensitive technique. The penetration of the electrons in the sample is demonstrated in Figure 12.19.

If we consider the Ewald sphere model, RHEED has a much larger incident wave vector value, and so the Ewald sphere will have a very large radius. In Figure 12.17, the large radius of the Ewald sphere, intercepting a reciprocal space rod will result in the diffraction spot being smeared out into a streak unless the rod is very thin (very high crystallinity and well-defined incident electron beam energy). If the surface has three-dimensional structures on

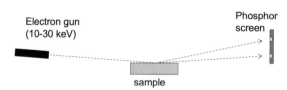

FIGURE 12.18
Schematic representation of RHEED.

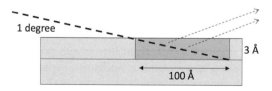

FIGURE 12.19
Penetration of electrons in RHEED for angle of incidence of 1°.

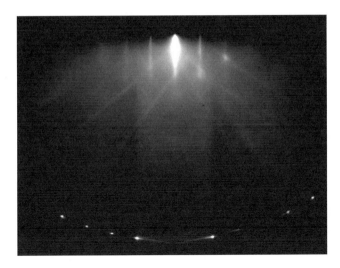

FIGURE 12.20
RHEED pattern from a clean, highly crystalline c-cut (0 0 0 1) sapphire surface. (Image courtesy of G. Bernhardt and R. J. Lad.)

it (such as islands or roughness), some electrons may be transmitted and diffracted through these structures resulting in additional diffraction features (typically sharp spots) in the image. Figure 12.20 shows an RHEED pattern from a highly crystalline surface.

RHEED is a powerful tool for monitoring thin film deposition, especially for highly crystalline films such as those grown in an MBE system (discussed in Section 9.4). If we have a partial film layer, then diffraction will be observed from both the partial layer and the first complete layer below it. These layers can destructively interfere resulting in a lower intensity than just a single layer. The amount of interference will change as the partial layer is completed, and so oscillations in the intensity of an RHEED diffraction spot can be used to monitor layer-by-layer film growth.

Consider a simple kinematic model of a partial layer forming on top of a completed layer (layer by layer growth). The amplitude (A_0) of the signal coming from the top layer will be proportional to the fractional coverage, σ, of the top layer. The amplitude (A_1) of the bottom layer will be proportional to how much of that surface is exposed, $(1 - \sigma)$, and to a phase factor, $e^{-i\Delta}$, that depends on the incident wave vector and the distance between the layers. The amplitude of the resultant electron wave will be

$$I(\sigma) = |A_0 + A_1|^2 \propto \sigma^2 + (1 - \sigma)^2 + 2\sigma(1 - \sigma)\cos(\Delta) \tag{12.23}$$

Further details about RHEED are available in various books (Brundle, Evans, and Wilson 1992, Vickerman and Gilmore 2009, Wang and Lu 2014).

12.4 X-Ray Reflectivity

Modern XRD instruments are also typically capable of performing X-Ray Reflectivity (XRR) measurements on films that have smooth interfaces. This technique can provide information about film thickness, density, and the roughness of the surface and interface.

Since the technique relies on reflection rather than diffraction, it can be used with amorphous as well as crystalline materials.

As in visible wavelength optics, when X-rays are incident below a critical angle, they will be totally reflected from a surface. This angle is typically less than one degree for most materials. As the angle of incidence is increased slightly above the critical angle, the reflectivity of the sample decreases rapidly. XRR experiments are thus restricted to a very narrow range of angles in approximately the 0°–5° range. The critical angle depends on the wavelength of the incident X-rays and the density of the materials. The density of the film can typically be determined from the measurement of the critical angle.

The large penetration depth of X-rays allows them to reach the interface between a film and the underlying substrate. If these two materials have different densities, then X-rays will be refracted and reflected at the interface. The reflected X-ray intensity will demonstrate constructive and destructive interference depending on the angle of incidence and reflection. An example of the measurement of intensity vs. angle is presented in Figure 12.21. The initial drop-off in intensity provides information about the film density. The period of the oscillations indicates the film thickness. The decay in the oscillations tells us something about the roughness of the interfaces.

This technique does require very smooth interfaces, typically less than a few nm of average roughness is best. If the films are less than a few nm in thickness, the X-rays may not have sufficient material to interact with to observe an effect. The upper thickness limit will vary with the materials but films up to several thousand nm in thickness can typically be measured. Additional information about XRR is available in the references (Daillant and Gibaud 2009, Sardela 2014) and in articles cited within those references.

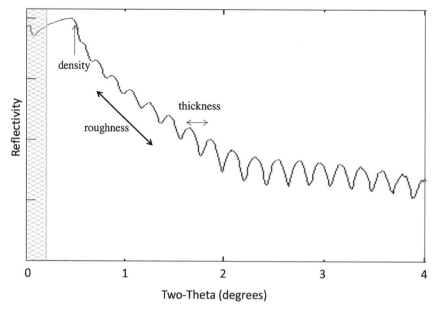

FIGURE 12.21
X-ray reflectance from a 40 nm film of Si_3N_4 on Si (Christensen 2016, unpublished).

12.5 Scattering Techniques

Various particles can be scattered from the surface region of a material to determine the positions of atoms in the material. For example, in ion scattering, ions (often He⁺) with energies of 0.5 keV–5 MeV are scattered off of a sample. The depth penetration of these ions varies with their incident energy with low energy ions being very surface sensitive and high energy ions being capable of penetrating a micron into the sample. The top atoms will shadow some lower atoms from contact with the incident beam and will block scattering from other lower atoms as indicated in Figure 12.22. By varying the angle of incidence and rotating the sample, atomic positions can be mapped out by examining the directions and energies of the scattered ions. Since the energy of the scattered ions carries elemental chemistry information about the target atoms from which they are scattered, the positions of atoms can be determined along with their elemental composition. We defer discussion of the chemical nature of scattering techniques to Chapter 13 where we examine chemical characterization.

The energy change of the ion during the collision with the target atom is derived from the conservation of energy and momentum. Letting E_i and M_i be the energy and mass of the incident ion, E_s be the energy of the scattered ion, M_t be the mass of the target atom, and $M_R = M_t/M_i$, the expression for energy is

$$\frac{E_s}{E_i} = \frac{1}{(1+M_R)^2}\left[\cos(\theta_s) + \sqrt{M_R^2 - \sin^2(\theta_s)}\right]^2 \tag{12.24}$$

where θ_s is the scattering angle from Figure 12.22.

High energy incident ions (MeV), typical in techniques such as Rutherford backscattering, can penetrate a micron into the sample. Reducing the energy of incident ions into the low keV range results in a surface-sensitive ion scattering technique called Low Energy Ion Scattering (LEIS). Almost all of the scattering will now take place from the top plane of atoms. The analyzer used in these experiments is typically sensitive only to charged

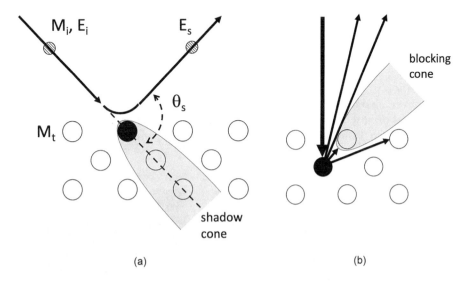

(a) (b)

FIGURE 12.22
Ion scattering. (a) Geometry and shadowing of lower atoms by top layer atom. (b) Blocking of scattered ions from lower atom by top layer atom.

particles, which is typical of surface-scattered ions. Ions that penetrate more deeply into the solid will typically emerge as scattered neutral atoms. Surface structural information can be obtained by considering the directions of the scattered ions. The references (Brundle, Evans, and Wilson 1992, Feldman and Mayer 1986) provide additional information on scattering theory and applications.

12.6 Stylus Profilometer

A stylus profilometer is a very common tool in a thin film laboratory. It is used to measure film thickness by measuring a step height from a bare substrate region to the thin film. The technique can measure changes in height over the range of about 200 Å to 65 µm with a vertical resolution of just a few Å. It can also be used to measure surface roughness with a horizontal resolution that depends on the radius of the tip being used. A wide range of solid samples may be studied if they are flat on a 10–100 µm scale. Soft films, however, may present difficulties due to the possible penetration of the film by the stylus tip.

The typical stylus profilometer experiment is represented schematically in Figure 12.23. A diamond tip (with a typical radius of curvature of 50 nm–1 mm) is placed at the end of a vertical arm that can be moved horizontally across the sample. Changes in the vertical position of the arm as it rides across the sample are recorded by electromagnetic sensors. By using a hard material, such as diamond, for the tip, there will be essentially no deformation of the tip, and so any changes in vertical position represent topographical changes in the sample. As the stylus reaches a step from the substrate to the film, the change in position will measure the thickness of the film. Some commercial units are restricted to doing a line scan. Others can combine multiple line scans into a map of the surface.

The radius of the stylus will clearly limit the resolution of the instrument both in the vertical and horizontal directions. As indicated in Figure 12.24, the stylus will not be able to penetrate into deep, narrow features that extend below the sample surface. The tip will also round off features that are sharper than the radius of curvature of the tip. Figure 12.24 shows both the actual surface topography and the measured topography from the stylus profilometer.

These resolution issues might suggest the use of as small a tip radius as possible. Small tip radii, however, also introduce another source of error into the measurement. The stylus must have a load on it to keep it in contact with the surface. These forces are typically in the 0.01–1 mN range. When these forces are applied over a very small area (tip size), they

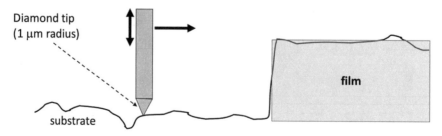

FIGURE 12.23
Schematic of stylus profilometer (scale greatly exaggerated!).

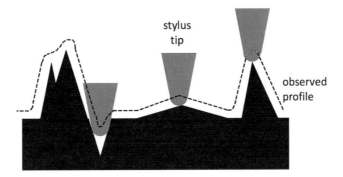

FIGURE 12.24
Sample geometry and observed profile for a stylus profilometer.

can result in very high pressures that may lead to having the stylus penetrate the film and not provide an accurate measurement of surface features.

The dynamic response of the profilometer/sample system can also lead to an error condition where a periodic signal will be observed that is not a result of the sample topography. Usually changing the speed of the horizontal motion of the stylus will remove this signal.

Figure 12.25 represents typical roughness data. Several different parameters are indicated that can be used to measure roughness. The total roughness, R_t, is the highest value above the mean surface minus the lowest value below the mean surface. The average roughness, R_a, is the average of the absolute values of the deviations from the mean surface. The root mean square roughness, R_q, is the square root of the average of the deviations from the mean surface squared.

Figure 12.26 presents a typical step profile from a thick film. Position measurements are averaged over the dark shaded region on the film and the light shaded region on the substrate. These are then compared to get the film thickness.

In Chapter 11, the confocal microscope was discussed. It can obtain information about surface roughness and topography which is similar to that of a stylus profilometer. One advantage of this method, when compared to the stylus profilometer, is that it is a noncontact method and so will work for soft samples that could be damaged by the force of the stylus on them. Additional information on measuring surface topography by stylus and other techniques is available in the references (Stout and Blunt 2000).

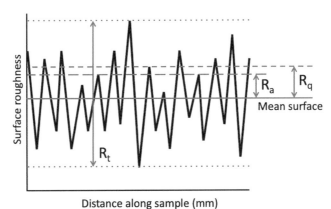

FIGURE 12.25
Representation of stylus profilometer roughness data showing different measurements of roughness.

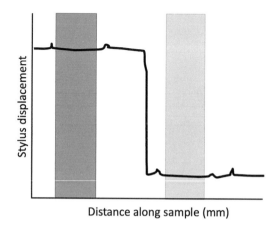

FIGURE 12.26
Simulated thickness measurement for a thick film with a stylus profilometer.

12.7 Quartz Crystal Microbalance

A Quartz Crystal Microbalance (QCM) can be used to indirectly measure the film thickness and deposition rate during thin film deposition. The measurement is not made directly on the sample, but rather in a location near the sample from which we infer what is happening on the sample. A typical deposition chamber geometry is shown in Figure 12.27. Film thicknesses can be estimated with about 3 Å resolution.

The QCM concept uses the piezoelectric properties of quartz crystals. A thin (0.3 mm) disk of quartz is cut with a particular crystalline orientation. Electrodes are deposited onto the crystal. When an alternating voltage is applied to these electrodes, the piezoelectric nature of the quartz causes the crystal to deform periodically. This vibration of the crystal has a resonant frequency in the 5–6 MHz range. As film material is deposited onto the crystal surface, the vibrational frequency shifts to lower values depending on the mass of the material added. The mass resolution is typically about 10^{-10} g.

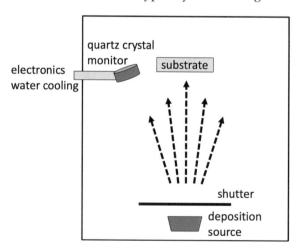

FIGURE 12.27
Deposition chamber showing the typical position of quartz crystal monitor.

The QCM measures the change in the vibrational frequency that is directly related to the mass of the deposited material. Since the exposed surface area of the quartz crystal is well known, the mass can be related to film thickness if the density of the film material is known since

$$m = \rho(Ad) \tag{12.25}$$

where m is the deposited mass, ρ is the density of the film material, A is the exposed area of the crystal, and d is the film thickness. It is common on many QCM instruments to have a table of bulk densities of various materials and to use these for the density in converting from the mass to the film thickness. This introduces potential error since film densities are often less than bulk densities. You can deposit up to about $5\,\mu m$ of material on most crystals before the mass gets to be so large that the frequency change is no longer proportional to the mass increase and the crystal should be replaced.

Another parameter that may need to be set on a QCM is the Z-ratio, which compares the elastic values of the thin films to the elastic values of quartz (the material of the oscillating crystal). It is defined as

$$Z = \left(\frac{\rho_q \mu_q}{\rho_f \mu_f} \right)^{1/2} \tag{12.26}$$

where ρ_q and μ_q are the density and shear modulus of quartz and ρ_f and μ_f are the density and shear modulus of the film. Since we know the properties for quartz, this expression can be simplified to

$$Z = \frac{8.84 \times 10^5}{\left(\rho_f \mu_f \right)^{1/2}} \tag{12.27}$$

where the film density is in g/cm³ and the shear modulus is in dynes/cm². Z-values for common materials have been tabulated and are often provided with the instrument. Some instruments may refer to the acoustic impedance, which is 1/Z.

A tooling factor may also be used with a QCM to correct for the fact that the substrate and the QCM are typically not located at the same place in the deposition chamber. The non-uniform distribution of depositing material coming from the source as described in Chapter 9 means that the QCM thickness and film thickness on the substrate will be different. By independently measuring the thickness deposited on the substrate and comparing this to the thickness reported by the quartz crystal monitor, the tooling factor is determined and applied to all measurements on this deposition system that use the same geometry.

Quartz crystal monitors are typically cooled since heat can alter the measurements. Most modern QCMs use quartz crystals that are cut in a particular orientation (A-T cut) that reduces the temperature dependence. Stress in the films deposited on the quartz can also alter the measurements.

A QCM can provide good reproducibility for monitoring deposition processes on a day-to-day basis. Accurate measurements, however, require calibration of the instrument by comparing the result with external measurements of film thickness and/or density. Further information on the theory and applications of QCM is available in the references (Steinem and Janshoff 2007).

12.8 Film Density Measurements

Our discussion of the QCM pointed out the need to independently measure the film thickness and/or density. Several of the techniques discussed so far can provide some information about film thickness. We will see that numerous chemical and optical techniques that we will discuss in subsequent chapters can also provide thickness information.

We discussed how XRR measurements can provide film density information. We will see that the optical properties of films depend on the film density, and so some optical techniques (Chapter 14) will also provide density information. Density can also be found from measurements of mass and volume. Usually, this involves measuring mass, area, and thickness. Mass may be possible to determine directly by weighing a sample on a microbalance (0.00001 mg resolution) or indirectly by measuring the frequency shift on a QCM. Thickness may be directly measured using a stylus profilometer, XRR, or an ellipsometer (to be discussed in Section 14.3). These measurements can be combined in order to determine the density of a film.

Measurements of surface area and porosity, which are typically related to density, can be obtained by introducing a known quantity of gas (often Ar or N_2) to a sample and observing the adsorption of the gas on the surface. Using a Langmuir or BET analysis (discussed in Chapter 7) the surface area of a porous film can be determined by examining the amount of gas adsorbed as a function of gas pressure.

References

Bock, F.X., T.M. Christensen, S.B. Rivers, L.D. Doucette, and R.J. Lad. 2004. Growth and structure of silver and silver oxide thin films on sapphire. *Thin Solid Films*. 468: 57–64.

Bowen, D.K. and B.K. Tanner. 2005. *High Resolution X-ray Diffractometry and Topography*. London: Taylor and Francis Ltd.

Brundle, C.R., C.A. Evans Jr., and S. Wilson, eds. 1992. *Encyclopedia of Materials Characterization: Surfaces, Interfaces, Thin Films*. Boston, MA: Butterworth-Heinemann.

Christensen, T.M. 1985. *Ellipsometry Studies of Metal Oxidation: Beryllium(0001) and Ni(111)*. Ph.D. Thesis, Cornell University.

Daillant, J. and A. Gibaud. 2009. *X-ray and Neutron Reflectivity*. Berlin: Springer-Verlag.

Ertl, G. and J. Küppers. 1974. *Low Energy Electrons and Surface Chemistry*. Weinheim: Verlag Chemie.

Feldman, L.C. and J.W. Mayer. 1986. *Fundamentals of Surface and Thin Film Analysis*. New York: North-Holland.

Ohring, M. 2002. *Materials Science of Thin Films*. 2nd ed. San Diego, CA: Academic Press.

Sardela, M., ed. 2014. *Practical Materials Characterization*. New York: Springer Science + Business Media.

Steinem, C. and A. Janshoff, eds. 2007. *Piezoelectric Sensors*. Berlin: Springer-Verlag.

Stout, K.J. and L. Blunt. 2000. *Three Dimensional Surface Topography*. 2nd ed. London: Penton Press.

Vickerman, J.C. and I.S. Gilmore, eds. 2009. *Surface Analysis – The Principal Techniques*. Chichester: John Wiley and Sons, Ltd.

Wang, G.-C. and T.-M. Lu. 2014. *RHEED Transmission Mode and Pole Figures: Thin Film and Nanostructure Texture Analysis*. New York: Springer Science + Business Media.

Waseda, Y., E. Matsubara, and K. Shinoda. 2011. *X-Ray Diffraction Crystallography: Introduction, Examples and Solved Problems*. Berlin: Springer-Verlag.

Problems

Problem 12.1 Using the grazing incidence X-ray diffraction pattern of face centered cubic Cu below, determine

a. The separation distance between (1 0 0), (1 1 0) and (1 1 1) type planes.

b. The lattice constant for Cu.

The data was collected using Cu X-rays with a wavelength of 1.54 Å. Recall that (2 0 0) and (2 2 0) planes are separated by only half the distance of (1 0 0) and (1 1 0) planes. You can check your answers by looking up the known lattice constant.

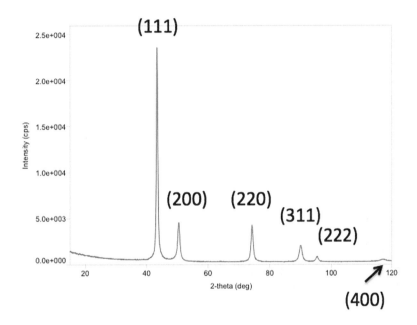

Problem 12.2 Consider X-ray diffraction from cubic crystals.

a. Show that the θ values corresponding to the two lowest peaks (θ_1 = lowest θ, θ_2 = next lowest θ) will obey:

$$\frac{\sin^2 \theta_1}{\sin^2 \theta_2} = \frac{h_1^2 + k_1^2 + l_1^2}{h_2^2 + k_2^2 + l_2^2}$$

b. Using information from Table 12.1, show that, for a fcc lattice, $\dfrac{\sin^2 \theta_1}{\sin^2 \theta_2} = 0.75$ and, for a bcc lattice $\dfrac{\sin^2 \theta_1}{\sin^2 \theta_2} = 0.50$.

c. Consider a metal element with diffraction peaks at $2\theta = 40, 58, 73 \ldots$ Is this element more likely to have a fcc or bcc structure?

Problem 12.3 The crystallite size of a film can be estimated using the Scherrer equation, size $= \dfrac{k\lambda}{(\text{FWHM})\cos(\theta)}$ where k is a shape factor with a value around 0.9, λ is the X-ray wavelength (in the same units as the size will be), FWHM is the full width at half maximum (in radians) for the peak being examined and θ is the angle of the peak. For a FWHM of 1.5° (remember to convert to radians), a peak position of $\theta = 47°$, and an incident X-ray wavelength of 1.54 Å, calculate the approximate crystallite size.

Problem 12.4 The Low Energy Electron Diffraction pattern for a cubic Rh(1 0 0) surface looks like the following:

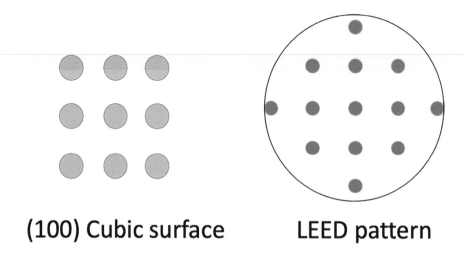

(100) Cubic surface   LEED pattern

We now deposit less than a monolayer of C atoms onto this Rh(1 0 0) surface and the LEED pattern changes to this:

LEED pattern

Draw a sketch of the Rh(1 0 0) surface and show the location of the C atoms on this surface.

Problem 12.5 By direct measurement on Figure 12.13 determine the ratio of the lattice spacings in the plane of the surface of BeO to Be.

Problem 12.6 Find the coverage where the RHEED oscillations predicted by Equation 12.23 have a minimum value. Explain why this result would be expected.

13

Characterization: Chemical and Elemental Techniques

Characterization of the elemental composition and/or chemistry of surfaces and thin films can seek to establish a range of information. The simplest goal is to identify which elements are present on the surface or in the thin film. We may also need to know how much of each element is present. Slightly more complex is determining how those elements are bound together. If Fe and O are both observed, for example, determining if they are bound together as an oxide. If they form an oxide, is it FeO, Fe_2O_3, Fe_3O_4, or some other compound? Finally, it may be desirable to know the spatial distribution of the chemical information both horizontally across the surface and as a function of depth. At the end of this chapter, we also examine vibrational techniques that provide chemical information about bonds and the bonding environment.

The ability of many of these techniques to provide some information on the chemistry of a sample as a function of location means that they are often able to provide structural information complementary to that discussed in Chapter 12. Many books (Brundle, Evans and Wilson 1992, Feldman and Mayer 1986, Frey and Khan 2015, Ohring 2002, Sardela 2014, Vickerman and Gilmore 2009) discuss a wide range of chemical characterization techniques including many of the techniques discussed in this chapter.

13.1 Auger Electron Spectroscopy

In Chapter 12, we used electrons as structural probes by diffracting them from crystalline surfaces. In Auger Electron Spectroscopy (AES), we typically use a beam of electrons (although incident X-rays can also be used) to remove electrons from the material under study. The energies of these secondary electrons provide us with chemical information. This technique easily identifies which elements are present and can be done in a quantitative manner. Some chemical state information is available indicating how elements are bound to one another, although this is not typically the technique of choice for collecting chemical state information. Since we are using an incident electron beam, it can be rastered over the surface as in the scanning electron microscope to produce elemental maps of the surface. Depending on the quality of the incident electron gun (and the cost of the instrument), the electron beam may have a diameter of 10 nm–1 mm at the sample.

In our discussion of the scanning electron microscope (Section 11.2.2), we discussed the interaction of electrons with materials in some detail. Those observations are valid for AES as well. The limited mean free path of electrons in solids makes this a surface-sensitive technique with most of the information coming from the top 3 nm. Similar to the SEM, this technique has charged particles both incident on and leaving the sample, and so charging of the material may be a problem. Conducting solid samples (metals, semiconductors, and

DOI: 10.1201/9780429194542-16

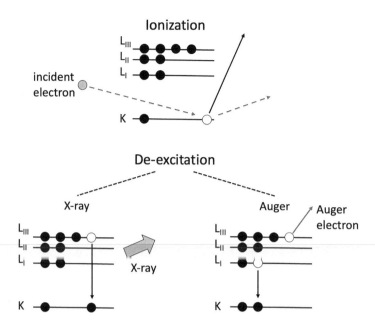

FIGURE 13.1
Atomic ionization leading to ejection of an X-ray or an Auger electron.

films) are best with this technique. The surfaces of the samples must be very clean. Since the technique only samples a few nm, a fingerprint on the sample will completely obstruct any information from the surface itself.

The need for cleanliness of the surface means that AES is typically performed in an ultra-high vacuum system. As discussed in Chapter 2, this means that there will be relatively little contamination incident on the sample over times scales needed for the measurements. Samples need to be compatible with the ultra-high vacuum environment, and so high vapor pressure materials are typically avoided.

The Auger electron emission process (named after the French scientist, Pierre Auger (pronounced O-jhay)) is shown in Figure 13.1. A primary electron with energy, E_P is incident on the sample. It interacts with atoms in the material causing the ejection of a secondary electron from an inner shell of the material atom. This leaves the resulting ion in an excited state with a missing electron in an inner shell. An electron from an upper level can drop to the inner shell to bring the ion back into the ground state, and energy can be conserved by one of two processes. An X-ray with wavelength characteristic of the energy gap of the electron that dropped down in the atom can be emitted. We will return to this mechanism when we explore Energy Dispersive Analysis of X-rays (EDAX) later in this chapter. The other mechanism for conserving energy is the emission of another electron (an Auger electron) from the atom in the sample. The Auger electron emission is favored for elements with low atomic number. The X-ray emission is favored for high atomic number elements. For a K-shell electron ejected, the cross-over point is around an atomic number of 35.

Note that this process requires an atom with at least three electrons. The Auger process will not occur in H or He, and so those elements cannot be detected using AES. Other elements can typically be detected at about a level of one atomic percent. The ionization cross-section (probability of creating a core hole) increases rapidly with energy reaching a maximum of a primary beam energy around 3X the energy of the ejected Auger electron.

The number of ejected electrons can thus be maximized by operating with an incident beam energy near that value. The cross-section falls off relatively slowly at higher energy, and so higher beam energies are typically acceptable.

The secondary electrons escaping from the sample, as well as any incident electrons that are scattered back from the sample, are collected and energy-analyzed. A schematic of an energy spectrum is provided in Figure 13.2. At the high-energy end of the spectrum, we observe an elastic peak from incident electrons with energy, E_P, that are scattered back from the sample with no energy loss. (This is often a useful peak to use in calibrating the instrument.) At slightly lower energies are peaks from primary electrons that are scattered back with some energy loss. The use of these peaks to obtain information about the sample will be explored later in this chapter. At lower energies (typically in the 50–2000 eV range) peaks from the Auger electrons are observed. Notice that a significant background of electrons is found across the spectrum with a large peak at very low energies.

Since the Auger electrons produce a relatively small peak on a high background, direct analysis of the peaks can be difficult. It is typical in AES analysis to take the derivative of the spectrum to reduce the effects of the slowly varying background and more clearly observe the Auger peaks. Typical $N(E)$ and derivative spectra from AES of a clean copper sample are shown in Figure 13.3.

This is the first spectroscopic technique that we have discussed, and so we explore the nature of spectroscopic analysis. Since each atom has electron energy levels with very well-defined energies, the Auger electrons from any specific element will be emitted only at very specific energies that are characteristic of that element. Each element has its own unique "fingerprint" of energies. The Cu signal in Figure 13.3 shows the multiple peaks, specific peak energies, and peak amplitudes that are typical of metallic elements. As a result, simple identification of which elements are found in a sample is quite easy with AES. An experienced user can often identify common elements by simple inspection of the spectrum.

Quantitative information is also available from analysis of the height of the peaks in the spectrum. This is complicated by effects both at the sample and at the detector. In the sample, the probability of emitting a particular Auger electron from one element differs from the probability of emitting a particular Auger electron from a different element. For

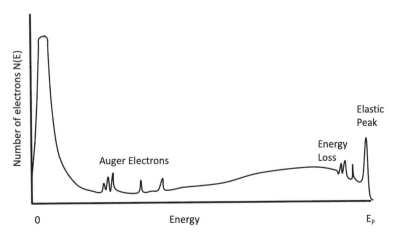

FIGURE 13.2
Schematic representation of an energy spectrum generated by an incident electron beam of energy E_P.

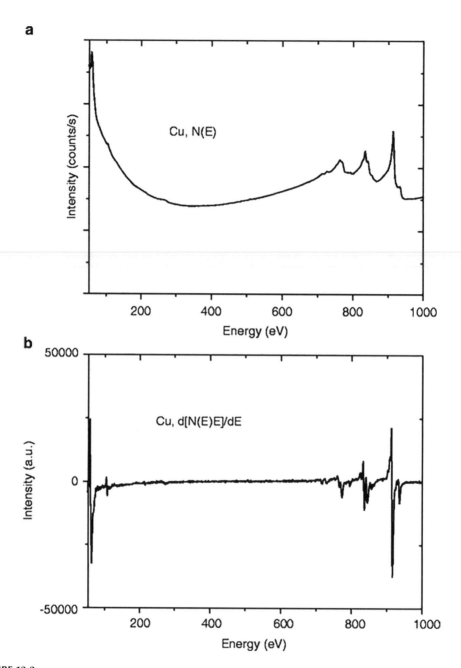

FIGURE 13.3
Typical (a) *N(E)* and (b) derivative AES spectra for a clean copper sample. (From Hofmann 2013 with permission of Springer.)

the same incident beam and the same surface concentrations, these two elements will produce Auger peaks with different heights. Furthermore, spectroscopic detectors (in AES and other spectroscopies) are more sensitive at some energies than at others. The combination of these effects means that we cannot simply compare the peak heights from two different elements to determine their relative concentrations.

A simple way of correcting these issues is to calculate a sensitivity factor for each common Auger peak and use this to correct the data. Such calculations have been done, and sensitivity factors are typically provided with the instrument to allow for quantitative analysis. The sensitivity factors depend on the incident electron beam energy and are typically calculated for incident beam energies of 3, 5, and 10 keV. As a result, much Auger data is collected at these three energies. Table 13.1 provides some typical sensitivity factors for 3 keV incident electrons to demonstrate a large range of factors, which emphasizes the need to account for these in analyzing Auger data.

The idea of using sensitivity factors for quantitative spectroscopic analysis can be demonstrated using a very simple method outlined in Davis (1978). We start by selecting a single peak for each element to represent the amount of that element found in the sample. The selected peaks are marked with an asterisk in Figure 13.4, which is a graph for a stainless steel sample that was fractured in vacuum to produce a very clean surface that would be typical of bulk stainless steel. The intensity of each peak is represented by the peak height, which can be measured by using a ruler to measure from the top of a peak to the bottom of the peak on the derivative graph. These values (measured before the graph was scaled for publication), along with the known sensitivity factors for these peaks are provided in Table 13.2.

The concentration (C) of each element (x) present in the sample can be found from:

$$C_x = \frac{\left(\text{height}_x / \text{sensitivity}_x\right)}{\displaystyle\sum_{\text{all elements } (i)} \left(\text{height}_i / \text{sensitivity}_i\right)} \tag{13.1}$$

The denominator is $(65/0.21) + (30/0.32) + (10/0.27) = 440$. We now find the corrected concentrations for each element:

$$C_{\text{Fe}} = \frac{65/0.21}{440} = 0.70.$$

$$C_{\text{Cr}} = \frac{30/0.32}{440} = 0.21$$

$$C_{Ni} = \frac{10/0.27}{440} = 0.08$$

We can compare these to the actual bulk concentrations that were: 0.702, 0.205, and 0.093.

This analysis was simplified by the fact that the peaks did not overlap with other peaks. Peak overlap can complicate analysis. For instance, in Figure 13.3, the Cr and O peaks overlap making it difficult to measure the peak height. Our simple analysis also neglects

TABLE 13.1

Approximate Sensitivity Factors for Selected Elements at 3 keV Incident Energy

Element	Atomic Number	Sensitivity Factor
Carbon	6	0.2
Oxygen	8	0.5
Sulfur	16	0.8
Iron	26	0.2
Platinum	78	0.02

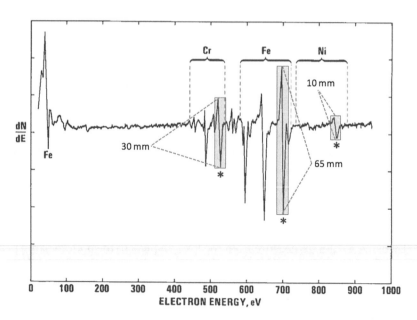

FIGURE 13.4
Auger spectrum of stainless steel sample fractured in vacuum showing the selected peaks and their peak-to-peak height. (Adapted from Davis 1978 with permission of Physical Electronics, USA.)

TABLE 13.2

Elements, Peak-to-Peak Heights, and
Sensitivity Factors for Data in Figure 13.4

Element	Height (mm)	Sensitivity
Fe	65	0.21
Cr	30	0.32
Ni	10	0.27

matrix effects of one element being in a solid of another element and variations in electron mean free path and backscattering. We are not considering changes in peak shape that might result from a different chemical bonding environment. Effects of surface roughness are also neglected.

Variations of the chemical composition with depth into the sample can be obtained by removing layers of the sample and then examining each newly exposed surface using AES. This is typically done using Ar^+ ion sputtering. A beam of Ar ions is directed at the sample either as a broad beam or as a rastered beam. Typical energies of the ions are 2–10 keV. Since Ar is a relatively heavy, inert gas, the collisions with the surface will lead to surface atoms being ejected from the sample as described in Sections 2.3 and 9.2.1. Depending on the rate of removal, AES analysis can be done during the removal process or by alternating with the removal process. A typical result is shown in Figure 13.5 of the native oxide layer that forms on stainless steel. The very surface of the sample (at sputter time=0) has extensive carbon contamination, which is typical of many samples which have been exposed to the air. As the sputter time increases, we enter into the oxide layer. Note that the Fe and Ni levels are suppressed inside the oxide since stainless steel forms an oxide

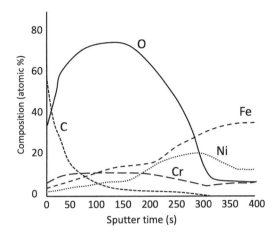

FIGURE 13.5
Experimental Auger depth profile of a stainless steel sample. (From Christensen unpublished.)

that is primarily Cr-based. As we get through the oxide, we see that Ni has accumulated at the interface region and see Fe rise dramatically toward a bulk level.

Three common difficulties exist with these depth profiles from sputter removal of the material. The first is simply that it is a destructive analysis. The second is that the sputtering process also causes some atoms from the surface layer to be driven down into the sample. This mixing means that the resultant profile is not necessarily indicative of the original sample. The last is that the horizontal axis is sputter time. We would prefer to know depth into the sample. Depth can be estimated by sputtering a known standard sample under the same conditions. For instance, a known thickness of SiO_2 on Si is often used as a standard for converting the sputter time to a thickness. The sample being studied, however, may have a different sputter rate than the standard, and so this method provides only an approximation of the depth into the sample.

The sputter profiling process, of course, damages the sample surface, but the electron beam may also cause damage to certain samples. Some polymer materials may experience bond breaking when exposed to an electron beam. The electron beam can also induce chemical reactions on the surface. Gases present in the chamber may break up on the surface leading to oxidation of metal surfaces or the deposition of C on surfaces. Maximally valent materials, such as Al_2O_3, can be reduced by exposure to an electron beam.

If the incident electron beam has sufficient spatial resolution and can be rastered over the surface as in an SEM, then we can perform Scanning Auger Microscopy (SAM). This allows for elemental mapping of the surface. When doing SEM, we count all electrons that reach the detector. When doing SAM, we use an energy analyzer to separate the electrons by energy. We then map only those electrons from a certain energy window. An example of SAM images obtained from a Cu(In, Ga)SeS sample is presented in Figure 13.6. Figure 13.6a shows the SEM image captured using all electrons. It shows that some sort of circular defect exists in the lower left corner of the image. Figure 13.6b plots the electrons in the energy range typical of In, which clearly shows that the defect is rich in In compared to the surrounding region. Figure 13.6c shows a similar plot for the energy range typical of Se, which shows minimal variation in Se across the sample. Further details on AES are available in various books (Brundle, Evans and Wilson 1992, Ertl and Küppers 1974, Feldman and Mayer 1986, Hofmann 2013, Sardela 2014, Vickerman and Gilmore 2009, Wolstenholme 2015) and their references.

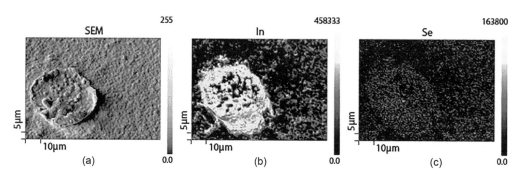

FIGURE 13.6
Scanning Auger images of a surface defect in Cu(In, Ga)SeS. (a) Image from all secondary electrons (b) image from In Auger electrons (c) image from Se Auger electrons. (Photo courtesy of National Renewable Energy Laboratory.)

13.2 Energy and Wavelength Dispersive X-Ray Analysis

In discussing the mechanism for the production of Auger electrons in the previous section, we noted that a competing process for de-excitation produces X-rays that also have energies characteristic of the elements present in the sample. This is the principle used in Energy Dispersive X-ray Analysis (EDAX or EDX) which is also known as Energy Dispersive Spectroscopy (EDS). This technique is often coupled with a SEM by adding an additional detector onto the SEM chamber. As a result, the discussion in Section 11.2.2 about the SEM and the interaction of electrons with materials is relevant here. The discussion in Section 13.1 about the ionization of surface atoms by an incident beam of electrons and the subsequent de-excitation is also of interest.

This technique, similar to SEM and AES, works with conducting solids. In this case, the detector is collecting X-rays that are emitted by atoms in the sample and are characteristic of the elements present. These X-rays are energy analyzed. EDX allows for elemental mapping like SAM. Spatial resolution is not as good as SAM since X-rays from much deeper in the sample can be detected. Electrons that have spread farther out in the sample will produce detectable X-rays broadening the spatial resolution to closer to 1 μm for thick samples.

Non-conducting samples can be analyzed, as in SEM, by applying a thin (about 10 nm) conducting layer to the surface. Typically C is used rather than Au so that it will not significantly interfere with the X-ray spectrum.

As with AES, identification of which elements are present in the sample is relatively simple. The quantitative analysis of the sample, however, is more complicated but can typically be done using computer analysis programs. The analysis is complicated by the poorer energy resolution of the X-ray detectors (70–130 eV), which often leads to overlaps of peaks. Quantitative analysis is typically accurate to within about 2 atomic %. Spectrometers that use Be windows in front of the detector are not able to detect elements with atomic numbers at or below 11 (Na) since the X-rays from those elements do not penetrate the Be window. Polymer windows are available that allow analysis of a wider range of energies.

Figure 13.7 shows an EDAX spectrum for a film deposited on glass. Multiple peaks are observed for each element as was observed in other spectroscopies. Notice the large peak widths compared to other spectroscopic techniques.

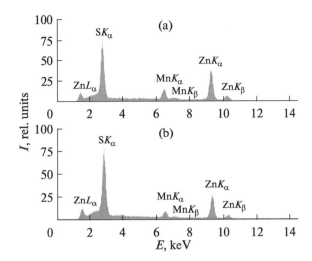

FIGURE 13.7
Experimental EDAX spectrum of $Zn_{1-x}Mn_xS$ showing (a) the blend from which the film was evaporated and (b) the resulting film on a glass substrate. (From Kurbatov et al. 2014 by permission of Springer Nature.)

By using a different detector, the technique of Wavelength Dispersive X-ray Analysis (WDX, Electron Microprobe) can provide better energy resolution of peaks (2–20 eV) as well as better sensitivity by about a factor of 10 compared to EDAX. The trade-off is a slower data collection speed.

EDAX is similar to X-ray Fluorescence (XRF) except that XRF uses an X-ray source to ionize the sample atoms. Since these X-rays penetrate deeply into the sample, the technique is more appropriate for bulk chemical analysis than for thin film or surface characterization. XRF can use either energy dispersive or wavelength dispersive energy analyzers. Additional information about these techniques is available in the references (Feldman and Mayer 1986, Goldstein et al. 2003).

13.3 X-Ray Photoelectron Spectroscopy

In the previous section, we examined using electrons as our incident beam and X-rays as our detected quantity. With X-ray Photoelectron Spectroscopy (XPS) we reverse this and use X-rays to excite the sample and then study the core electrons that are ejected from atoms in the sample. As suggested in Section 13.2, Auger electrons can also be emitted from the sample from these incident X-rays and they will be collected and analyzed along with the core electrons.

XPS (also referred to as Electron Spectroscopy for Chemical Analysis, ESCA) provides rapid identification of what elements are present in a sample through a distinctive energy fingerprint for each element. It can be readily quantified using peak-fitting routines to obtain the relative abundances of the elements. The key feature of XPS is the ability to identify chemical states. The peaks associated with Si in an elemental form are measurably shifted from the Si peaks associated with SiO_2 for example. This allows the identification of chemical state information. The surface sensitivity of this technique comes from the detected electrons rather than the incident X-rays. The X-rays penetrate deeply into the sample and cause core electrons to be ejected throughout the sample. Only those electrons close enough to the surface, however, will escape and be detected without energy loss. These are typically from about the top 30 Å of the sample. Electrons from deeper in the sample undergo energy loss before reaching the surface and provide a broad background signal observable in Figure 13.9. Since charged particles are escaping from the surface, some charging effects are observed, but not as significantly as with AES. XPS studies of insulators, metals, and semiconductors are all common with some insulators presenting complications from charging. With the high surface sensitivity, samples must be very clean. This again makes XPS an ultra-high vacuum technique and the samples must be compatible with ultra-high vacuum. Recent progress in extraction optics using differential pumping has allowed for near ambient pressure XPS measurements to be accomplished relaxing the requirement that the sample must be UHV compatible.

X-rays cannot be focused as well as electrons, and so the sampling region for XPS is often in the 30 μm–10 mm range (although diameters of 150 nm have been reported). This limits the usefulness of mapping although some XPS systems are capable of mapping. The detection limit of XPS is comparable to AES at about 1 atomic percent. XPS does not detect H or He due to the very low ionization cross-sections of those elements. A typical experimental arrangement is represented schematically in Figure 13.8.

The process of producing photoelectrons begins with X-rays of known wavelength incident on the material. The X-rays are often from a Mg (1253.6 eV) or Al (1486.6 eV) source. The X-ray energy can be described in terms of the frequency, v, of the X-ray radiation as

$$E = hv \tag{13.2}$$

where h is Planck's constant. The X-rays deliver energy to the atoms in the sample ionizing them by ejecting core electrons from the atoms. The core electrons are bound into the atoms by a binding energy, E_B. Those electrons that are close enough to the surface to escape from the sample, exit with a kinetic energy, E_K, which is approximately the difference of the incident energy and the binding energy.

$$E_K = hv - E_B \tag{13.3}$$

The energy analyzer then separates the electrons according to their kinetic energies and detects the number of electrons at each kinetic energy. It is binding energy, however, that tells us about the chemistry of the sample, and so we typically plot the number of electrons vs. the binding energy. Historically, the data were recorded on strip chart recorders that plotted against the detected kinetic energies. As can be noted from Equation 13.3, the binding energy has the opposite sign of the kinetic energy. Plots of binding energy, then, were

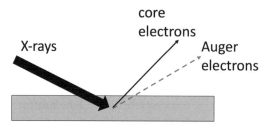

FIGURE 13.8
Basic interactions of X-ray photoelectron spectroscopy.

labeled with the binding energy decreasing from left to right (since the data was really kinetic energy increasing from left to right). This convention is still followed with XPS data, and the horizontal scale is typically decreasing binding energy.

For any given element, photoelectrons from different electron orbitals can be released leading to many different peaks with widely differing binding energies as shown in Figure 13.9 for gold. These peaks will have different intensities since the ionization cross-section for each type of electron will be different. The orbitals are identified initially with the quantum number, n, with the most tightly bound electrons in the $n=1$ state and the binding energies generally becoming less for higher values of n. The quantum number, l, is associated with a spectroscopic notation such that $l=0$ is labeled as the s state, $l=1$ is the p state, $l=2$ is the d state, and $l=3$ in the f state. States can thus be labeled: 1s, 2s, 2p ... Atomic physics predicts a splitting in atomic levels beyond the s level. The p, d, and f states will exhibit a splitting associated with the quantum number $j=l\pm\frac{1}{2}$. The intensities of these doublet peaks are expected to have a ratio of $(2j_1+1)/(2j_2+1)$. A $p_{3/2}$ peak should be twice the intensity of a $p_{1/2}$ peak. A $d_{5/2}$ peak should be 1.5 times the amplitude of a $d_{3/2}$ peak. The ionization cross-sections, doublet splitting intensities, and the energy dependence of the detector lead to a need to use sensitivity factors in a process similar to that described for AES to get quantitative results. Table 13.3 provides approximate binding energies and cross-sections (at two common X-ray excitation energies) as well as the intensity ratios of the doublet peaks for Au. Note that the cross-sections depend on the incident X-ray energy.

The direct dependence on binding energy, coupled with high-energy resolution of the electron analyzer, allows very small differences in the bonding environment of an atom in the sample to be detected. These changes in binding energy are characteristic of different compounds that the atom can be bound into. Figure 13.10 shows some data from two polymers, polymethyl methacrylate (PMMA) and polystyrene (PS). The data associated with the electrons emitted from the C 1s level and the O 1s level are shown. PMMA has C bound in

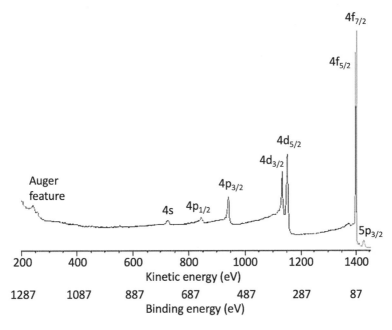

FIGURE 13.9
Experimental XPS spectrum of Au showing peaks from several atomic levels and peak splitting.

TABLE 13.3

Selected Au XPS Peaks with Binding Energies, Cross-Sections at Two Common X-Ray
Excitation Energies, and Doublet Intensity Ratios

Peak	Binding Energy (eV)	Cross-Section (Mbarns) at Mg Kα (1253 eV)	Cross Section (Mbarns) at Al Kα (1487 eV)	Doublet Intensity Ratio
$4f_{7/2}$	84	0.42	0.25	1.33
$4f_{5/2}$	88	0.42	0.25	1
5s	110	0.01	0.01	
$4d_{5/2}$	335	0.36	0.27	1.5
$4d_{3/2}$	353	0.36	0.27	1
$4p_{3/2}$	547	0.16	0.12	2
$4p_{1/2}$	643	0.16	0.12	1
4s	763	0.05	0.04	

Source: Cross-section data from Eletta Sincrotrone Trieste (2016).

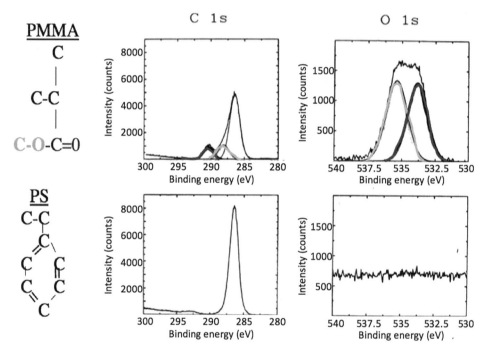

FIGURE 13.10

XPS of PMMA (top row) and PS (bottom row). The first column shows the molecular structure. The second column shows the C 1s peaks. The third column shows the O 1s peaks. (Adapted from Green, Christensen, Russell and Jerome 1989 with permission of the American Chemical Society.)

three forms. Singly bonded C atoms bound to other C atoms account for three of the five C atoms in PMMA. One C atom is singly bonded to an O atom and one C atom is doubly bonded to an O atom. This gives us three bonding environments for C and two bonding environments for O. The data shows this as well. The C 1s signal can be fit with three peaks. The tallest peak has an intensity three times the other two peaks. The O 1s signal can be fit with two equal peaks. The locations of these peaks correspond to other measurements of C and O in these bonding environments. Polystyrene has only C atoms and they are bonded either in single bonds or in a C ring structure. The energy difference between these two bonding states was not large enough to be resolved by the instrument used in this study, and so only a single C peak is observed in the data and no oxygen is detected.

Each of the PMMA graphs in Figure 13.10 shows both the raw data and the fits to the data. The need to fit the experimental data with theoretically based peaks adds to the complexity of the analysis. A great deal of discussion in the literature has examined the shape of the theoretical peaks considering, for example, whether they should be Gaussian, Lorentzian, or some combination of the two. Other questions include whether it is sufficient to measure peak height or whether peak area should be used. With most analysis done by computers today, peak area is typically used. This then raises the question of where is the bottom of the peak? The signal on either side of the peak is often not at the same level. The simplest solution is to use a linear background, but determining the endpoints for starting and ending the background is not obvious. More sophisticated models exist that use non-linear backgrounds based on arguments put forth by Shirley or Tougaard (Repoux 1992).

If additional information on the chemistry below the surface is desired, some depth profiling can be accomplished in two ways. One way is to use Ar ion sputtering in exactly the same manner as was described in Section 13.1 where it was used with AES. A second way is to make use of the significant difference in mean free paths of X-rays and electrons in the sample. This provides a non-destructive method for determining a depth profile of the top 100 Å using angle-resolved XPS. This involves rotating the sample. Since the X-rays excite atoms throughout the sample, the electrons that will be detected depend only on those that escape from the sample in the direction of the detector. Figure 13.11 shows

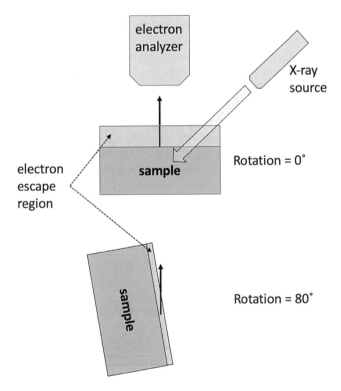

FIGURE 13.11
Angle-resolved XPS allows for depth profiling of the surface region showing un-rotated sample and sample rotated 80°.

how samples that are perpendicular to the detector will have electrons escape from much deeper in the material than samples that have been rotated so that electrons leaving in the direction of the detector can only have come from the surface region. By varying the angle of the sample to the detector, various depths can be probed down to about 10 nm.

While most samples are stable under X-ray radiation at these energies, some polymer samples may be altered by exposure to X-rays. Further details on XPS are available in various books (Brundle, Evans and Wilson 1992, Ertl and Küppers 1974, Feldman and Mayer 1986, Hofmann 2013, Sardela 2014, Vickerman and Gilmore, 2009) and their references.

13.4 Ultraviolet Photoelectron Spectroscopy

Instead of using X-rays as the source of energy to eject photoelectrons, Ultraviolet Photoelectron Spectroscopy (UPS) uses ultraviolet light incident on the sample. The lower energy (<50 eV) of the incident light means the photoelectrons will be emitted from the valence levels in UPS rather than the core levels typical of XPS. While XPS can, in principle, examine the valence levels as well, the cross-sections for photoionization are quite small for the higher energy incident photons. Equation 13.3 will still govern the process. Like XPS, this is a surface-sensitive technique due to the short mean free paths of the photoelectrons. UPS is often used to study the electronic structure of the occupied states of the valence bands of materials. It can also be used to examine adsorbed molecules and to calculate the work function of materials. The energy resolution of UPS is much better than XPS with resolutions in the meV range possible.

Although UPS is good for examining valence levels, it is somewhat arbitrary to claim a sharp distinction between the core and valence levels. In some materials, core levels can be found with relatively low binding energies. As a general guide, however, a binding energy around 10 eV is often a good divider between core and valence binding energies. The photoemission spectrum in UPS is sensitive to the incident photon energy, especially at lower energies. Most laboratory UPS systems use a single energy source (often He I radiation from a gas discharge lamp). Tunable synchrotron sources are also used.

The analysis of UPS spectra, which involve valence states, is typically more complicated than XPS analysis. The valence states will experience hybridization where orbitals from different atoms combine in various ways leading to a more complicated spectrum. The emitted photoelectrons have energies that are similar to the valence electrons leading to a spectrum that includes significant interactions in these states with considerable electron-electron correlation. Like XPS, the UPS experiment can be performed in an angle-resolved manner leading to detailed mapping of the band structure of the occupied states of the material.

A typical UPS spectrum builds on the discussions in Section 6.1. Of particular interest is the simple picture of Figure 6.12 showing the relationship of the Fermi energy and work function in a metal to the filled states in the system. We recall that the Fermi energy was defined as the energy of the highest filled state and the work function was the additional energy needed to escape from the solid. If we make Figure 6.12 somewhat more realistic by including the density of states for the filled states, we get Figure 13.12, which can then be directly related to the UPS spectrum through Equation 13.3. Further details on UPS are available in various books (Brundle, Evans and Wilson 1992, Ertl and Küppers 1974).

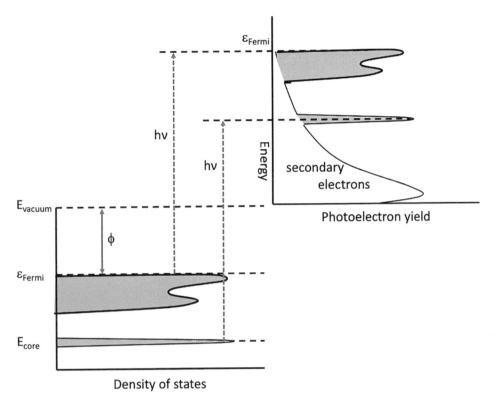

FIGURE 13.12
Relationship of the energy levels and density of states to the photoelectron spectrum.

13.5 Near-Edge X-Ray Absorption Fine Structure

Instead of measuring the emitted photoelectrons as is done in XPS, the absorption of X-rays in a solid can be measured by looking at the emitted photons or Auger electrons resulting from transitions caused by the X-ray absorption. Figure 13.1 examines this process when it is initiated by an electron beam, but the same processes occur through X-ray absorption as well. As described in Section 13.1, the cross-section for Auger electron emission is higher for low Z elements and also provides greater surface sensitivity, so this mechanism is more commonly measured. Near-Edge X-ray Absorption Fine Structure (NEXAFS), also called X-ray Absorption Near Edge Structure (XANES), is a technique that makes use of this to examine the elemental composition and chemical state of materials. Surface properties and buried monolayers can be detected. The X-rays need to be scanned over a range of energies (synchrotron radiation) while the electron yield or emitted photon yield are measured to indicate the absorption of the X-rays. Typically the range of X-rays of interest is the first 30–40 eV above the absorption edge as indicated in Figure 13.13.

Since the energy of the absorption edge is specific to each element as shown in Figure 13.13, this technique can be used for chemical identification in a manner similar to AES or XPS. The fine structure above the absorption edge can be used to determine the bonding environment of the target atom. By using a linearly polarized incident beam, NEXAFS can be used to probe directional bonds. For instance, bonds perpendicular or parallel to the

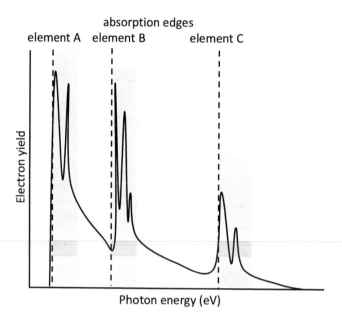

FIGURE 13.13
Representation of a NEXAFS spectrum showing different absorption edges for different elements and fine structure (shaded) above the absorption edges.

surface can be explored by using two different incident polarizations. This technique is complementary to XPS and UPS in the sense that it probes the unoccupied states in the band structure while the photoelectron spectroscopies probe the occupied states.

In Figure 13.10, we examined the XPS spectra of polystyrene and noted that the carbon-carbon single bonds could not be distinguished from the bonds in the carbon ring structure since they were too close in energy. NEXAFS is able to resolve these bond differences and provide more information about the bonding environment. Additional detail about NEXAFS is available in the references (Brundle, Evans and Wilson 1992, Stöhr 2003).

13.6 Secondary Ion Mass Spectrometry

The techniques discussed so far have been unable to detect H and He directly. Secondary Ion Mass Spectrometry (SIMS) provides elemental composition information that includes H and He and has much greater sensitivity with some elements detectable to the parts per billion range. SIMS can analyze a wide range of materials, including insulators, and is a surface-sensitive technique with most of the information coming from the top 30 Å of the sample. As a result, samples must be clean and compatible with an ultra-high vacuum environment. SIMS is typically used to provide elemental composition information in both a qualitative and quantitative manner. Since it is mass spectroscopy rather than energy spectroscopy, it can also provide information about the presence of different isotopes of an element.

A typical SIMS configuration is presented in Figure 13.14. An incident ion beam (often Ar, O, Ga, or Cs) is directed at the sample with an energy of 1–40 keV. The width of the

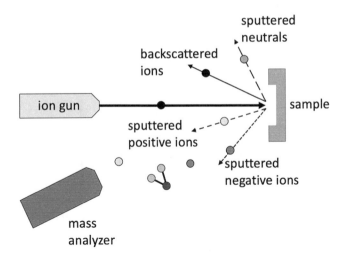

FIGURE 13.14
Schematic of SIMS experiment.

beam varies from system to system but is typically in the 50 nm–2 mm range. The incident beam of ions sputters material from the sample surface. Typically a larger area is sputtered with only material ejected from the central region of the crater being analyzed to avoid effects from the crater walls. This process results in sputtered neutral atoms/molecules as well as sputtered positive and negative ions/ionized molecules leaving the surface. Backscattered ions are also present in the gas phase during the measurement. A mass analyzer and detector (typically a quadrupole, magnetic sector, or time of flight mass analyzer) are placed near the sample to detect charged species and separate them by mass. The mass analyzer distinguishes the mass-to-charge ratio of the detected species. The sputtered material is mainly neutral atoms that are not detected in SIMS. About 0.01%–1% of the sputtered products are charged. These secondary ions are detected.

The number of atoms of a particular element emitted for each incident ion (sputter yield) is highly dependent on the element. As a result, sensitivity factors must be determined for each element and used in a manner similar to that described in Section 13.1 for AES. The yield is further found to depend strongly on what other elements are found in the sample. These matrix effects of the elements influencing one another make the quantitative analysis of SIMS data a much more complex process.

SIMS can detect the positive and/or negative ionized species produced by the sputtering process. Data are typically plotted as the number of ions detected as a function of the mass/charge ratio. For example, a singly charged Ar ion (mass=40 amu) would show up at position 40/1=40 on the graph. A doubly ionized Ar ion would show up at position 40/2=20 on the graph. Major peaks from SIMS data (Wikipedia 2007) for both positive and negative ions from a polytetraflouroethylene (PTFE) sample are shown in Figure 13.15.

Although the technique is surface-sensitive, it involves sputtering away the surface layers for analysis. This readily allows for depth profiling by continuing to sputter deeper into the film. The same issues, discussed in Section 13.1, of mixing caused by the sputtering process are present in SIMS depth profiling. Clearly, the technique is destructive of the sample surface. Some materials may be altered by the sputtering process through chemical reactions or the decomposition of compounds found in the sample. Further details on SIMS are available in various books (Brundle, Evans, and Wilson 1992, Feldman and Mayer 1986, Sardela 2014, Stevie 2016, van der Heide 2014, Vickerman and Gilmore 2009).

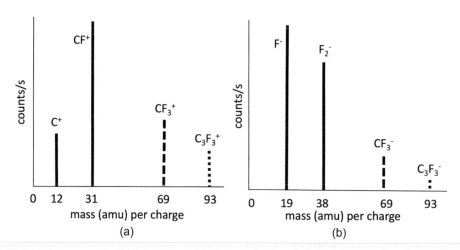

FIGURE 13.15
Major peaks in static SIMS spectra from PTFE (polytetrafluoroethylene). (a) Positive ion spectrum. (b) Negative ion spectrum.

A related technique, secondary neutral mass spectroscopy (SNMS) examines the neutral species that are emitted. In order to analyze neutral species, they must first be ionized before entering the mass analyzer. Molecular species may break up during the ionization process and have components show up at several different masses. These "cracking patterns" are well known for common compounds.

13.7 Scattering Techniques

The topic of scattering techniques was introduced in Chapter 12 with an emphasis on the determination of atomic positions. As was indicated there, elemental information about the target atoms in the sample is also available. This section explores several scattering techniques.

Like SIMS, Rutherford Backscattering Spectroscopy (RBS) makes use of a beam of incident ions, but studies the scattering of those ions to learn about the sample. This technique produces an elemental composition depth profile of the sample down to a depth of about 1 μm. It is capable of detecting the full range of elements including H. By providing a depth profile, it can also be used to measure the thickness of thin films. A wide range of samples can be studied as long as they are not volatile.

A typical RBS geometry is demonstrated in Figure 13.16. An incident beam of high-energy (often around 2 MeV) ions is incident on the sample. He$^+$ ions are commonly used since they are chemically inert and very low mass. The beam size tends to be large (about 1 mm) so the technique does not have a good lateral resolution. While most of the incident ions end up implanted into the sample, some collide with atoms in the sample and are scattered back out of the sample. Those ions are energy analyzed and detected. The energy of the scattered ions depends on the elemental mass of the atom they collided with, the angle of the collision and how deep in the sample the collision occurred. Typically,

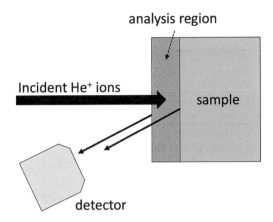

FIGURE 13.16
Schematic of RBS experiment.

collisions near the surface will result in scattered ions with higher energy. Collisions with heavier elements also result in scattered ions with higher energy. This technique has a depth resolution that can be around 20 Å in some cases. Measurements are limited for low atomic number elements by a weak signal. High atomic number elements produce ample intensity but do not have a good mass resolution in the spectra.

The number of scattered ions detected over the energy range is plotted as shown in Figure 13.17, which simulates a TaSi film on a Si surface. Three distinct features are visible. At high energies is a feature that must arise from the heavier of the elements found near the surface. This is Ta in this case. As we move further down in energy, we reach a feature arising from the lighter element, Si, at the surface. Continuing further down in energy, we see an increase in signal as we move into the Si substrate with its higher concentration of Si. The height of each region provides information about the concentration of each element in the material (after correcting for sensitivity factors). The width of the regions associated with the surface film provides information about the film thickness. The value at which a peak begins and ends provides information about which element was scattered from and how deep in the material it was located.

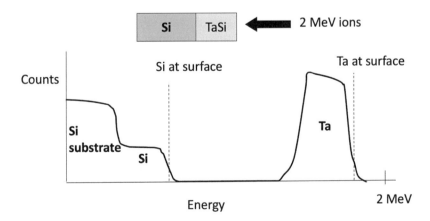

FIGURE 13.17
Simulated RBS spectrum for TaSi film on Si substrate.

Rather than examine the backscattering events, the ion beam can be directed onto the sample at glancing incidence and forward scattering events can be detected. If the mass of the incident ion is greater than the mass of the target atom, the incident energy of the ion will be elastically transferred to the target atom in a recoil collision. Both the incident ion and the target atom will be scattered in a forward direction. Typically, it is the target atom that is detected. Since this technique, Elastic Recoil Detection Analysis (ERDA), is best for low mass target atoms, it is useful for examining hydrogen or deuterium in films and surfaces. The technique has approximately constant sensitivity for all light elements (<50 amu) and can be sensitive to <0.1 atomic percent concentrations. Additional information about RBS is available in various books (Brundle, Evans and Wilson 1992, Feldman and Mayer 1986, Vickerman and Gilmore 2009).

As introduced in Chapter 12, reducing the energy of the incident ions into the low keV range results in a surface-sensitive ion scattering technique called Low Energy Ion Scattering (LEIS) or Ion Scattering Spectroscopy (ISS). Most of the scattering now takes place in the top plane of atoms. The analyzer used in these experiments is typically sensitive only to charged particles, which is typical of surface-scattered ions. Ions that penetrate more deeply into the solid will typically emerge as scattered neutral atoms. Chemical information is obtained since the energy of the scattered ion will depend on the relative masses of the incident and target particles as well as the incident energy. The choice of incident ion depends on the mass of the target atoms being studied. He$^+$ ions work well for target atoms with masses up to about 50 amu (Vanadium). Ne$^+$ ions work from atomic masses of around 30–100 amu (Ruthenium). Heavier noble gases are used for heavier target atoms. Further details are available in the references (Brundle, Evans and Wilson 1992, Vickerman and Gilmore 2009).

13.8 Fourier Transform Infrared Spectroscopy

Chemical information is also available using optical techniques. In this chapter, we describe optical techniques that are used primarily for chemical characterization. The next chapter will examine other optical techniques. One of the most common optical techniques for providing chemical characterization is Fourier Transform Infrared Spectroscopy (FTIR). FTIR can be used for studying a wide range of materials including solids, liquids, and gases and is capable of detecting a wide range of elements including H. It is most commonly used for examining organic materials as well as oxides and hydrides. Since the technique relies on the absorption of infrared (IR) radiation, the main requirement is that the material being studied must have a bond that absorbs in this part of the spectrum. As a result, this technique is not good for metals. FTIR involves transmitting infrared light through a sample or, more common in surface and thin film measurements, by reflecting from the sample surface.

Early IR spectrometers relied on a monochromator to scan the wavelengths of light coming from the sample after exposure to IR light. This process was relatively slow. Modern instruments, pictured schematically in Figure 13.18, make use of an interferometer allowing the collection of data from all wavelengths rapidly (often less than 1 s) since the only moving part is the mirror in the interferometer that moves a few mm.

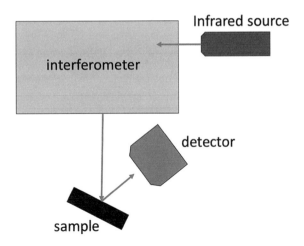

FIGURE 13.18
FTIR schematic for reflection measurements.

A broad spectrum IR source emits light into an interferometer. The light leaving the interferometer is incident on the sample. Certain wavelengths of the light incident on the sample are absorbed by interaction with chemical bonds in the sample. The remaining light is detected and the intensities are sent to a computer for Fourier analysis to convert the interference pattern into a spectrum. Data is typically plotted as % transmission vs. frequency. Instrumental effects can be removed by running a background scan with no sample in place.

The basic concept of IR spectroscopy involves molecular bonds that can typically be excited into various vibrational modes by absorption of light with energies typical of the IR part of the spectrum. Figure 13.19 shows some examples of vibration modes for a simple linear molecule consisting of an atom of element A with an atom of element B on either side of it. Figure 13.19a shows a stretching vibrational mode where the two B atoms oscillate symmetrically about their equilibrium positions. Figure 13.19b shows the case where the atom of element A oscillates back and forth between the two atoms of B. Figure 13.19c shows a bending vibration where the two B atoms oscillate together perpendicular to the line of the molecule. Many other vibrational modes are possible.

The energy absorbed will depend on the type of bond, the vibrational mode that is excited and the elements that form the bond. FTIR is particularly sensitive to dipole-active vibrations. Typically, single bonds of O, N, or C with H in a stretching vibration require more energy than double bonds between C, N, and O atoms in a stretching vibration that requires more energy than single bonds between C, N, and O atoms in a stretching vibration. Some typical ranges of bond excitations are shown in Figure 13.20 for H, C, N, and O atoms.

Typical data of an FTIR spectrum of 1,4-butanediol ($C_4H_{10}O_2$) are presented in Figure 13.21. Characteristic absorptions from the presence of OH and CH_2 bonds suggest that the molecule could be best represented as HO CH_2 CH_2 CH_2 CH_2 OH. This technique is not inherently surface-sensitive. Operating at glancing incidence can improve the surface sensitivity. Additional information about FTIR spectroscopy is available in the references (Brundle, Evans, and Wilson 1992, Martin 2007, Smith 2011, Vickerman and Gilmore 2009).

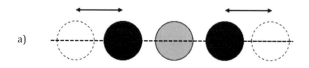

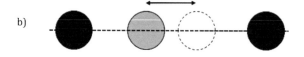

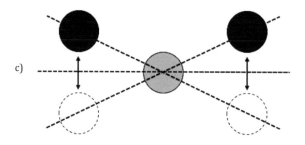

FIGURE 13.19
(a) Stretching vibration (b) oscillation vibration (c) bending (wagging) vibration.

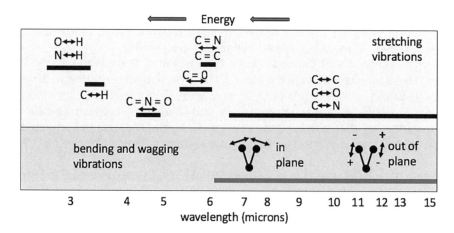

FIGURE 13.20
Common IR bond excitations as a function of wavelength.

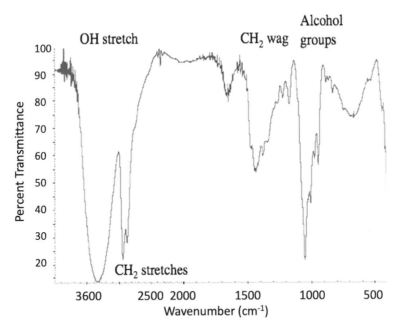

FIGURE 13.21
FTIR spectrum of 1,4-butanediol. (Adapted from Garcia and Catterton 2003.)

13.9 Raman Spectroscopy

A complementary technique to FTIR is Raman spectroscopy, which explores molecular bonds based on a change in polarizability of the molecule. This technique will only be useful for studying materials that have a bond that can be excited in a Raman interaction such as many carbon-based bonds. The Raman principle, however, is much different from IR absorption.

Most IR light incident on a sample is scattered elastically with no change in wavelength upon reflection (Rayleigh scattering). Some radiation, however, may interact with the vibrational energy levels in the molecule in order to either gain or lose energy resulting in a small change in the wavelength of the outgoing radiation relative to the incoming radiation. Typically, the inelastically scattered Raman signal represents about 1 part in 10^7 of the incident radiation so a high-intensity laser source is needed. If the outgoing radiation loses energy, the process is referred to as a Stokes process. A gain in energy is an anti-Stokes process. Often, only the Stokes Raman scattering is reported.

The instrument scans a laser onto the sample. All wavelengths of scattered light are passed into the spectrometer with a filter to block out the Rayleigh scattered peak. This allows the much smaller Raman scattered light to be observed without saturating the detector.

Raman spectroscopy can be used for molecular identification, especially for organic molecules. Phonons and plasmons can also be detected. Polarized laser light can provide information on crystal orientation in the sample. Samples that exhibit fluorescence at the wavelength of the exciting laser may mask the Raman signal. The sample also needs to be able to withstand exposure to the powerful lasers, which are typically used.

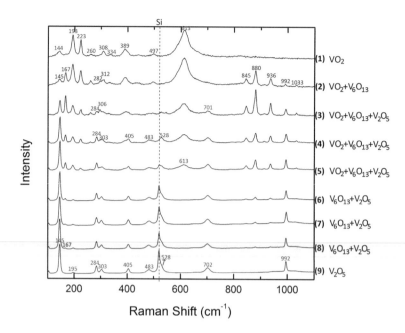

FIGURE 13.22

Raman spectra of vanadium oxide thin films on Si prepared by oxidation of vanadium thin films at 400°C for (1) 10 minutes, (2) 30 minutes, (3) 1 hour, (4) 1.5 hours, (5) 2 hours, (6) 2.5 hours, (7) 3 hours, (8) 3.5 hours, and (9) 4 hours. (From Zhang et al. 2016 with permission from Elsevier.)

Figure 13.22 shows various Raman spectra for different thin film oxides of vanadium deposited on a Silicon substrate. The horizontal axis is the Raman shift in wavenumber from the elastic peak.

Molecules near the surface can experience a significant amplification in the Raman signal on the order of 10^{14} times enhancement. This surface-enhanced Raman signal comes from an amplification of the electric field at the surface related to the excitation of surface plasmons. Further details about Raman spectroscopy are available in various books (Brundle, Evans, and Wilson 1992, Vandenabeele 2013, Vickerman and Gilmore 2009).

13.10 Electron Energy Loss Spectroscopy

Figure 13.2 indicated a variety of electrons emitted or scattered by a material when exposed to an incident electron beam. Of interest now are the primary electrons that are scattered from the surface with only small losses in energy from the primary energy. These energy losses typically arise from interactions with electrons in core levels in the surface atoms, surface and bulk plasmons, phonons, and vibrational states of adsorbed atoms and molecules. These interactions cause the incident electrons to leave the surface with characteristic reductions in energy depending on the interaction they experience. These experiments can be done in either transmission (for very thin samples) or in reflection.

Some of these features, such as core level interactions or plasmons, result in energy losses in the 1–100 eV range. These features can be observed in a standard AES experiment. An example is shown in Figure 13.22 for plasmons and core interactions in a Bi film deposited on Si. The energy loss of the plasmon features follows the predictions of Chapter 6. The presence of the core features allows chemical identification much like with AES. These features can also be used to examine the density of states of the sample since the core electrons are typically being excited into higher energy levels.

Phonon interaction with low energy (<50 eV) incident electrons, however, result in much smaller energy losses, typically 0.1–1 eV. Observation of these energy loss features requires much higher energy resolution, which typically requires a monochromatized incident electron beam and a high energy resolution detector. Energy can be lost by the interaction of the incident electron with a phonon in the solid. If the electron interacts with two phonons, a second peak will be observed at twice the energy loss.

When examining molecules adsorbed on surfaces, EELS, like FTIR, primarily detects dipole-active modes of vibration. Molecules adsorbed on conducting surfaces will induce image charges below the surface. As a result, dipoles that are oriented perpendicular to the surface will produce a strong dipole and be easier to detect. Dipoles oriented along the surface will induce image charges that form a quadrupole structure, which will be harder for EELS to detect. These orientations are demonstrated in Figure 13.23.

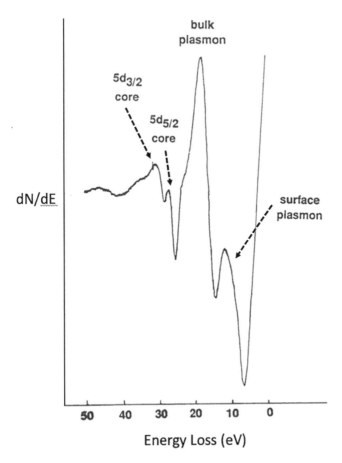

FIGURE 13.23
Energy loss spectrum of Bi film showing plasmon and core level interactions. (From Christensen unpublished.)

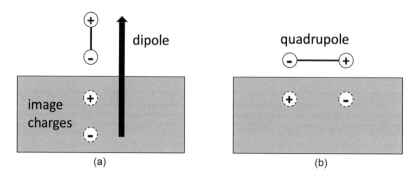

FIGURE 13.24
Induced image charges and vibrational modes for different orientations of adsorbed molecules. (a) Dipole perpendicular to the surface produces a dipole-active mode. (b) Dipole parallel to the surface produces a quadrupole mode.

The amount of energy loss can be indicative of the bonding site for atoms adsorbed to surfaces as demonstrated in Figure 13.24. An atom bonded to an on-top site will vibrate normal to the surface with a relatively low energy. If the atom is bonded in a bridge site, the bonds to the two bridging atoms form a higher energy bond. The vibration is still normal to the surface. If the atom is bonded asymmetrically to the bridge site, two peaks will appear in the energy loss spectrum due to the two different bond lengths. Stretching modes of these two bonds will have components in the direction normal to the surface, which can be detected. The longer bond will have the lower energy vibration. When molecules are bonded to the surface, they will exhibit energy loss features from the bonds to surface atoms and from the internal vibrations of the molecule. The orientation on the surface of complicated molecules with multiple dipole-active vibrational modes can often be determined by noting which modes are stronger and thus oriented more normal to the surface. Additional information about EELS is available in various books (Brundle, Evans, and Wilson 1992, Brydson 2001, Ertl and Küppers 1974, Vickerman and Gilmore 2009).

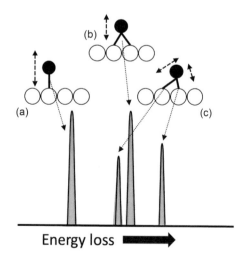

FIGURE 13.25
Energy loss for atoms bonded to a surface. (a) Atom bonded to on-top site. (b) Atom bonded to bridge site. (c) Atom bonded asymmetrically to bridge site.

References

Brundle, C.R., C.A. Evans Jr., and S. Wilson, eds. 1992. *Encyclopedia of Materials Characterization: Surfaces, Interfaces, Thin Films.* Boston, MA: Butterworth-Heinemann.

Brydson, R. 2001. *Electron Energy Loss Spectroscopy.* New York: Taylor and Francis.

Davis, L.E. et al. 1978. *Handbook of Auger Electron Spectroscopy.* 2nd ed. Eden Prairie, MN: Physical Electronics.

Eletta Sincrotrone Trieste. *Atomic Calculation of Photoionization Cross Sections and Asymmetry Paramters.* vuo.elettra.eu/services/elements/WebElements.html (accessed 3 Aug 2016).

Ertl, G. and J. Küppers. 1974. *Low Energy Electrons and Surface Chemistry.* Weinheim: Verlag Chemie.

Feldman, L.C. and J.W. Mayer. 1986. *Fundamentals of Surface and Thin Film Analysis.* New York: North-Holland.

Frey, H. and H.R. Khan, eds. 2015. *Handbook of Thin-Film Technology.* Berlin: Springer Verlag.

Garcia, A.D. and A.J. Catterton. 2003. 1,4-Butanediol (BD) – forensic profile. *Microgram Journal* vol 1: 44–54. Washington D.C.: U.S. Drug Enforcement Agency. https://www.dea.gov/pr/microgram-journals/2003/2003_1-2.pdf (accessed 26 July 2016). https://www.thevespiary.org/rhodium/Rhodium/chemistry/mjournal_v1_pg7.html (accessed 16 May 2022).

Goldstein, J.I., D. Newbury, D. Joy, C. Lyman, P. Echlin, E. Lifshin, L. Sawyer, and J. Michael. 2003. *Scanning Electron Microscopy and X-Ray Microanalysis.* 3rd ed. New York: Springer.

Green, P.F., T.M. Christensen, T.P. Russell, and R. Jerome. 1989. Surface interaction in solvent-cast polystyrene/poly(methylmethacrylate) diblock copolymers. *Macromolecules* 22: 2189–2194.

Hofmann, S. 2013. *Auger- and X-Ray Photoelectron Spectroscopy in Materials Science: A User-Oriented Guide.* Springer Series in Surface Sciences, vol 49. Berlin: Springer.

Hofmann, S. 2013. Qualitative analysis (principle and spectral interpretation). In: *Auger- and X-Ray Photoelectron Spectroscopy in Materials Science.* Springer Series in Surface Sciences, vol 49. Berlin: Springer.

Kurbatov, D., A. Klymov, A. Opanasyuk, et al. 2014. Studying the elemental composition and manganese distribution in $Zn_{1-x}Mn_x$ Te and $Zn_{1-x}Mn_x$ S films using the μ-PIXE and EDAX methods. *J. Synch. Investig.* 8: 259–262.

Martin, P. 2007. Infrared spectroscopy. *Vacuum Technol. Coat. Mag.* 8(9): 38–42.

National Renewable Energy Laboratory. *Field Emission Auger Electron Spectroscopy with Scanning Auger Microscopy.* https://www.nrel.gov/materials-science/auger-electron.html (accessed 16 May 2022).

Ohring, M. 2002. *Materials Science of Thin Films.* 2nd ed. San Diego, CA: Academic Press.

Repoux, M. 1992. Comparison of background removal methods for XPS. *Surface Interface Anal.* 18: 567–570.

Sardela, M., ed. 2014. *Practical Materials Characterization.* New York: Springer Science+Business Media.

Smith B.C. 2011. *Fundamentals of Fourier Transform Infrared Spectroscopy.* Boca Raton, FL: Taylor and Francis Group, LLC.

Stevie, F.A. 2016. *Secondary Ion Mass Spectrometry: Applications for Depth Profiling and Surface Characterization.* New York: Momentum Press, LLC.

Stöhr, J. 2003. *NEXAFS Spectroscopy.* Berlin: Springer-Verlag.

Vandenabeele, P. 2013. *Practical Raman Spectroscopy: An Introduction.* Chichester: John Wiley & Sons, Ltd.

van der Heide, P. 2014. *Secondary Ion Mass Spectrometry: An Introduction to Principles and Practices.* Hoboken, NJ: John Wiley & Sons, Inc.

Vickerman, J.C. and I.S. Gilmore, eds. 2009. *Surface Analysis – The Principal Techniques.* Chichester: John Wiley and Sons, Ltd.

Wikipedia. 2007. *Static Secondary-Ion Mass Spectrometry.* http://en.wikipedia.org/wiki/Static_secondary_ion_mass_spectrometry (accessed 16 May 2022).

Wolstenholme, J. 2015. *Auger Electron Spectroscopy: Practical Application to Materials Analysis and Characterization of Surfaces, Interfaces, and Thin Films.* New York: Momentum Press, LLC.

Zhang, C., Q. Yang, C. Koughia, F. Ye, M. Sanayei, S.-J. Wen, and S. Kasap. 2016 Characterization of vanadium oxide thin films with different stoichiometry using Raman spectroscopy. *Thin Solid Films* 620: 64–69.

Problems

Problem 13.1 The intensity of the Auger electron signal does not depend only on surface atoms. Atoms from below the surface contribute depending on the mean free path of the Auger electrons. The intensity for Auger electrons from element X, will depend on the concentration of X atoms as a function of depth, $C(z)$ and the electron mean free path, λ_{mfp}. $I_X = K \int_0^\infty C_X(z) e^{-z/\lambda_{mfp}} dz$ where K is a constant. What would be the intensity for a pure X material in terms of K and λ_{mfp}?

Problem 13.2 The energy of a particular Auger electron is $E_{Auger}=E_K-E_L-E_L$ and the momentum can be described as $p_{Auger} = \hbar k$ where k is the wave vector. We can make a reasonable estimate of these values using the Bohr model of the atom where binding energy is $E_n = \dfrac{me^4Z^2}{2\hbar^2 n^2}$ and the K-shell correspond to $n=1$ and L shell corresponds to $n=2$. Show that $k = \dfrac{1}{\sqrt{2}r_1}$ where r_1 is the radius of the K-shell given by $r_n = \dfrac{\hbar^2 n^2}{mZe^2}$.

Problem 13.3 In XPS, a simple approximation for the binding energy of the photo-electrons is the Bohr model, $E_n = -13.58\dfrac{Z^2}{n^2}$ eV.

a. Calculate the binding energy (expressed as a positive number) for the 1s and 2s levels of Ca. Compare your results with the known values of 4038 eV and 438 eV for the 1s and 2s levels respectively. Is the Bohr model a reasonable estimate?

b. Using the known values of the binding energies, determine the kinetic energy for photoelectrons from these two levels if the incident X-rays are from a Mg Kα source with energy of 1.25 keV. (If you get a negative number, that photoelectron will not be ejected from the atom.)

c. Repeat the kinetic energy calculation for a Al Kα source with energy of 1.49 keV. Note that the energy where you detect the electrons will change depending on the energy of the source. While the kinetic energy of photoelectrons changes with source energy, the kinetic energy of Auger electrons does NOT change. XPS systems will often have two sources so that peak overlaps between XPS and Auger peaks can be resolved by moving the XPS peaks.

Problem 13.4 Using Auger Electron Spectroscopy with 5 keV incident electrons, we measure peak intensities for the following elements. The corresponding sensitivity factors are also provided. What is the surface composition in atomic percent?

Element	Intensity (Arb Units)	Sensitivity Factor
Fe	5200	0.22
Ni	3300	0.26
O	2400	0.43
C	2000	0.14
S	1400	0.75

Problem 13.5 Using Rutherford Backscattering Spectroscopy, which film would be easier to measure the thickness of: a Pt film on Si or a Si film on Pt? Why?

Problem 13.6 XPS can be used to measure the thickness of thin films on substrates by examining the attenuation of the substrate signal. The attenuation will depend on the thickness of the film (d) as well as the mean free path (λ_{mfp}) of photoelectrons in the film at the energy of the substrate peak (E_{sub}) and the film peak (E_{film}). For simplicity in this calculation assume that these two mean free paths are the same.

The intensity of the substrate signal can be shown to be $I_{sub} = I_{sub}^0 \, e^{-\left(d/\lambda_{mfp}\cos(\theta)\right)}$ where I^0 is the intensity of the bare material and θ is the angle between the normal and the escape direction of the photoelectrons. Similarly the film will have an intensity $I_{film} = I_{film}^0 \left[1 - e^{-\left(d/\lambda_{mfp}\cos(\theta)\right)}\right]$. Notice that these two equations have the correct limiting behaviors as d goes to 0 or infinity.

a. Show that $\dfrac{I_{film}}{I_{sub}} = \dfrac{I_{film}^0}{I_{sub}^0}\left[e^{\left(d/\lambda_{mfp}\cos(\theta)\right)} - 1\right]$

b. Solve this for the film thickness in terms of the intensities, angle, and electron mean free path.

c. In angle-resolved XPS, the angle θ will change, while d and λ_{mfp} are constants. Plot $\ln\left(1 + \dfrac{I_{film}/I_{sub}}{I_{film}^0/I_{sub}^0}\right)$ vs. $\dfrac{1}{\cos\theta}$ for $\lambda_{mfp} = 4\,\text{nm}$ and for both $d=10$ and $3\,\text{nm}$.

14

Characterization: Electrical, Magnetic, and Optical Techniques

This chapter explores electrical, magnetic, and optical techniques for studying surfaces and thin films. In exploring the characterization of the electrical and magnetic properties of materials, we will not be considering device characterization but rather techniques that focus on fundamental materials properties introduced in Chapter 6. Properties such as resistivity and conductivity, magnetoresistance, various magnetic properties, and hysteresis loops in ferromagnetic materials will be examined. Much of the discussion from Chapter 6 is relevant to understand how these techniques work and the types of measurements that may be of interest. Optical characterization will examine how the interaction of light with materials can help us understand the optical/dielectric properties of surfaces and thin films, film thickness, surface roughness, and other properties. Many of the techniques included in this chapter are discussed in books (Brundle, Evans, and Wilson 1992, Frey and Khan 2015, Hudson 1998, Ohring 2002, Prutton 1994, Sardela 2014, Somorjai 2010, Zangwill 1988) that broadly cover characterization of surfaces and films.

14.1 Electrical Characterization

14.1.1 Hall Effect

The Hall effect can be used to determine the concentration and sign of charge carriers in a sample. While not a surface-specific measurement, it can be performed in thin films, especially when using an insulating substrate. A standard Hall measurement is demonstrated in Figure 14.1. A current is produced by applying a voltage at opposite ends of a sample. A magnetic field is applied across the sample perpendicular to the direction of current flow.

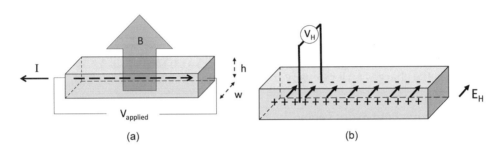

FIGURE 14.1
Hall effect measurement. (a) Applied voltage and magnetic field. (b) Resulting charge separation, Hall field and Hall voltage measurement.

DOI: 10.1201/9780429194542-17

Moving charges in a magnetic field are deflected by the Lorentz force resulting in the build up of charges on the front and back surfaces of the sample. The figure shows the direction of motion of negative charge carriers.

The charge carriers will be deflected until the force from the electric field associated with the Hall voltage (qE_H) balances the Lorentz force (qv_dB)

$$qE_H = qv_dB \tag{14.1}$$

Instead of the drift velocity (v_d), it is often easier to use the current density,

$$j = (8/3\pi)qnv_d \tag{14.2}$$

where n is the charge carrier density. The factor of ($8/3\pi$) arises from a more detailed analysis of collisions of charge carriers.

Substituting this into Equation 14.1 we get:

$$qE_H = (3\pi/8)(j/n)B \tag{14.3}$$

We define the Hall Coefficient as

$$R_H = \frac{E_H}{jB} = \frac{3\pi}{8}\frac{1}{nq} \tag{14.4}$$

Figure 14.1a provides dimensions for the sample, which will be helpful for expressing this in terms of measurable parameters.

We can now express

$$E_H = V_H/w \tag{14.5}$$

and

$$j = I/wh. \tag{14.6}$$

These can be substituted into Equation 14.4 to yield

$$R_H = \left(\frac{V_H}{I}\right)\frac{h}{B} \tag{14.7}$$

If the sample dimensions and applied magnetic field are known, we can measure the Hall voltage arising from a current to determine the Hall Coefficient. Once known, Equation 14.4 provides the charge carrier density in the material.

14.1.2 Resistivity: Four-Point Probe

Four-point probes are used to measure the resistivity of bulk and thin film materials which was introduced in Section 6.1.5. The analysis of the measurements is different depending on the thickness of the film relative to the spacing of the probes and the position of the probes relative to the edges of the sample.

Measuring resistance using a standard two-point probe (like a multimeter) introduces several other resistances in series with the sample resistance. Of particular concern are the contact resistance where the probes touch the sample and the lead resistance in the wires, probes, and the instrument itself. The resistance of the substrate can influence the measurement, especially for thin films. When measuring semiconductors, the work function difference between the sample and probes may also need to be considered.

In a four-point probe configuration, shown in Figure 14.2, we have four equally spaced probes. Current is passed between the outer two probes and the voltage, which arises from the resistance of the sample, is measured between the inner two probes. This eliminates the lead resistance since no current is flowing through the leads where the voltage is being measured. Since the geometry of the probes is well known, we can often relate the measured resistance to the resistivity of the sample.

We begin with the simple definition of electrical resistance (R) of a wire. Ohm's Law states

$$R = V/I \tag{14.8}$$

where V and I are the voltage and current measured between two ends of the wire. This will depend on both the geometry of the wire (how long it is and how thick it is) as well as on the properties of the material that the wire is made up of. The resistivity, ρ, includes all properties of the material itself and is defined such that

$$R = \rho \left(L/A \right) \tag{14.9}$$

where L is the length of the wire and A is the cross-sectional area of the wire.

This can be generalized for a differential resistance

$$dR = \rho \left(dx/A \right) \tag{14.10}$$

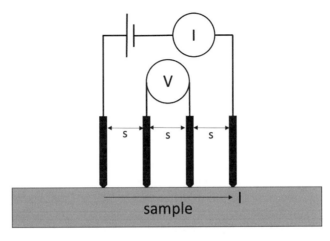

FIGURE 14.2
Geometry of the four-point probe.

where we find the resistance difference, dR, between two points separated by a small distance dx.

In bulk samples, the sample thickness is much greater than the distance (s) between the probes and we can assume that the current comes out of the outer probes in a spherical manner. With this assumption, it can be shown that the bulk resistivity is given by:

$$\rho = 2\pi s (V/I) \tag{14.11}$$

where s is the spacing between the probes, as indicated in Figure 14.2.

If you are dealing with films where the thickness of the film is greater than about 5/8 of the probe spacing, you can usually use this bulk formula for analyzing the film resistivity. For a probe spacing $s = 1.25$ mm, this would be films thicker than 0.78 mm $= 781$ μm $= 781,000$ nm $= 7,810,000$ Å. Very few thin films will be close to this large thickness.

In thin film samples where the film thickness (t) is small compared to the spacing between the probes, the assumption is that the current comes out of the probe tips in rings rather than spheres. The resistivity now can be expressed as

$$\rho = (\pi/\ln(2)) t (V/I) \tag{14.12}$$

Now the resistivity will vary with film thickness but is independent of the spacing between the probes. $\pi/\ln 2 = 4.532$ so you will sometimes see this factor in the equations rather than writing out the constants.

Sheet resistance depends on the film resistivity and thickness

$$R_S = \rho/t \tag{14.13}$$

The units of sheet resistance are typically given as Ohms per square. Some people will set the current through the probes to 4.532 mA so that the voltage measured in mV can simply be read out directly to give a sheet resistance.

The geometry of the resistance measurement is also distorted if you measure too close to the edge of the sample. Typically, it is recommended to measure at least 20 times the probe spacing from the edges. For 1.25 mm spaced probes, this would be a distance of 25 mm (1 in) from the edges. This is not always possible. Often a correction factor term is introduced into the equation for resistivity to take into account the proximity of the edges.

Very low-resistance materials (like metals) may require higher currents to achieve a measurable voltage difference between the inner probes. The current, however, is best kept under about 10 mA to avoid excessive heating of the sample. Very high-resistance materials may require lower currents to keep the voltage from being too high between the inner probes. Typically this voltage should be less than about 200 mV.

An alternative to the linear four-point probe is the Van der Pauw method. The placement of small measurement probes needs to be close to the edges of the sample but otherwise is arbitrary. Several standard symmetric configurations, however, are typically used to simplify the mathematical analysis. Two measurements are required to measure current and voltage in two different configurations as indicated in Figure 14.3. This can be continued around the sample for a total of four measurements. Often a total of eight measurements is made with the polarity of the probes reversed for the additional four measurements.

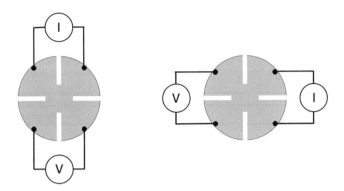

FIGURE 14.3
Van der Pauw resistivity measurement showing two measurement configurations for a cloverleaf sample design.

14.2 Magnetic Characterization

Many techniques for characterizing magnetic properties of materials are presented in Zhu (2005).

14.2.1 Magneto-Optical Kerr Effect

In Chapter 6, we discussed the magnetic properties of ferromagnetic materials and noted the importance of the hysteresis curve. One way of measuring this, which works well for thin films and surfaces, is using Magneto-Optical Kerr Effect (MOKE). This technique looks at the change in polarization of light reflected from a sample in the presence of a magnetic field. In some sense, this is an ellipsometry measurement (to be described in Section 14.3) but with a magnetic field present. The presence of the magnetic field causes an additional polarization rotation beyond that induced by the material itself. The Kerr effect in reflected light is similar to the Faraday effect in transmitted light. The dielectric constant, ε in a material is actually a tensor and the off-diagonal terms cause a change in the polarization of light upon reflection due to the anisotropy of the material.

Three geometries for the direction of the magnetic field relative to the surface and incident light are pictured in Figure 14.4. The polar geometry has the magnetic field perpendicular to the surface. The longitudinal geometry has the magnetic field parallel to the surface and in the plane of incidence of the light. The transverse geometry orients the magnetic field parallel to the surface and perpendicular to the plane of incidence of the light.

A typical experiment uses monochromatic light (often a laser) of known linear polarization incident on the surface. The reflected light passes through an analyzing prism that is set to extinguish the light reflected from the surface with no magnetic field. As the magnetic field is increased, the intensity of light transmitted through the prism is monitored. A plot of light intensity (proportional to magnetization) vs. applied magnetic field results in the hysteresis curve for the sample. For very thin films, the effect is rather small with only about 10^{-3} degrees of rotation for a monolayer of ferromagnetic film. When performing measurements on very thin films, the technique is sometimes referred to as Surface Magneto-Optical Kerr Effect (SMOKE). If the incident light is well focused, the beam can be scanned over the sample to map variations in the magnetic properties. Further details are available in the references (Brundle, Evans, and Wilson 1992).

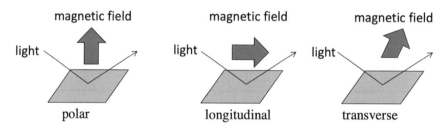

FIGURE 14.4
Magneto-optic Kerr effect geometries showing the applied magnetic field and the incident and reflected light path relative to the sample surface.

14.2.2 Spin-Polarized Electron Techniques

In earlier chapters, we have examined many characterization techniques that made use of incident beams of electrons. These electrons have a quantum mechanical property known as spin, which can be used to probe the magnetic properties of materials. For electrons, which are part of a class of particles known as fermions, the spin can take on two possible values that are often referred to as spin up (+½) and spin down (−½). Electrons in the two spin states will scatter differently from a material depending on the magnetic properties of the material. If we take our incident beam of electrons and create an uneven spin distribution so that one spin state is preferred (known as spin polarization) and then have a spin-sensitive detector so that we know the spin state of the detected electrons, we can perform a range of experiments that will help us understand the magnetic properties of surfaces and thin films.

In Section 11.2.4, we introduced low-energy electron microscopy. Spin-Polarized Low Energy Electron Microscopy (SPLEEM) uses an incident beam of spin-polarized electrons scattered from a sample and detected in a multichannel detector as indicated in Figure 14.5. The result is an electron intensity map of the surface resulting in an image much like a Scanning Electron Microscope (discussed in Section 11.2) but with contrast that depends on the magnetic properties of the material as well as the topography. The SEM topographic image is typically subtracted out to yield an image that depends only on magnetic properties. This can allow imaging in real time of magnetic domains with a spatial resolution of around 10 nm. Incident electron energies are typically less than 100 eV (much lower than traditional scanning electron microscopes) making this a very surface-sensitive technique.

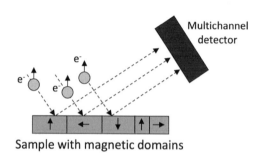

FIGURE 14.5
Spin-polarized low energy electron microscopy showing the spin-polarized incident electrons scattering from the surface.

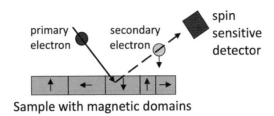

primary electron   secondary electron   spin sensitive detector

Sample with magnetic domains

FIGURE 14.6
Scanning electron microscopy with polarization analysis.

Highly crystalline materials and strongly oriented thin films are ideal for this technique since the detected electrons are diffracted with high concentration in the specular beam. Less crystalline or amorphous materials can be imaged but require signal acquisition over longer times. The observed magnetic contrast in the images arises from two processes: (1) interactions between the spin-polarized incident electrons and the spin-polarized electrons in the magnetic material and (2) differences in the inelastic mean free path of incident electrons with different spin directions (parallel or anti-parallel) relative to the spins in the surrounding material that they are traveling through.

Another imaging technique is Scanning Electron Microscopy with Polarization Analysis (SEMPA). In this case, the incident beam is not spin-polarized, but the spin of the detected secondary electrons is determined as indicated in Figure 14.6. Since the spin of the secondary electrons depends on the magnetic properties of the region of the sample that they are emitted from, this technique can image magnetic domains. Incident energies of the primary electrons tend to be more typical of scanning electron microscopy (>10 keV). Although the secondary electrons may have significantly lower energies. Spatial resolution of the images is about 10 nm.

14.2.3 Magnetic Force Microscopy

In Section 11.2, we discussed atomic force microscopy. A variation on this technique, Magnetic Force Microscopy (MFM), uses a magnetized tip to detect magnetic forces and to map out the magnetization of the sample. If the sample has magnetic domains with the magnetization perpendicular to the surface, then a tip magnetized perpendicular to the surface will detect the difference in forces between the domains magnetized up or down as shown in Figure 14.7. If the sample domains are magnetized parallel to the surface, then a probe with a parallel magnetization will detect the differences. The result is an image of the magnetic structures on the surface of the sample with spatial resolution better than 100 nm. If more than just an image is desired, the data can be quantified by detailed calibration of the MFM system.

MFM measurements can be done in the air, and the sample can have a nonmagnetic film coated over the magnetic elements as long as the magnetic field can be sensed above the sample. Fortunately, magnetic forces tend to be relatively long range. A standard AFM image can be formed by scanning with tip close to the surface. An MFM scan can then be obtained with the tip around 20 nm from the surface where magnetic forces are measurable but atomic forces are negligible. MFM does not, however, detect the magnetization of the sample directly, which is usually the property of interest. As with AFM, the measurements can be done by simply measuring the cantilever and tip deflection as a function of position or can be done by oscillating the cantilever and tip near the resonant frequency. For the oscillating measurements, the interaction of the magnetic sample, through the

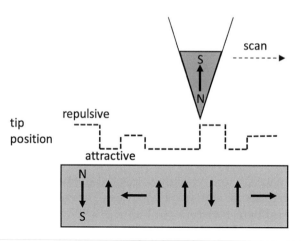

FIGURE 14.7
Magnetic force microscope operation.

gradient in the magnetic force, will shift the resonant frequency. This shift in frequency is then detected.

MFM probes often involve coating an AFM tip with 15–40 nm of a magnetic material such as a Co-Cr alloy. Depending on the application and how strong a magnetic interaction is desired between the tip and sample, tips of different magnetic coercivity may be desired. Hard magnetic materials can be examined using a high coercivity tip where a lower coercivity tip may be more appropriate for magnetically soft materials that could be altered by the presence of a stronger magnetic field.

14.2.4 Brillouin Light Scattering

Brillouin Light Scattering (BLS) will be discussed again in Chapter 15 since it is typically used as a method for detecting phonons in materials. A schematic diagram of the experimental configuration is provided in Figure 14.8. With careful attention to experimental detail, the technique can be used to explore magnons (see Chapter 6) as well. When an incident beam of light is reflected from the sample, a photon could lose some energy through the creation of a magnon. Similarly, an existing magnon could interact with a photon and give it additional energy as it leaves the sample. The result is that most of the light

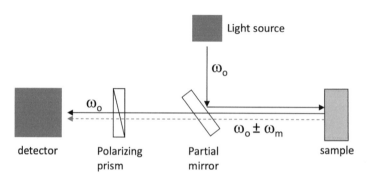

FIGURE 14.8
Schematic diagram of Brillouin light scattering experiment.

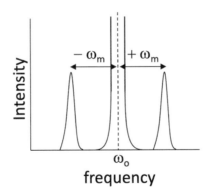

$-\omega_m$ $+\omega_m$

ω_o

frequency

FIGURE 14.9
Low-intensity shifts in frequency from the interaction of incident light with magnons.

is reflected at the same wavelength as the incident light, but a very small amount of light will be shifted up and shifted down in wavelength by the energy of magnons ($\hbar\omega_m$) in the sample, as indicated in Figure 14.9. A Fabry-Perot interferometer geometry is used to make the high-resolution frequency measurement that is required. Typical incident frequencies are in the 10^6–10^7 GHz range but the shifts in frequency are only about 1–500 GHz, so the experiment requires detecting a very small signal that is very close in the spectrum to a very large peak. The measurements provide details about the coupling between magnetic layers in multilayer samples as well as information about magnetic anisotropies.

14.2.5 Magnetometers

A vibrating sample magnetometer (VSM) can be used to examine the magnetization of a bulk or thin film sample as a function of the applied magnetic field. This will trace out a hysteresis loop, as discussed in Section 6.2, which can be used to characterize ferromagnetic materials. In a thin film sample, the magnetization component in the plane of the sample may not be the same as the magnetization in the direction perpendicular to the film. This can be explored using the VSM by rotating the sample relative to the applied magnetic field.

The basic design of the VSM is shown in Figure 14.10. We apply a uniform magnetic field around the small sample vibrating vertically with typical vibration frequencies of around 100 Hz. This motion creates a time-varying change in the magnetic flux detected by the pickup coils as an induced electric field (Faraday's Law of Induction) that produces a current in the coils. The time-varying signal is detected using a lock-in amplifier. The amplitude of this signal is proportional to the magnetic moment of the sample. By varying the current in the large electromagnet, we can vary the uniform incident magnetic field and trace out a hysteresis loop.

A Superconducting Quantum Interference Device (SQUID) Magnetometer is a much more sensitive type of magnetometer. It is based on two superconducting Josephson junctions in a ring structure as shown in Figure 14.11. A constant bias current is applied to the device and the voltage across the two junctions is measured as indicated in Figure 14.11. This voltage will oscillate as the magnetic flux through the ring changes allowing for accurate measurements of very small magnetic effects. Oscillating the sample produces a changing magnetic field in the ring.

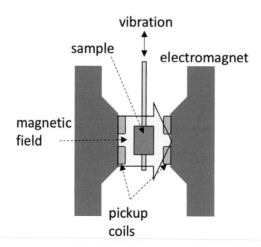

FIGURE 14.10
Vibrating sample magnetometer schematic.

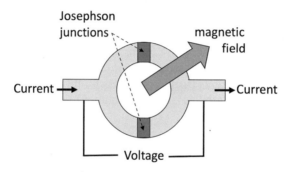

FIGURE 14.11
Configuration of a SQUID magnetometer with the magnetic field coming out of the page.

SQUID magnetometers typically allow for measurement over a range of sample temperatures from around 4–400 K and a large range of applied magnetic fields. Magnetization and magnetic susceptibility of the sample can be determined allowing for very sensitive determination of hysteresis curves.

14.2.6 Ferromagnetic Resonance

The magnetization of ferromagnetic materials can be examined using the interaction of microwave radiation with magnetic spin waves through Ferromagnetic Resonance (FMR). The precession frequency of the spin waves depends on the magnetization of the material and the strength of the applied magnetic field. A typical apparatus has a fixed frequency of microwave radiation (in the range of 1–35 GHz) that interacts with the spin waves of the sample while the applied magnetic field is swept through a range (0–1 T depending on the magnet) as indicated in Figure 14.12. When the precession frequency of the spin waves and the microwave frequency are equal, a strong absorption of microwave intensity will be detected. The resonant frequency can be related back to the magnetization of the sample if the applied magnetic field is known. Data is often collected in a derivative mode rather than the absolute microwave intensity.

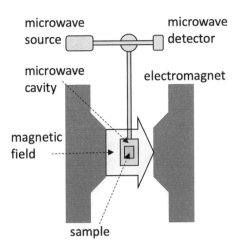

FIGURE 14.12
Schematic of a typical ferromagnetic resonance experiment.

The width of the resonant absorption peak can be directly related to the relaxation processes going on in the material. The anisotropy of the material can be determined by rotating the sample relative to the applied magnetic field direction. For a multilayer system of magnetic films, the interlayer exchange coupling can also be measured.

14.2.7 X-Ray Magnetic Circular Dichroism

The absorption of X-rays in a material depends on the energy of the incident X-rays and the material leading to absorption features at particular energies for specific materials. We discussed one example of this type of measurement in Section 13.5 when considering near-edge X-ray absorption fine structure. In magnetic materials, the absorption also depends on the magnetic properties of the material. In X-ray Magnetic Circular Dichroism (XMCD), X-rays are used to probe the magnetic spin states of the material in the presence of an applied magnetic field. Right or left circularly polarized X-rays are incident on the sample as indicated in Figure 14.13. These X-rays transfer the angular momentum information to photoelectrons that are excited to empty valence levels. This results in a difference in absorption of X-rays between the two polarizations of incident radiation for magnetic materials. The detected signal can be an energy spectrum of transmission of the incident X-rays. Fluorescence and emission of electrons can also be used as in NEXAFS. The surface sensitivity of the technique will be influenced by which detection scheme is used. Since X-rays have a large penetration depth compared to thin film thicknesses, the transmission technique will average over the film and substrate. Emission of low energy electrons, on the other hand, is spatially limited to a few nm by the short mean free paths of electrons in solids.

This technique is not dependent on the crystalline nature of the sample and so works well for amorphous and less crystalline materials. Spin and orbital magnetic moments can both be determined in an element-specific manner. The orbital moment information helps to probe anisotropy in the magnetic structure at thin film interfaces. Studies of exchange bias have also been conducted. XMCD has revealed the magnetic properties of ultra-thin ferromagnetic films. Often the temperature of the sample is also varied with cryogenic measurements being of interest.

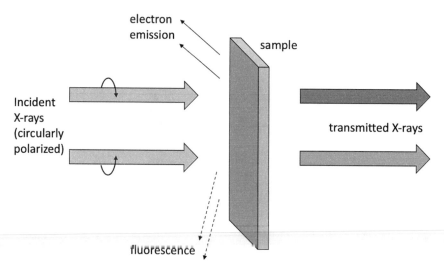

FIGURE 14.13
General schematic of X-ray magnetic circular dichroism experiment.

14.3 Optical Characterization

The optical characterization techniques discussed in this chapter provide information about the optical properties and structure of thin films and surfaces. Techniques that use optical phenomena to study sample chemistry (FTIR, Raman spectroscopy) were discussed in Chapter 13.

When using light as a probe, we start by considering the simple model of light incident upon a single layer of sample as shown in Figure 14.14. The incident light can be reflected from the sample in a specular manner with the angle of incidence, θ_i, equal to the angle of reflection, θ_r. If the surface is not perfectly smooth, light can also be scattered from the surface at non-specular angles. Some light may also enter the sample after being refracted at the interface with the angle of refraction, ϕ_r, given by Snell's Law

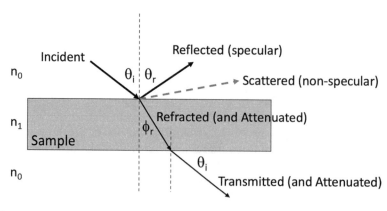

FIGURE 14.14
Interaction of incident light with a sample.

TABLE 14.1

Several Optical Techniques that Rely on Different Aspects of Light to Characterize Surfaces and Thin Films

Technique	Specular	Non-Specular	Intensity	Polarization	Interference
Reflectance	×		×	×	
Ellipsometry	×				
Scattering		×	×		
Interferometry	×		×		×

$$n_0 \sin \theta_i = n_1 \sin \phi_r \tag{14.14}$$

where n_0 is the refractive index of the ambient medium and n_1 is the refractive index of the sample. We will discuss the nature of these refractive indices in more detail shortly. As the light passes through the sample, it may lose intensity through various attenuation processes. Finally, some light may make it through the sample, be refracted at the interface and be transmitted from the sample at the same angle that it was incident.

Light carries a remarkable amount of information. We can examine the intensity of the light and the polarization (orientation of the electric field vector of the electromagnetic wave). Both of these can be done as a function of wavelength of the light. We can examine the specular and non-specular scattering of light. We can compare light coming from different locations using interference techniques. Table 14.1 shows how several different optical techniques sample these various aspects of light.

14.3.1 Reflectance

A simple reflectance experiment where light from a source is reflected from a sample and detected is shown in Figure 14.15. The wavelength of the incident light and the angle of incidence can both be changed to gather more information than is possible in a single wavelength, single angle system. Light can be focused and rastered across the sample to determine the uniformity of the sample by mapping. Absorption features in the reflectance spectrum can provide information about the chemistry. Samples should be highly reflective and flat with a minimal non-specular scattering of light. The intensity of the reflected light is measured and compared to the intensity of the incident light in order to determine the reflectivity of the sample. Notice that the geometry of the technique would allow measurements to be made during thin film deposition for a sufficiently high angle

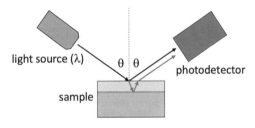

FIGURE 14.15
Schematic representation of a general reflectance measurement.

of incidence. This technique can provide information about film thickness and quality by examining the interference of light reflected from the top and bottom of the film.

When the wavelength of the incident light is varied, reflectance is a version of ultraviolet-visible spectroscopy that can also be performed in transmission mode to obtain chemical information by examining absorption from atomic and molecular bonds. As with other types of spectroscopies discussed in previous chapters, optical spectroscopies require knowledge of how the instrument behaves at different wavelengths. The light source does not emit equal intensities at all wavelengths and so the spectral emission of the source must be known. The detector does not give the same response for the same intensity of light incident at different wavelengths and so we also need to know the spectral response of the detector. The final measurement is a combination of the source, sample, and detector. The behavior of the source and detector needs to be removed if quantitative knowledge is needed. If only reproducibility or differences from one sample to another are needed, then this more detailed knowledge may not be necessary. Some reflectivity systems use a two-beam approach where the sample and a standard reference are measured and compared simultaneously. Other systems use a single beam but sequentially measure the sample and a standard reference.

When measuring thin film systems, the interference of light reflected from the film surface and from the film-substrate interface will cause variations in the reflected intensity that are a function of the wavelength of the incident light as well as the angle of incidence. Many reflectance systems vary the wavelength of the light resulting in an interference pattern such as that pictured in Figure 14.16.

The interference pattern depends on the thickness of the films. Figure 14.17 shows the normal incidence interference patterns for several different thicknesses of SiO_2 films on Si. Notice that the 100 nm film does not display much variation compared to the thicker films. This sets a lower limit on accurate thin film thickness measurements that can be extended by having a larger wavelength range. Many instruments will have an option to work in the UV part of the spectrum to measure thinner films.

The interference pattern also depends on the optical properties of the film and substrate. Figure 14.18 demonstrates this dependence for SiO_2 and Si_3N_4 films of the same thickness on a Si substrate.

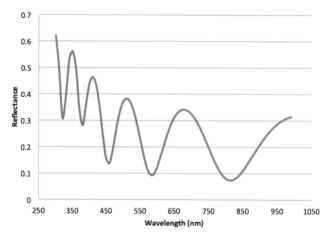

FIGURE 14.16
Model reflectance for 700 nm of SiO_2 on Si at normal incidence.

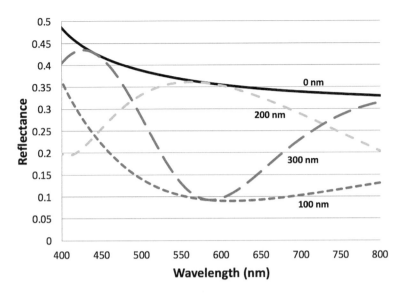

FIGURE 14.17
Normal incidence reflectance models for several thicknesses of SiO_2 on Si.

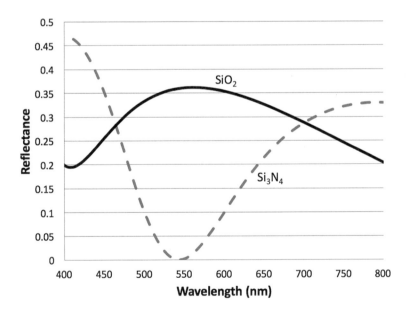

FIGURE 14.18
Reflectance models for 200 nm thick films of SiO_2 or Si_3N_4 on Si substrates.

14.3.2 Ellipsometry

Ellipsometry is widely used for determining the dielectric/optical properties of bulk materials, the thickness and dielectric/optical properties of thin films, and information about surface roughness and material porosity. While typically operated in a reflection mode, it differs from reflectance in that the polarization rather than the absolute intensity of the light is measured. As a reflection technique, samples must be flat and specularly reflecting.

Simple ellipsometers often operate at a single wavelength of incident light and at a single incident angle. Both wavelength and incident angle, however, can be varied to provide more information about the sample. Most spectroscopic ellipsometers operate in visible wavelengths. Some extend into the ultraviolet and infrared ellipsometers have also been developed. The lateral sampling region on the sample varies with different light sources and can range from about 30 μm to 1 cm in diameter. Depth information depends on the sample's absorption of light. Metallic samples with high absorption may only reflect light from down to about 100 Å into the material. Transparent samples may provide information to depths well beyond the thin film range.

Many designs of ellipsometers are available with slightly different operating principles. We focus our attention on the rotating analyzer ellipsometer, which is one of the most common of the automated ellipsometer designs. Figure 14.19 shows a schematic of a simple rotating analyzer ellipsometer. The light source emits unpolarized light of a single wavelength. This light passes through a polarizing prism (often set to 45°) that produces linearly polarized light incident on the sample. Upon reflection, interaction with the sample changes the polarization state of the light to a more general, elliptical polarization (hence the name "ellipsometry"). The elliptically polarized light passes through an analyzing prism rotating about the optical axis. The resulting sinusoidal variation in intensity is detected and sent to a computer. The intensity variation is analyzed to determine the polarization state of the reflected light.

Since the initial, linear polarization state of the light is known, what is typically examined is a comparison of the reflected polarization state to the initial state. The data is reported as a pair of angles, ψ and Δ. Δ is a measure of the relative change in phase upon reflection and ψ is a measure of the relative change in amplitudes upon reflection.

Optical measurement science was one of the earliest branches of experimental science to develop to a level of high precision. Making good ellipsometry measurements is not difficult to do. The complications arise in trying to interpret how these changes in the polarization of the light provide information about the sample. This requires developing models to convert the change in polarization to material properties.

Ellipsometry models are typically built on a simple layer-by-layer concept where each layer will be described by a thickness, an index of refraction, n, and an extinction coefficient, k. Sometimes the absorption index, κ, is used where $k = n\kappa$. Figure 14.20 shows the simplest possible case of a thin film on a substrate. Light is incident on the sample from the ambient medium (typically air). Some light is reflected and some refracted into the film layer. The light in the film layer travels through the film to the film-substrate interface where some is reflected back toward the surface and some light is refracted into the substrate. We assume that the substrate is infinitely thick, and so no light will be reflected back

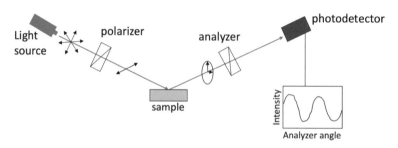

FIGURE 14.19
Schematic diagram of a rotating analyzer ellipsometer.

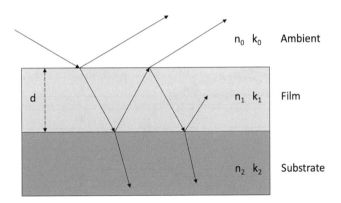

FIGURE 14.20
Optical layer model of a single film on a substrate.

out of the substrate. The light that was reflected from the film-substrate interface reaches the surface where some is refracted out of the sample and some is reflected back down into the film again.

If there is no film layer and the optical properties of the ambient medium are known, then we have two model parameters (n_2, k_2) and two experimentally determined parameters (ψ, Δ) and the system is solvable for the optical properties of a surface. As discussed in Section 6.3.2, however, even a bare surface may have a surface layer with different optical properties from the bulk. This situation is comparable to a surface with a thin film on it.

We continue to assume that the optical properties of the ambient medium are known. The presence of a film layer leaves us with five unknown parameters (n_1, k_1, d, n_2, and k_2). Our measurement, however, only yields two parameters (ψ and Δ). This system cannot be solved. If we can measure the optical properties of the substrate before film deposition (and assume that they do not change), then we still have three unknown parameters and only two measurable parameters. In some cases, however, we have some additional information about the film. For instance, SiO_2 or Si_3N_4 films on Si are commonly measured with ellipsometers. These films are known to be transparent ($k_1 = 0$) for visible light. Now we only have two unknowns (n_1 and d) and so the model can be solved exactly.

Even in these solvable systems, however, there is a complication arising from the periodic behavior of the ellipsometric parameters. Transparent films will cycle through the same values of ψ and Δ after roughly an increase in thickness corresponding to a half wavelength of the incident light. This behavior is demonstrated in Figure 14.21, which shows the periodicity for several different refractive indexes of transparent films growing on Si substrates. The circled data point on the right side of the figure could correspond to a thickness of 142, 425, or 708 nm of SiO_2.

More complicated systems, however, cannot be solved in this simple model. Additional information can be obtained by varying the light wavelength and angle of incidence. Another option is to produce a series of samples where some parameter is changed. For instance, this could be a series of samples grown under identical conditions with only the film thickness changing. More difficult to produce might be a series of films of identical thickness where the film morphology is changed by varying the substrate temperature. The film properties are often a function of temperature and so measurements could also be made on the same sample at different temperatures. The possible experimental parameters that could be varied are indicated in Figure 14.22.

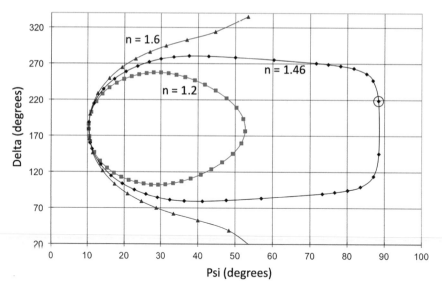

FIGURE 14.21

Ellipsometric cycles of transparent films grown on SiO_2 assuming incident light with $\lambda=633\,nm$.

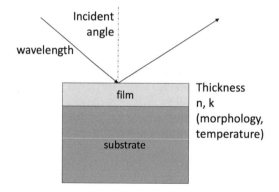

FIGURE 14.22

Experimental parameters that can be varied in ellipsometry of a film/substrate system.

Consider the variation of wavelength of the incident light. While the thickness of a thin film is constant, the optical properties of the film are functions of wavelength. Figure 14.23 shows the ellipsometric parameters (ψ and Δ) plotted against the wavelength of the incident light. Notice that operation at a single wavelength generates only two points on this graph. Turning the data into an optical model of the sample involves determining which possible models intersect those two points. There are an infinite number of these solutions and so we cannot adequately model the sample. Measurement of ψ and Δ as functions of wavelength changes the process to one of trying to find models that will fit two curves rather than just two points. This will lead to a much smaller number of physically reasonable models and is often sufficient for determining the properties of the sample.

Another option is to vary the angle of incidence of the light on the sample. Ellipsometry achieves the minimum uncertainty when operated at the principal angle where $\Delta=-90°$, which for many materials is around 70°. This is typically very close to the pseudo-Brewster angle where $\tan(\psi)$ is minimized. Some samples are very sensitive to variations in the

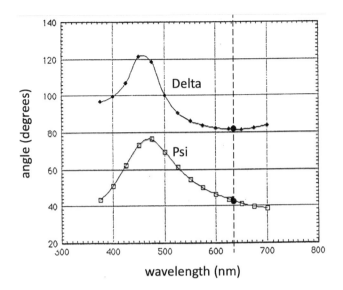

FIGURE 14.23
Ellipsometric parameters for SiO_2 film on Si plotted vs. wavelength. The points at the HeNe laser wavelength are indicated by the larger circles at 633 nm.

angle of incidence and so this variation can provide additional information to help restrict models. If incident angle variation is used in conjunction with wavelength variation, it adds a third axis to Figure 14.23 with an incident angle axis coming out of the page. This means that we will now be trying to find models that will fit two surfaces rather than just two curves. This will further restrict the number of physically reasonable solutions that can model the sample.

Once optical properties, n and k, are determined, the next step is often relating these to other material and structural properties of the sample. Effective medium approximations, as described in Section 6.3, are frequently used in modeling ellipsometry data.

Ellipsometry has been used to examine samples located inside vacuum systems for film deposition and materials characterization. The windows in the chamber that the incident and reflected light pass through, however, will typically change the polarization of the light and this needs to be taken into account. If studies are done during deposition, the temperature of the sample may change leading to changes in the optical properties of the sample. Accurate positioning of the sample on the optical axis can also be more difficult inside a vacuum system.

Mapping can be done with ellipsometry either by moving the sample under the beam in a raster pattern or by using a broad field of illumination and having a position-sensitive optical detector (such as a CCD camera). Measurements are frequently made as a function of time as well to examine film growth, oxidation, and corrosion processes. Additional information about ellipsometry is available in several books (Azzam and Bashara 1989, Brundle, Evans and Wilson 1992, Sardela 2014, Tompkins 1993) and the references within them.

14.3.3 Light Scattering

Light scattering experiments examine the roughness of surfaces by examining the light that is diffusely scattered away from the specular beam. The intensity of the reflected light is measured, but now at multiple exit angles rather than just at the angle of incidence

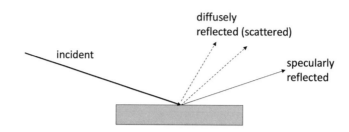

FIGURE 14.24
Specularly and diffusely scattered light.

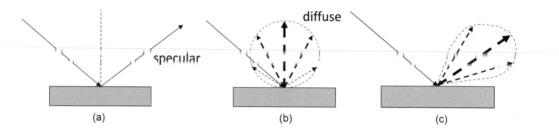

FIGURE 14.25
Light scattering: (a) specular reflection (b) totally diffuse scattering, and (c) mixture of specular reflection and diffuse scattering.

as shown in Figure 14.24. Measuring the polarization of the scattered light (non-specular ellipsometry) can also provide roughness information.

Figure 14.25 models the scattering of light with different types of samples. If the sample is very flat, the light will be primarily reflected as a specular beam as in Figure 14.25a. If the sample is randomly rough, the reflected light may be totally diffusely reflected as in Figure 14.25b. In this extreme case, there is no evidence of a specular reflection. The light is emitted with a maximum intensity normal to the surface. Most common is the case represented in Figure 14.25c where there is a strong specular reflection but diffuse scattering around that angle.

Careful measurement of the angular distribution of the scattered light can provide a measure of the surface roughness. The total integrated scattering is directly related to the root mean square roughness, R_q, discussed in Chapter 12 as well as to the reflectance of the surface and the angle of incidence and wavelength of the light. Higher reflectance, shorter wavelengths, and more normal incidence scatter more total light. The total integrated scattering, I_S, measured as a percentage of the incident light intensity, can be expressed as

$$I_S = R_0\left[1 - e^{-\left(4\pi R_q \cos\theta / \lambda\right)^2}\right] \tag{14.15}$$

where R_0 is the ideal reflectance of the smooth surface, θ is the angle of incidence, and λ is the wavelength of the incident light.

14.3.4 Interferometry

If light reflected from the top of a film and from a bare substrate region are combined, they will interfere with one another since they have traveled different path lengths (dependent on the thickness of the film) and have reflected from materials with different optical properties. A schematic representation of this interference experiment is presented in Figure 14.26. Often a Fizeau plate, which is a partially transparent but highly reflective coating on the bottom of a planar reference plate, is brought near the sample. The differences in the light wave between the film and the Fizeau plate and the substrate and the Fizeau plate lead to an interference pattern that can be observed from above the Fizeau plate. This interference pattern can be used to determine the thickness of a film if the optical properties are known. Thicknesses in a range of 1–2000 nm can be measured. The figure shows the substrate surface being exposed, which is necessary for non-transparent films.

Several different designs are available. Two beam interferometry results in bright and dark bands called fringes arising from the constructive and destructive interference of light from the top and bottom of the film. For transparent films, fringes of equal thickness will produce an image like a contour map of the film. If the film is perfectly flat, the surface will appear uniform. When fringes are observed, they represent a difference in thickness of one-half the wavelength of the incident light. A multiple beam system allows for better resolution by producing narrower fringes. Figure 14.27 show some examples of fringes. Fringes of equal thickness are demonstrated for a symmetric wedge-shaped film in Figure 14.27a. A film shaped like a spherical cap is demonstrated in Figure 14.27b.

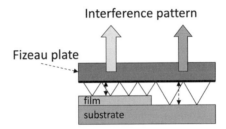

FIGURE 14.26
Schematic diagram of interferometer experiment.

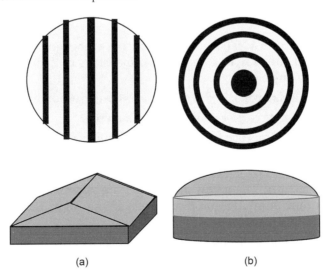

(a)   (b)

FIGURE 14.27
Fringes of equal thickness. (a) Symmetric wedge-shaped film. (b) Spherical cap film.

A Tolansky interferometer measures the relative fringe shift, using a single wavelength of light and achieves about 10–30 Å resolution in film thickness. An interferometer using Fringes of Equal Chromatic Order (FECO) uses white light, which results in an independent fringe pattern for each wavelength. Measuring the shift in fringes in the resultant spectrum allows a resolution of about 1 Å for film thickness.

References

Azzam, R.M.A. and N.M. Bashara. 1989. *Ellipsometry and Polarized Light*. Amsterdam: North Holland Elsevier Science Publishers.
Brundle, C.R., C.A. Evans Jr., and S. Wilson, eds. 1992. *Encyclopedia of Materials Characterization: Surfaces, Interfaces, Thin Films*. Boston, MA: Butterworth Heinemann.
Frey, H. and H.R. Khan, eds. 2015. *Handbook of Thin-Film Technology*. Berlin: Springer Verlag.
Hudson, J. 1998. *Surface Science: An Introduction*. New York: John Wiley & Sons, Inc.
Ohring, M. 2002. *Materials Science of Thin Films*. 2nd ed. San Diego, CA: Academic Press.
Prutton, M. 1994. *Introduction to Surface Physics*. Oxford: Clarendon Press.
Sardela, M. ed. 2014. *Practical Materials Characterization*. New York: Springer Science+Business Media.
Somorjai, G.A. 2010. *Introduction to Surface Chemistry and Catalysis*. 2nd ed. New York: John Wiley & Sons, Inc.
Tompkins, H.G. 1993. *A User's Guide to Ellipsometry*. Boston, MA: Academic Press, Inc.
Zangwill, A. 1988. *Physics at Surfaces*. Cambridge: Cambridge University Press.
Zhu, Y. ed. 2005. *Modern Techniques for Characterizing Magnetic Materials*. Boston, MA: Kluwer Academic Publishers.

Problems

Problem 14.1 We perform a Hall measurement using a magnetic field with $B = 145$ mT and a current of 20 mA. Our sample has dimensions of 2 cm in length, 1 cm in width and 0.1 cm in height.

   a.   We measure a Hall voltage of 50 mV. What is the corresponding Hall coefficient?

   b.   What is the charge carrier density in this material?

   c.   How would you improve the experiment to gain confidence in these values?

Problem 14.2 Consider making four-point probe measurements on a series of Cu films deposited on glass substrates. You set the current through the four-point probe to be 4.532 mA. The following voltages are recorded from the four-point probe. 50 nm film: 56 mV, 100 nm film: 22 mV, 150 nm film: 11.3 mV, 200 nm film: 8.5 mV, 250 nm film: 6.8 mV.

   a.   What are the sheet resistances (in Ohms) for each of the films?

   b.   What are the resistivities (in μΩ cm) for each of the films?

   c.   What does the data suggest about the dependence of resistivity on film thickness?

Problem 14.3 Ellipsometry measurements for transparent films cycle through the same (ψ, Δ) values for thicknesses that vary by $\Delta d = \dfrac{\lambda}{2\sqrt{n_{film}^2 - \sin^2(\phi)}}$. Calculate this thickness interval for SiO_2 films ($n_{film} = 1.46$) for an ellipsometer using a HeNe laser ($\lambda = 633\,nm$) and an angle of incidence, $\phi = 70°$.

Problem 14.4 When performing spectroscopic ellipsometry or reflectance measurements, we are interested in how the index of refraction varies with wavelength. For transparent films we can use the Cauchy model where $n(\lambda) = A + \dfrac{B}{\lambda^2} + \dfrac{C}{\lambda^4}$.

For SiO_2: we will use $A = 1.447$, $B = 4808$ and $C = 0$ with the wavelength measured in nm. Determine the index of refraction for SiO_2 at several common wavelengths: Hg vapor lamp (filtered): 436 nm, Na vapor lamp: 589 nm, HeNe laser: 633 nm, IR: 1000 nm. (Note that the index of refraction of a film will depend on the density of the film which can vary with deposition parameters and film thickness.)

15

Characterization: Thermodynamic, Thermal, and Mechanical Techniques

This chapter begins by examining methods for measuring thermodynamic properties such as surface tension or adsorption/desorption energies, which were introduced in Chapter 5. We then explore the measurement of thermal properties such as heat conduction. The chapter concludes with an examination of mechanical properties. Mechanical properties of films and surfaces have been referred to at various places in this book, but not explicitly developed. We explore the characterization of elastic waves/phonons and then examine mechanical properties such as friction, stress, strain, adhesion, hardness, elasticity, etc. and ways of characterizing these properties. Many of these techniques are discussed in general references (Frey and Khan 2015, Hudson 1998, Ohring 2002, Prutton 1994, Somorjai 2010, Zangwill 1988).

15.1 Thermodynamic Characterization

A wide range of thermodynamic properties of surfaces and thin films, discussed in Chapter 5, can be measured either directly or indirectly. In the following sections, we explore some examples of these measurements.

15.1.1 Surface Tension

Surface tension is a fundamental thermodynamic property of surfaces. We describe here four methods that have been used for measuring this quantity as a sample of many possible experiments that could be done.

If you are working with crystals that cleave easily, it is possible to apply a carefully controlled force to separate the material and create a new surface area as represented in Figure 15.1. The surface tension will be the work done per unit of new surface area exposed. Measurements of the force, displacement, and new surface area created will provide a value for the surface tension of the crystal. This measurement assumes that there is no shear deformation or plastic deformation of the crystal and so it is best performed at low temperatures.

Another experiment uses very thin, polycrystalline, wires with a load applied to one end. The surface tension will seek to cause the wire to contract in order to minimize the amount of surface. The application of a load, however, will produce a stress field leading to diffusion to cause a lengthening of the wire through diffusional creep. If a load is applied that produces zero creep, the two effects are balanced and the surface tension can be measured. These experiments are typically performed at temperatures slightly below the melting point so that diffusion will be fast. If the presence of grain boundaries is neglected,

DOI: 10.1201/9780429194542-18

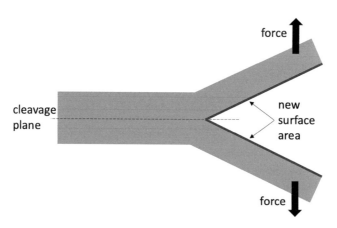

FIGURE 15.1
New surface area created in cleaving a crystal by a controlled force

FIGURE 15.1
New surface area created in cleaving a crystal by a controlled force.

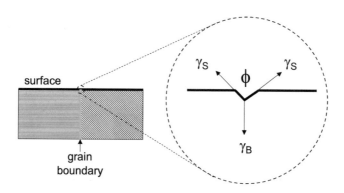

FIGURE 15.2
Surface tensions at the intersection of a grain boundary with the surface.

$$F_{\text{applied}} = \pi r \gamma \tag{15.1}$$

where r is the radius of the wire.

If the material can be shaped into a sharp tip it may be possible to use field emission to determine the surface tension. When a high voltage is applied to the tip, the electric field will tend to sharpen the tip by removing atoms from the tip. Surface diffusion, however, will try to minimize the surface area by rounding the tip. Again these effects can be balanced so that no change in the tip is observed and the surface tension can be related to the high voltage at this point of balance.

A final technique uses grain boundaries in material that reach a surface. Figure 15.2 shows two grains and the grain boundary between them. The enlargement examines more closely the region where the grain boundary reaches the surface. A triangular indentation in the surface will form where the angle, ϕ, depends on the value of the surface tension. Direct observation of this angle can be used to determine the surface tension.

15.1.2 Surface Adsorption and Desorption

In Chapter 7, the subject of adsorption and desorption from surfaces was discussed in detail both to examine submonolayers of gases adsorbed on surfaces and in the process of

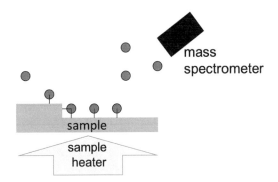

FIGURE 15.3
Schematic diagram of temperature-programmed desorption experiment. Note that the atom at the step is more strongly bonded to the surface.

thin film growth. Adsorption and desorption information can be obtained in a variety of experiments.

One common approach uses temperature-programmed desorption (TPD), which measures the desorption energy of atoms and molecules bound to a surface. The technique is schematically represented in Figure 15.3. A mass spectrometer is placed near a surface. The sample temperature is rapidly increased, and we observe the number and type of desorbed molecules at each temperature.

Plots of the number of molecules (N) vs. temperature (T) or of dN/dT vs. T can be used to determine the desorption energy for each molecular species. The change in pressure in the chamber can also be used as a measure of the number of molecules escaping from the surface. These experiments can be conducted at various initial surface coverages to explore the existence of multiple binding sites. A molecule may prefer to adsorb at a particular type of binding site on the surface. As the coverage increases, however, other types of binding sites may need to be used to accommodate the larger number of molecules on the surface. These two sites will typically have different desorption energies that will be indicated by different peaks in a TPD spectrum. An example of this is presented in Figure 15.4.

If a series of experiments are conducted with the sample exposed to different amounts of gas in order to get different coverages, the resulting TPD curves can be integrated since the area under the curve should represent the coverage. A plot of the coverage vs. the initial exposure provides a measure of the sticking coefficient of the surface.

TPD results often fall into two categories. If the molecule desorbs without dissociation, the desorption rate is typically proportional to the surface coverage. In this case, the peak temperature and the width of the TPD peak are independent of the initial coverage on the surface. If the molecule typically experiences dissociative desorption, the peak temperature and the width of the TPD peak will change with initial coverage. In this case, the desorption rate typically is proportional to the square of the surface coverage.

The presence of steps and kinks on surfaces, described in Section 3.6, provides additional adsorption sites on a surface that will have different desorption energies. As a result, the TPD spectrum from a flat surface and from a similar surface with steps and kinks on it will look quite different. Comparison of the spectra can be used to determine the presence of the steps and kinks and the desorption energies associated with them. This is also indicated in Figure 15.4.

Sources other than temperature can be used to stimulate the desorption of molecules from a surface. Electron beams can be used. These produce two effects on the surface. One is just a heating effect but the other is to excite the electrons involved in specific bonds

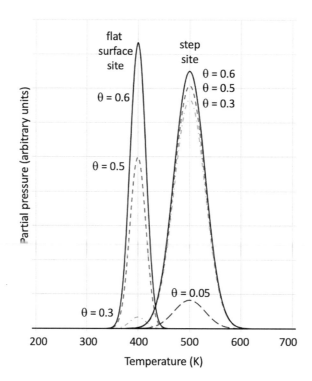

FIGURE 15.4
Simulated TPD spectra. At low coverages only the more strongly bound step sites are occupied. At higher coverages, the flat surface sites are also occupied.

into a repulsive state where the bonds will be broken. Electron Stimulated Desorption Ion Angular Distributions (ESDIAD) experiments make use of these effects. This is a rather complicated process and obtaining desorption energies in this manner is not easy. This technique, however, can be used to produce images that indicate whether the molecules are oriented on the surface. For example, if water molecules are randomly oriented in the plane of a surface, the distribution of H^+ ions emitted in an ESDIAD experiment will be the same in all directions. If the H_2O molecules are oriented in particular directions due to the crystallography of the surface and/or the presence of other adsorbates, the H^+ ions will come off the surface in particular directions.

Using light for photostimulated desorption produces some heating of a surface but can induce surface chemistry to occur that may lead to desorption. The frequency of the light can be varied to observe at what frequency (energy) desorption begins.

Desorption energies can also be measured using beams of atoms and molecules incident on a surface. The typical experiment is represented in Figure 15.5. The incident beam is chopped at a known frequency to obtain a periodic signal. The desorbed species are detected, using a mass spectrometer, a short time later depending on their lifetime (τ) on the surface, which is a function of the surface temperature (see Equation 7.1). An Arrhenius plot of $\ln(\tau)$ as a function of $1/T$ will have a slope that will be proportional to the desorption energy.

The adsorption and desorption of atoms from a surface can also be examined using molecular beam relaxation spectroscopy. Some atoms (noble gases) or molecules incident on the surface will have minimal interaction with the surface and will scatter elastically such that the angle of incidence is equal to the angle of scatter. Other atoms or molecules will interact

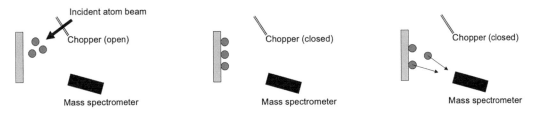

FIGURE 15.5
Desorption experiment using a chopped beam of incident atoms.

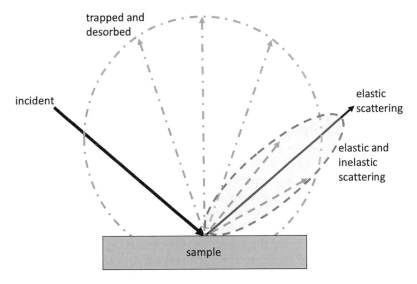

FIGURE 15.6
Scattering directions from interactions of atoms or molecules incident on a surface.

so strongly with the surface that they lose all information of their initial momentum and are desorbed from the surface with no recollection of the direction of incidence. This is presented in Figure 15.6. Examination of the non-specular molecules can yield a desorption rate. Repeating the experiment as a function of surface temperature allows the construction of an Arrhenius plot with the desorption energy available from the slope.

15.1.3 Differential Scanning Calorimetry and Thermogravimetric Analysis

Although more commonly used for bulk samples and powders, Differential Scanning Calorimetry (DSC) and Thermogravimetric Analysis (TGA) can be used with small pieces of film/substrate samples. For DSC, the substrate should be a material that will not experience any chemical reaction or phase change over the temperature range of interest (or it should be well characterized.) In DSC, a small sample (<100 mg) is placed in a pan with a thermocouple to carefully monitor the sample temperature. An identical, empty, pan is placed in an identical DSC cell and monitored for temperature changes. By comparison of these temperatures, heat flow can be detected. Model data is shown in Figure 15.7 indicating exothermic and endothermic reactions. Common exothermic events include crystallization and oxidation. Common endothermic events include melting and evaporation.

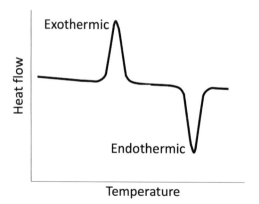

FIGURE 15.7
Representation of DSC data showing exothermic and endothermic reactions.

When working with thin films, DSC can be used to look for reactions of a film with the substrate or with other films in a multilayer structure. Melting or phase changes in the film can also be observed. If a gas (air, nitrogen, oxygen) is introduced into the DSC cell, reactions of the gas with the film such as oxidation can be observed.

TGA measures the change in the weight of small (<200 mg) samples. One typical experiment involves using TGA during exposure of the film to gases at elevated temperatures (up to about 1500°C) for examining reactions of the gas with the film. For instance, the weight gain of a film during oxidation can be measured as a function of time and temperature. As with DSC, a primary issue is that the substrate is also exposed to the gas and so this is restricted to systems where the substrate will be inert and only the film will react. TGA experiments can also be conducted in vacuum or inert gas environments to look for reactions in the film or between the film and substrate. In vacuum, these experiments can be combined with a mass spectrometer to identify volatile species which are created in the reaction. Additional information about DSC and TGA is available in the literature (Gabbott 2008, Höhne, Hemminger, and Flammersheim 2003).

15.1.4 Diffusion

Surface diffusion rates and energies can be determined using several methods. A considerable amount of older work was done using field ion microscopy. A sample shaped as a sharp tip is formed consisting of a series of atomic planes and steps. A high voltage is applied to the tip to ionize atoms and remove them from the sample. Atoms on the step edges or atoms on top of the flat areas will ionize more readily and be removed from the sample and directed toward a phosphor screen. More recent work uses a scanning tunneling microscope to directly image atoms and vacancies in motion on the surface. Care must be taken that the STM tip is not interacting with the sample to change the diffusion.

Another technique for observing diffusion is laser desorption. This is demonstrated in Figure 15.8. A layer of adsorbed atoms is produced on the surface. A focused laser heats a small region on the sample causing the adatoms in that region to evaporate from the surface. Other adatoms then diffuse into this region. Sometime later the laser is pulsed again and a mass spectrometer detects the quantity of adsorbed atoms that had diffused into the region during that time. From this, the diffusion properties of the adsorbed gas on this surface can be determined.

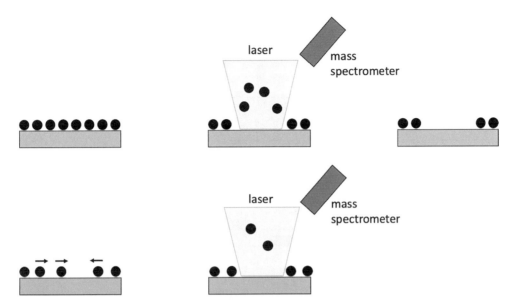

FIGURE 15.8
Laser desorption. Each image is at a successive time.

15.2 Thermal Characterization

15.2.1 Micro-Thermal Microscopy

The direct measurement of thermal properties of thin films is difficult due to the presence of the substrate. The thermal properties of the sample can be dominated by the more massive substrate unless measurements can be made in a very localized manner. Micro-thermal microscopy (Price et al. 1999) makes use of some principles of the scanning probe microscopies discussed in Chapter 11. It resembles atomic force microscopy in design, but the tip is replaced with a resistive heating element that also functions as the temperature sensor. This allows the technique to provide information on a very localized scale about the thermal conductivity of the sample as well as providing some topographic information with about a 10 μm resolution due to the much larger size of the tip. Phases in the film or surface can be identified from their thermal properties and phase transitions can be examined on a μm scale. The lower the thermal conductivity of the sample, the more localized the information will be.

The micro-thermal microscope can be operated in several different modes. If the temperature of the tip is held constant (DC mode) then the measurements will provide a mixture of thermal conductivity and topographic information. If the temperature is modulated (AC mode), then the same information can be collected with an improvement in depth resolution. The tip can be held still over one region of the sample and the temperature of the tip can be increased. This will provide more detailed, localized thermal information and can be used to detect temperatures at which phase transitions occur. The instrument can also be used as a thermo-mechanical probe where a constant force is applied by the tip to the sample. The tip is then heated and the deflection of the tip is monitored as the film softens.

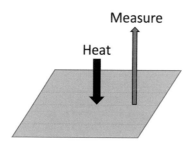

FIGURE 15.9
Schematic of photothermal analysis measurement.

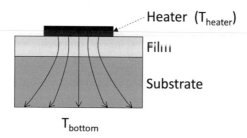

FIGURE 15.10
Cross-plane thermal analysis measures the difference between T_{heater} and T_{bottom}.

15.2.2 Photothermal Analysis

Another way of examining the thermal conductivity of a thin film or surface is using photothermal analysis. In this technique, a small region of the sample is heated with a laser. The laser is often modulated to produce a periodic signal. The flow of heat from this region is then measured as a function of distance from the heated region as indicated in Figure 15.9. The heat can be sensed by a variety of different techniques such as an infrared scanner, a thermocouple, or by measuring the thermoreflectance of the sample. This technique is best for films that are more than 1 μm thick in order to reduce the effects of the substrate, but films down to 100 nm in thickness can be measured using a very short heating pulse (<10 ns). In this technique, a substrate with low thermal conductivity is preferred.

15.2.3 Cross-Plane Thermal Analysis

The thermal conductivity through the sample can also be measured (Dames 2013). In cross-plane thermal analysis, a heater is placed on top of the film and the temperature is sensed at both the top and bottom of the sample. Since the substrate is included in the measurement, this technique prefers a substrate with very high thermal conductivity so that the temperature drop will primarily arise from the film. The geometry for this measurement is presented in Figure 15.10. The temperature drop across the sample is used to determine the thermal conductivity. This measurement can be performed in a DC mode or the temperature can be modulated with a driving current of frequency, ω. For the modulated case, it can be shown that measuring the voltage at the bottom at a frequency of 3ω will provide information about thermal properties.

15.3 Mechanical Characterization

15.3.1 Brillouin Light Scattering

The elastic properties of a material can be determined from knowledge of elastic waves in materials. In Chapter 14, we introduced Brillouin Light Scattering (BLS) where light interacts with a particle in the sample and is scattered at a different frequency than the incident light as indicated in Figure 14.8. In that chapter, the particle of interest was a magnon. Much more common is to use the technique for studying phonons, which are the particles associated with elastic waves in the solid as discussed in Section 4.2. If an existing phonon transfers energy to the incident light, the scattered light will have higher frequency (anti-Stokes component). If a phonon is created and takes energy from the light, the scattered light will have lower frequency (Stokes component). The resulting spectrum is indicated in Figure 15.11. The spectrum is similar to that discussed for Raman spectroscopy (Chapter 13), although the frequency shift is smaller. The position of this shifted peak provides the natural frequency of excitation of phonons in the solid. These measurements can be used to determine the sound velocities of phonons in materials and consequently the elastic constants. The measurements can be done as a function of sample temperature.

15.3.2 Stress

Forces arising from the interaction of a film with the substrate often result in stresses in the film and in the substrate. Focusing our attention on the film, stress is an internal force per unit area applied to a film and typically results in some change in the lattice spacing of the film that leads to strain. Strain is defined as the change in the dimension in the direction of the stress divided by the unstressed dimension. Two situations are common. If the film is being stretched so that the lattice spacings are larger than they normally would be, the film is in tensile stress. The film would prefer to be smaller but is stretched by the interaction with the substrate. If the film is being compressed so that the lattice spacings are smaller than they would normally be, the film is in compressive stress from the interaction with the substrate and would prefer to have a larger lattice spacing. These conditions are indicated in Figure 15.12.

One direct way of measuring stress is to use a flexible substrate that allows the film/substrate system to bend. For instance, Si wafer substrates can be purchased with a thickness of only 50 μm making them fairly flexible. A film in tensile stress would prefer to be smaller and can achieve this by curving in a concave manner so that the film is

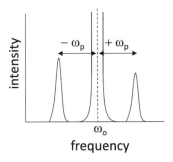

FIGURE 15.11
Brillouin light scattering spectrum from creation or annihilation of phonons.

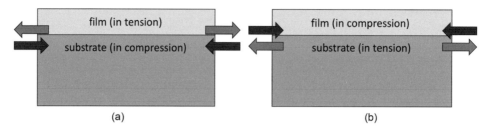

FIGURE 15.12
Forces (arrows) creating (a) Film in tension (b) Film in compression.

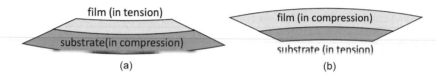

FIGURE 15.13
Curvature of films on flexible substrate for (a) Film in tension (b) Film in compression.

closer to a normal lattice spacing. Similarly, a film in compressive stress would prefer to be larger and can achieve this by curving in a convex manner. These curvatures are indicated in Figure 15.13.

Thermal expansion differences between the substrate and film can also lead to stresses in the system. Thin films are often deposited at elevated temperatures to get desired properties. As the materials cool, the film and substrate will contract at different rates since the thermal expansions of the two materials are different. This can lead to stresses in the film and substrate.

The curvature of the film/substrate system can be measured in many different ways. If the curvature is assumed to follow a known shape, then a measurement of the deflection of the sample at the center is sufficient. More complete information could be obtained by using a stylus profilometer (discussed in Section 12.5) to make a continuous measurement across the sample. If the stress is anisotropic, it may be necessary to make measurements in multiple directions across the sample. Optical methods using the reflection of a laser off of the curved surface can also provide a precise measurement of the curvature. If the film is reflective, an optical flat can be placed over the curved surface and the curvature can be determined from interferometry (discussed in Section 14.3). One could also place a flat electrically conducting disk over the sample to make a measurement of how the capacitance varies across the sample to determine the curvature. Several of these methods are schematically represented in Figure 15.14.

Direct measurements of strain in the lattice are also possible yielding information about the stresses that are being applied. By directly measuring the lattice spacing and comparing it to the known lattice spacing for the relaxed material, the deformation of the lattice can be determined. These experiments are often done using diffraction techniques. X-ray diffraction (discussed in Section 12.1) can be used in grazing incidence configuration to examine the stress both perpendicular to the plane of the surface and in the plane of the surface. Electron diffraction techniques (Sections 12.2 and 12.3) can provide information in the plane of the surface and some limited information perpendicular to the surface. Further details about stress in thin films are available in the references (Freund and Suresh 2003).

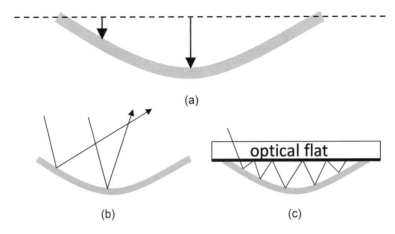

FIGURE 15.14
Curvature measurements: (a) Physical measurement at multiple points or profilometer (b) Reflected light from multiple points (c) Optical interference measurement.

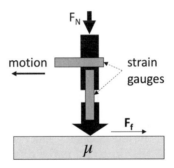

FIGURE 15.15
Friction measurement geometry and terms.

15.3.3 Friction

The force of friction (F_f) depends on the magnitude of a force applied normal to a surface (F_N) where two objects are interacting and a friction coefficient (μ), which contains all of the materials-related properties of the two objects.

$$F_f = \mu F_N \tag{15.2}$$

The direction of the force is opposed to the relative motion of the two objects. Static friction represents a force that must be overcome to initiate motion. Kinetic friction, which is of more interest in many industrial applications, is the force that opposes the motion once it has begun. Typically the friction coefficient for static friction is larger than that for kinetic friction.

The characterization of friction typically involves the determination of μ from measurements of the applied normal force and the frictional force. A very common tool for measuring these forces is a strain gauge. A typical experimental configuration is shown in Figure 15.15.

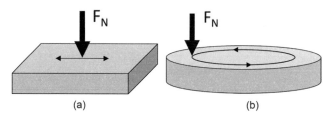

FIGURE 15.16
Friction measurement configurations: (a) pin on flat (b) pin on disk.

Two very common geometries for friction and wear measurements are the pin on flat and the pin on disk configurations as represented in Figure 15.16. The pin on flat configuration oscillates the sample or pin in a linear manner relative to one another. The pin on disk rotates the sample under a stationary pin. The pin on disk will keep a constant velocity between the sample and pin during the entire experiment for a thin tip. For wider tips, the pin on disk design will have different linear velocities on the inner and outer edges of the track leading to different wear rates. The normal force should not be great enough that the pin penetrates the film. The pin on flat design allows for easier side-by-side comparisons of multiple tracks with different experimental parameters. The measurement depends on the shape of the pin tip, which is often chosen to be spherical.

15.3.4 Microindentation – Nano-indentation

When discussing the atomic force microscope (Section 11.2.4) we observed that an increasing repulsive force occurs when the tip gets very close to the surface. If we continue to move the tip toward the sample, the tip could make direct contact with the surface and could even penetrate the surface. These indentation experiments, analogous to large-scale mechanical testing of materials, can be used to generate a stress-strain curve for a surface or thin film and allow us to measure the hardness, elastic modulus, plastic deformation, stress relaxation, and other mechanical properties of a film or surface.

The measurement involves applying a force (typically normal to the surface) and measuring the displacement of the tip into the material. The tip is made of a very hard material, such as diamond, so that it will not deform during the measurement. A common tip geometry is the Vickers pyramid with a radius of curvature around 50 nm. To get the most accurate measurements, the size of the tip should be greater than or equal to the scale of the surface roughness.

The applied force can typically be determined with about 0.3 mN resolution. The penetration of the tip into the surface can be measured with about 2 Å resolution. These measurements can be made on both the nano and micro scales in terms of how far the tip will penetrate into the sample. For nanometer scale indentations (nano-indentation), the applied forces will typically be in the 10–500 mN range. Microindentation, where the penetrations are in the micrometer range, requires larger applied forces in the 0.1–20 N range. The experimental arrangement is presented in Figure 15.17.

A schematic representation of a stress-strain curve is presented in Figure 15.18. Once the tip begins to be displaced into the sample, increasing the applied force causes a greater displacement. The shape of the load part of the graph gives information about the hardness of the sample. If the force is held constant, the tip may continue to be displaced further into

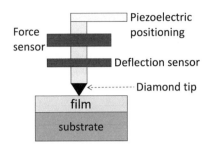

FIGURE 15.17
Micro or nano-indentation apparatus.

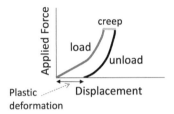

FIGURE 15.18
Stress–strain curve showing three stages of experiment as force is increased (load), held constant (creep) and decreased (unload).

the sample in a process described as creep. When the force is reduced, the tip may not follow the same stress-strain curve as during the indentation stage. The shape of the unloading curve gives the elastic modulus of the material. Once zero applied force is reached, the tip may still be displaced into the sample indicating that plastic deformation of the sample has occurred.

The hardness of a material is a measure of the material's resistance to surface penetration by the indenter tip. On the nano and micro scales, this can be used to examine the hardness of different phases within a sample or different surface crystal orientations. The hardness is defined as the applied force (load) divided by the contact area of the indenter tip. One problem with measuring this is that the contact area is very small in nano-indentation and is very difficult to measure. It can be determined indirectly by examination of the shape of the indentation portion of the stress-strain curve.

Nano-indentation measurements can be complicated by several experimental factors. If the system under study is not held at a constant temperature over the duration of the experiment, thermal drift may complicate the interpretation of the results. During the measurement, material from the region surrounding the indentation area can pile up at the edges of the indentation area or sink into the indentation area altering the measurements. The models applied in interpreting the stress-strain curves typically assume that surface roughness is on a scale much smaller than the contact area. Tip rounding may change the actual contact area compared to the expected contact area.

15.3.5 Adhesion

Measurements of the adhesion (Mittal 1995, 2001) of a film to a substrate are particularly challenging in that the separation of the two components may not occur at the

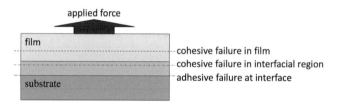

FIGURE 15.19
Possible cohesive and adhesive failure locations if a film is pulled away from a substrate.

film-substrate interface. Several possibilities are indicated in Figure 15.19. If there is inter-diffusion of the film and substrate, the separation may occur somewhere in this poorly defined interfacial region. The separation could also occur somewhere in the interior of the film in which case it is cohesive failure rather than adhesive failure. The failure will occur at the weakest location in the film substrate system, which may not be where you want to measure. Chemical analysis of the separated surfaces (using AES, XPS, EDX ...) can help to identify the location of the failure.

We focus on direct adhesion measurements where a force is applied to a film-substrate system to test when separation occurs. Other adhesion tests can be done under accelerated testing conditions to determine the time required for some other agent (oxidation, liquid penetration ...) to separate a film from a substrate without directly applying a force.

Adhesion tests are dependent on a range of material properties and test configurations. Material properties include the mechanical properties (particularly ductility) of the film and substrate, the thickness of the film, the presence of stresses in the film and substrate, and the presence or development of cracks in the film. Since film properties can change with time, especially when exposed to air or heat, the results may depend on the time since the film was grown. Results will also depend on how the test is applied. The location and direction of the applied forces and the rate at which the applied force is increased can produce different results.

A common set of adhesion tests are the pull-off tests. The most simple is the adhesive tape test in which a piece of household adhesive tape is attached to the film and then pulled off to see if the film comes off with the tape or stays attached to the substrate. This test is actually quite reproducible as well as being simple and inexpensive. It is a qualitative test and is typically not calibrated. Two different practitioners may pull the tape at a different angle and with a different speed. If the test shows that something was removed with the tape, it is not clear whether this was adhesive or cohesive failure.

A more calibrated set of adhesion pull-off tests can be conducted by gluing test blocks to the film and substrate. The blocks can then be pulled using a known force and at a known angle. Often these tests are done with the blocks pulled apart normal to the surface as shown in Figure 15.20a. If shear forces are of greater interest, the blocks can be pulled parallel to the surface to examine adhesion under shear forces as in Figure 15.20b.

One disadvantage of these tests is the need to apply an adhesive to the surface of the film. If the film is thin, this may change the properties of the film. A test that avoids this is an impact deceleration test. In this case, only the substrate is glued to a test block. The film/substrate/block assembly is shot using an air gun at a target with an opening in it that is large enough so that the film and substrate can pass through. The attached block, however, is stopped by the target. If the assembly was initially shot with sufficient force, the film may detach and keep going when a high enough initial velocity is reached. This concept is shown in Figure 15.21.

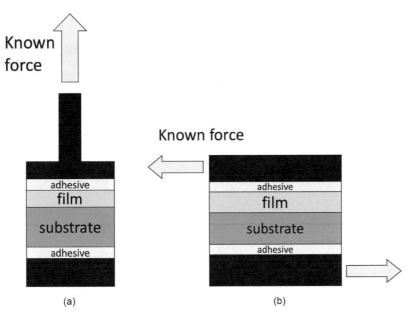

FIGURE 15.20
Pull-off tests: (a) normal force (b) shear force.

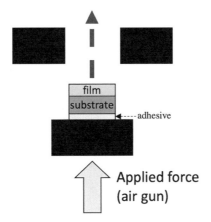

FIGURE 15.21
Deceleration adhesion test schematic.

Another set of adhesion measurements are the scratch tests. In these tests, a stylus of known radius is dragged across a film while the applied force on the stylus is increased until the stylus completely removes the film. It can be difficult to determine when complete removal of the film occurs since cohesive failure may occur rather than adhesive failure. The radius of the tip and the speed with which it is moved horizontally over the surface can also change the results. A typical experimental configuration is shown in Figure 15.22. Typical data is presented in Figure 15.23. For measurements on a small area, a microindenter can be used to perform these tests. Failure can be detected using microscopic observation, observing the frictional forces (which will typically change upon failure), acoustic emission of elastic waves from cracking in the film, and observing sudden changes in the depth of the tip.

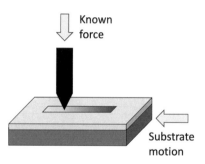

FIGURE 15.22
Scratch test configuration.

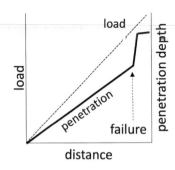

FIGURE 15.23
Representative data for scratch test.

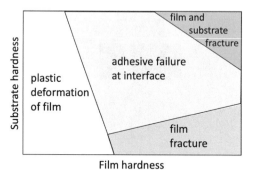

FIGURE 15.24
Scratch test failure mechanisms as a function of substrate and film hardness.

Film failure may occur by cracking rather than by adhesive failure. This typically depends on the relative hardness of the film and substrate as shown in Figure 15.24. If the film has low hardness, failure may occur by plastic deformation of the film where the film is permanently compressed, but not removed. For moderate values of film hardness, adhesive failure at the film-substrate interface may occur. If the film is much harder than the substrate, the deformation of the substrate under the film may cause fracture to occur through the entire thickness of the film. If both the film and substrate are very hard, the fracture may occur throughout the film/substrate system.

References

Dames, C. 2013. Measuring the thermal conductivity of thin films: 3 Omega and related electrothermal methods. In: *Annual Review of Heat Transfer*, vol 16. G. Chen, V. Prasad, and Y. Jaluria, eds. Danbury, CT: Begell House, Inc.

Freund, L.B. and S. Suresh. 2003. *Thin Film Materials: Stress, Defect Formation and Surface Evolution.* Cambridge: Cambridge University Press.

Frey, H. and H. R. Khan, eds. 2015. *Handbook of Thin-Film Technology.* Berlin: Springer Verlag.

Gabbott, P. ed. 2008. *Principles and Applications of Thermal Analysis.* Oxford: Blackwell Publishing Ltd.

Höhne, G.W.H., W.F. Hemminger and H.-J. Flammersheim. 2003. *Differential Scanning Calorimetry,* 2nd ed. Berlin: Springer-Verlag.

Hudson, J. 1998. *Surface Science: An Introduction.* New York: John Wiley & Sons, Inc.

Mittal, K.L. ed. 1995. *Adhesion Measurement of Films and Coatings.* Ultrecht: VSP.

Mittal, K.L. ed. 2001. *Adhesion Measurement of Films and Coatings,* vol 2. London: CRC Press.

Ohring, M. 2002. *Materials Science of Thin Films.* 2nd ed. San Diego, CA: Academic Press.

Price, D.M., M. Reading, A. Hammiche, and H. M. Pollock. 1999. Micro-thermal analysis: scanning thermal microscopy and localized thermal analysis. *Int. J. Pharmaceut.* 192: 85–96.

Prutton, M. 1994. *Introduction to Surface Physics.* Oxford: Clarendon Press.

Somorjai, G.A. 2010. *Introduction to Surface Chemistry and Catalysis.* 2nd ed. New York: John Wiley & Sons, Inc.

Zangwill, A. 1988. *Physics at Surfaces.* Cambridge: Cambridge University Press.

Problems

Problem 15.1 In a temperature-programmed desorption experiment with a second order reaction, the peak temperature (T_P) depends on the initial coverage (σ_0). If the temperature is increasing linearly with time as $T = T_0 + \alpha t$, an equation for the activation energy (E_a) is $\dfrac{E_a}{RT_P^2} = \dfrac{v_2 \sigma_0}{\alpha} e^{-E_a/RT_P}$ where v_2 is a frequency factor and R is the universal gas constant.

a. Show that this equation can be rewritten as $\ln\left(\sigma_0 T_P^2\right) = -\dfrac{E_a}{RT_P} + \ln\left(\alpha - \dfrac{E_a}{v_2 R}\right)$.

b. In an experiment, different initial coverages will result in different T_P values. How would you use the experimental data and this equation to determine the activation energy, E_a?

Problem 15.2 Temperature-programmed desorption with first order reactions are independent of the coverage. If the temperature is increasing linearly with time as $T = T_0 + \alpha t$, an equation for the activation energy (E_a) is $\dfrac{E_a}{RT_P^2} = \dfrac{v_1}{\alpha} e^{-E_a/RT_P}$ where v_1 is a frequency factor and R is the universal gas constant. Solve this equation numerically for a gas desorbing from a surface with $\alpha = 30$ K/s, $v_1 = 10^{19}$ s^{-1}, and $T_P = 500$ K.

Problem 15.3 In a thermogravimetric analysis experiment on a thin Ni film, we expose the film to oxygen at high temperature and observe a weight gain from the incorporation of oxygen into the Ni film as it forms NiO.

a. What is the mass of a Ni film with dimensions of 1 cm×1 cm×500 nm if the density of the film is 8.9 g/cm³?

b. How many moles of Ni does this correspond to given that the atomic mass of Ni is 28 g/mole?

c. If we fully oxidize the Ni film by forming NiO, what mass change should we see given that the atomic mass of O is 8 g/mole?

Problem 15.4 When doing a diffusion measurement such as the laser desorption experiment, it is helpful to determine several values to optimize the experiment.

a. Assume that the diffusing element forms a uniform complete monolayer on the surface and that the atoms are in a simple cubic lattice with lattice constant of 0.5 nm. If the laser illuminates a circular region with a diameter of 1 mm, what is the maximum number of atoms that we could desorb?

b. If the chamber has a volume of 0.25 m³ and the desorbed gas has a temperature of 400 K, use the ideal gas law (Equation 2.1) to estimate the maximum partial pressure of the gas that would be observed.

c. If the diffusing element has a diffusion constant, $D=2.0\times10^{-15}\,\text{m}^2/\text{s}$ use Equation 4.23 to estimate how long it would take for an atom to diffuse 1.0 μm and 1 mm.

Problem 15.5 When measuring average film stress (σ_f) by the curvature of a thin Si wafer substrate (Figure 15.13), the stress is typically determined by some form of the Stoney formula. A simple form of this relationship is $\sigma_f = M_S \dfrac{d_s^2}{6 d_f R}$ where M_S is the biaxial modulus of the Si wafer substrate, d_s is the thickness of the substrate, d_f is the thickness of the film and R is the measured radius of curvature of the film/substrate system after deposition. For Si(1 0 0) wafers with $M_S = 1.8\times10^{11}$ Pa and $d_s=380$ μm, assume that a film with thickness $d_f=600$ nm is deposited. Determine the average film stress for $R=1$ cm and $R=1$ m.

Appendix 1: Physical Constants and Unit Conversions

Physical Constants

Quantity	Symbol	Value
Speed of light	c	$2.998 \times 10^9 \, \text{m/s}$
Planck's constant	h	$6.626 \times 10^{-34} \, \text{J/s}$
		$4.136 \times 10^{-15} \, \text{eV/s}$
Elementary charge	e	$1.602 \times 10^{-19} \, \text{C}$
Electron mass	m_e	$9.109 \times 10^{-31} \, \text{kg}$
Proton mass	m_p	$1.673 \times 10^{-27} \, \text{kg}$
Neutron mass	m_n	$1.675 \times 10^{-27} \, \text{kg}$
Avogadro's number	N_A	$6.022 \times 10^{23} \, \text{mol}^{-1}$
Boltzmann's constant	k_B	$1.381 \times 10^{-23} \, \text{J/K}$
		$8.617 \times 10^{-5} \, \text{eV/K}$
Gas constant	R	$8.314 \, \text{J mol/K}$

Unit Conversions

Electron volt	eV	$1.602 \times 10^{-19} \, \text{J}$
Atomic mass unit	u	$1.660 \times 10^{-27} \, \text{kg}$
Ångstrom	Å	$10^{-10} \, \text{m}$
		$0.1 \, \text{nm}$

TABLE A1.1

Pressure Unit Conversion Factors

	mbar	Pascal (N/m²)	Atmosphere	torr (mm Hg)	micron (μm Hg)	psi (lb/in²)	dyne/cm²	molecules/m³
1 mbar =	1	100	9.87×10^{-4}	0.75	750	0.0145	1000	2.65×10^{22}
1 Pa =	0.01	1	9.87×10^{-6}	7.5×10^{-3}	7.5	1.45×10^{-4}	10	2.65×10^{20}
1 atm =	1010	10,100	1	760	7.6×10^{5}	14.69	1.01×10^{6}	2.69×10^{25}
1 torr =	1.333	133.3	1.31×10^{-3}	1	1000	0.0193	1333	3.53×10^{22}
1 μm =	1.33×10^{-3}	0.133	1.31×10^{-6}	0.001	1	1.93×10^{-5}	1.333	3.53×10^{19}
1 psi =	68.94	6.89×10^{3}	0.068	51.71	5.17×10^{4}	1	6.89×10^{4}	1.83×10^{24}
1 dyne/cm² =	0.001	0.10	9.87×10^{-7}	7.50×10^{-4}	0.75	1.45×10^{-5}	1	2.65×10^{19}
1 molecule/m³ =	3.77×10^{-23}	3.77×10^{-21}	3.72×10^{-26}	2.83×10^{-23}	2.83×10^{-20}	5.47×10^{-25}	3.77×10^{-20}	1

Appendix 2: Acronyms and Abbreviations

AC	Alternating Current
AES	Auger Electron Spectroscopy
AFM	Atomic Force Microscopy
ALD	Atomic Layer Deposition
ATR	Attenuated Total Reflection
bcc	Body Centered Cubic
BET	Brunauer-Emmett-Teller Equation
BLS	Brillouin Light Scattering
CLSM	Confocal Scanning Laser Microscope
CVD	Chemical Vapor Deposition
DC	Direct Current
DSC	Differential Scanning Calorimetry
EBIC	Electron Beam Induced Current
ECR	Electron Cyclotron Resonance
EDAX, EDS, EDX	Energy Dispersive (X-Ray) Spectroscopy
EELS	Electron Energy Loss Spectroscopy
EPMA	Electron Probe Microanalysis
ERDA	Elastic Recoil Detection Analysis
ESCA	Electron Spectroscopy for Chemical Analysis (see XPS)
ESDIAD	Electron Stimulated Desorption Ion Angular Distribution
fcc	Face Centered Cubic
FECO	Fringes of Equal Chromatic Order
FIB	Focused Ion Beam
FMR	Ferromagnetic Resonance
FTIR	Fourier Transform Infra-Red (Spectroscopy)
GIXD	Grazing Incidence X-Ray Diffraction
hcp	Hexagonal Close Packed
HEIS	High Energy Ion Scattering
HREELS	High Resolution Electron Energy-Loss Spectroscopy
ICP	Inductively Coupled Plasma
ISS	Ion Scattering Spectrometry
LCAO	Linear Combination of Atomic Orbitals
LDOS	Local Density of States
LECVD	Laser Enhanced Chemical Vapor Deposition
LEED	Low Energy Electron Diffraction
LEELS	Low Energy Electron-Loss Spectroscopy
LEEM	Low Energy Electron Microscopy
LEIS	Low Energy Ion Scattering
LPCVD	Low Pressure Chemical Vapor Deposition
MBE	Molecular Beam Epitaxy
MEIS	Medium Energy Ion Scattering Spectrometry
MFM	Magnetic Force Microscopy
mfp	Mean Free Path
MOCVD	Metalorganic Chemical Vapor Deposition (see OMVPE)

MOKE	Magneto-Optical Kerr Effect
NEXAFS	Near-Edge X-Ray Absorption Fine Structure
OMVPE	Organometallic Vapor Phase Epitaxy (see MOCVD)
PCVD	Photoassisted Chemical Vapor Deposition
PECVD	Plasma Enhanced Chemical Vapor Deposition
PEEM	PhotoElectron Emission Spectroscopy
PL	Photoluminescence
PLD	Pulsed Laser Deposition
PVD	Physical Vapor Deposition
QCM	Quartz Crystal Monitor
RBS	Rutherford Backscattering Spectrometry
RF	Radio frequency
RHEED	Reflected High Energy Electron Diffraction
SAM	Scanning Auger Microscopy
SEM	Scanning Electron Microscopy
SEMPA	Secondary Electron Microscopy with Polarization Analysis
SERS	Surface Enhanced Raman Spectroscopy
SIMS	Secondary Ion Mass Spectrometry
SMOKE	Surface Magneto-Optical Kerr Effect
SNMS	Secondary Neutral Mass Spectrometry
SLEEM	Spin-Polarized Low Energy Electron Microscopy
SQUID	Superconducting Quantum Interference Device (Magnetometer)
STM	Scanning Tunneling Microscopy
TEM	Transmission Electron Microscopy
TGA	Thermogravimetric Analysis
TPD	Temperature Programmed Desorption
UHV	Ultra-High Vacuum
UHVCVD	Ultra-High Vacuum Chemical Vapor Deposition
UPS	Ultraviolet Photoelectron Spectroscopy
VASE	Variable Angle Spectroscopic Ellipsometry
VSM	Vibrating Sample Magnetometer
WDS, WDX	Wavelength Dispersive (X-Ray) Spectroscopy
XANES	X-Ray Absorption Near Edge Structure
XMCD	X-Ray Magnetic Circular Dichroism (XMCD)
XPS	X-Ray Photoelectron Spectroscopy (see ESCA)
XRD	X-Ray Diffraction
XRF	X-Ray Fluorescence
XRR	X-Ray Reflectance

Appendix 3: Basic Vacuum Technology

Chapter 2 introduces the vacuum environment and explores the aspects of that environment which are important in surface and thin film science. In this appendix, we briefly examine vacuum technology to understand how a vacuum can be contained, obtained, and measured. The references (Borichevsky 2016, Delchar 1993, Hablanian 1997, Hata 2008, Hoffman and Thomas 1998, Lafferty 1997, O'Hanlon 2003, Roth 1990, Tompkins 1997) describe the details of vacuum technology much more completely.

We start by describing the elements of a generic vacuum chamber, as shown in Figure A3.1. The chamber is typically made of non-magnetic stainless steel since it is a strong, relatively inexpensive, material with a reasonably low reactivity to other gases. The empty chamber has two main sources of gas. The first is outgassing from materials in the chamber. This arises from the vapor pressure of these materials. We typically want to use low vapor pressure materials and keep them very clean to minimize outgassing. The second source is leaks in the chamber. Since our goal is to do something inside this chamber, we often attach instruments, mechanical, fluid, and electrical feedthroughs, and other elements to our chamber. These connections will sometimes not seal effectively leading to small leaks of gas into the chamber. The chamber will also have a pump to produce vacuum and a pressure gauge to measure the vacuum.

A variety of technologies are available to seal parts that are attached to the chamber. For relatively poor vacuums rubber O-rings can often be used. For better vacuums, a special type of rubber material (viton) is often used. In the high and ultra-high vacuum ranges, metal seals, such as copper gaskets, are typically used. The details of these technologies are discussed in the references.

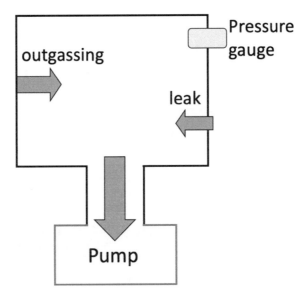

FIGURE A3.1
Generic vacuum system.

When designing a vacuum system, we need to consider four sources of gas which will be pumped: (1) the gas in the volume of the chamber, (2) the gas adsorbed on the walls of the chamber which can outgas, (3) gas entering the chamber from leaks, and (4) gas created by the process that we are performing. Our ability to create a vacuum depends on (1) how fast our pumps can remove gas, (2) how fast gas is entering the volume of the chamber, and (3) the conductance of the path between our chamber and the pump.

Not surprisingly, the conductance gets better for shorter and wider paths. We typically try to place the pump as close to the chamber and with as wide a connection as possible. This will minimize the pumping speed required of the pump.

Different types of pumps operated most effectively in different pressure ranges. As a result, it is common to need multiple pumps on a single vacuum system. Figure A3.2 shows the operating ranges of some common types of pumps.

Pumps can be broadly classified as either gas transfer pumps or entrapment pumps. The gas transfer pumps actually remove the gas from the chamber and exhaust it into the outside environment. The entrapment pumps, leave the gas in the vacuum system but trap it somewhere such that it will not interfere with our process. Gas transfer pumps can be further distinguished as positive displacement pumps which use a change in volume to move gas or kinetic pumps which pump by transferring momentum to the gas particles. Positive displacement pumps, such as rotary vane mechanical pumps, compressive dry pumps, and non-compressive (roots blower) pumps, are typically most effective at pressures near atmosphere and down to 10^{-2} to 10^{-4} torr. Kinetic pumps, such as turbomolecular and diffusion pumps, work at lower pressures. Entrapment pumps use a variety of physical processes to contain the gas. Adsorption (cryosorption) pumps use large surface areas which are typically cooled to very low temperatures to trap large quantities of gas onto these surfaces. Getter pumps (bulk getter materials, sublimation pumps, sputter ion pumps) use chemical reactions to trap reactive gases in the pump. Cryocondensation pumps work at liquid He temperatures to provide cold surfaces for gas to condense onto. The details of how these pumps operate are discussed in the references.

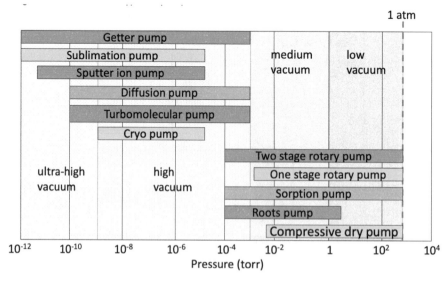

FIGURE A3.2
Operating ranges of typical pumps.

In addition to having pressure ranges in which they are most effective, many of these pumps are also more effective at pumping certain gases. Turbomolecular pumps, for example, are less effective on very low mass gases such as H_2 or He. Oil diffusion pumps work slightly better for low mass gases. Getter pumps work well for reactive gases (O, N, CO, ...) but do not pump inert gases (Ar, Ne ...) very well. As mentioned in Chapter 2, this can lead to a change in the composition of the gas in a vacuum chamber as it pumps down from atmosphere to high or ultra-high vacuum.

As with pumps, the gauges that we use to measure the pressure in a vacuum system have different effective pressure ranges. These are indicated in Figure A3.3. We can divide pressure gauges into two categories. One group are direct gauges which measure the pressure in the system directly as a force per unit area. The other group are indirect gauges which measure some property that is typically related to the number density of gas molecules in the chamber which is proportional to pressure according to the ideal gas law. Direct gauges include mechanical gauges where a piston is pushed to deflect a needle on the gauge or capacitance manometers where the pressure moves a diaphragm that changes the capacitance in the gauge. Indirect gauges often rely on thermal conductivity (thermocouple or thermistor gauges), resistance (Pirani gauge), ionization (hot or cold cathode ionization gauges), or momentum transfer (viscosity gauges). Again, the details of the operation of these pumps are described in the references.

Gauges can also be dependent on the type of gas that is being measured. While many gauges are relatively insensitive to gas composition, ionization gauges are quite sensitive to the type of gas being measured.

Vacuum technology is a well-developed practice and so the creation and measurement of acceptable vacuum environments in which to conduct our processes is typically a matter of learning about and following best practices within the constraints of our budgets.

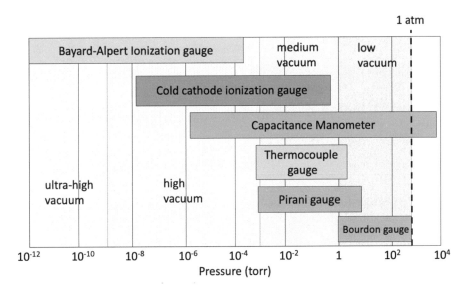

FIGURE A3.3
Operating ranges of typical pressure gauges.

References

Borichevsky, S. 2016. *Understanding Modern Vacuum Technology*. Ipswich, MA: Blue Dasher.

Delchar, T.A. 1993. *Vacuum Physics and Techniques*. London: Chapman and Hall.

Hablanian, M.H. 1997. *High Vacuum Technology, A Practical Guide*. 2nd ed. New York: Marcel Dekker, Inc.

Hata, D.M. 2008. *Introduction to Vacuum Technology*. San Francisco, CA: Pearson.

Hoffman D.M., B. Singh and J.H. Thomas III, eds. 1998. *Handbook of Vacuum Science and Technology*. San Diego, CA: Academic Press.

Lafferty, J. M. ed. 1997. *Foundations of Vacuum Science and Technology*. New York: John Wiley and Sons, Inc.

O'Hanlon, J.F. 2003. *A User's Guide to Vacuum Technology*. 3rd ed. Hoboken, NJ: John Wiley and Sons Inc.

Roth, A. 1990. *Vacuum Technology*. 3rd ed. Amsterdam: Elsevier.

Tompkins, H.G. 1997. *The Fundamentals of Vacuum Technology*. 3rd ed. New York: American Vacuum Society.

Index

Note: **Bold** page numbers refer to tables; *italic* page numbers refer to figures.

Abbe limit *229–30*
absorption index 120
adhesion *339–42*
adsorption isotherm 138–9
amorphous materials 54
anisotropy energy 117
arc vaporization *see* cathodic arc deposition
Arrhenius plot 70–1, 215, 330–1
atomic force microscope *243–6*
atomic layer deposition *222–3*
atomic spacing 2
Auger electron spectroscopy *273–280*

backscattered electrons *234–6*
basis *35–6*, 42
body-centered cubic lattice *38–9*, 50
bonds
    covalent 44–5
    ionic 44–5
    metallic 44–6
    Van der Waals 44, 135, 159
Bragg's Law *249–51*
Bravais lattice *37–9*
Brillouin light scattering *310–11*, *335*
Brillouin zone *43–4*, 63
Bruggeman model *126–7*
Brunauer-Emmett-Teller isotherm 139

capillarity model 161–7
cathodic arc deposition *200–1*
charging *236*
chemical adsorption (chemisorption) *135–6*, *144*, 159
chemical potential *74–5*, 81
chemical vapor deposition *209–21*
    chemical reactions 209–12
Clausius-Mossotti equation 125
cleaning surfaces 27–9
clusters–atomic *160–6*
commensurate structures *142–3*
condensation coefficient 136
conduction band *105–6*
conductivity 106–7
confocal microscope *231–2*
coverage *136–9*, 329

critical nucleus *156–7*, *162–7*
cross-plane thermal analysis *334*
Curie temperature *117*
cyclotron frequency 22

DC sputter deposition *188–94*
Debye length 20–1
Debye temperature 261
Debye-Waller factor 261
defects in crystals *46–8*, *51–2*
degree of ionization 20–1
deposition process atomic *151–2*
depth of field **213**
depth profile *278–9*
desorption 136, 159, *328–31*
diamagnetic materials *114–15*
dielectric constant 120, *127*
differential scanning calorimetry *331–2*
diffusion *68–74*; surface 73–4, *160*
diffusion coefficient 68, 70, 73, 213
diffusion energy 70–1
diffusion measurement *332–3*
dislocations *47*, 198–9
dispersion relation *63–6*, *102–4*
domains (magnetic) *115–16*
drift velocity 23, 304

effective medium approximation *124–7*
elastic waves *61–5*
electron cyclotron resonance 25
electron energy loss spectroscopy *296–8*
electron impact ionization 19, 195, 200
electron stimulated desorption ion angular
        distribution 330
Ellingham diagram 84, *211–12*
ellipsometry *317–21*
energy barrier 4–5
    adsorbed atoms 145
    chemical vapor deposition 215–16
    desorption 6, 159, 163
    diffusion 69, 72, 74, 165, 172–3
    ionization 18
    nucleation 156
energy dispersive X-ray analysis *280–1*
enthalpy 81, 83

entropy 81, 83, 91
epitaxial films 198–9
equation of state 4, 9–10, 139
eutectic *88–9*, 152–3
evaporation 179–88
 point source *181–2, 184*
 surface source *182*, 183–5
Ewald sphere *254–6, 259–60*, 262
exchange energy *117*
exciton *112–13*
extensive thermodynamic variables 81
extinction coefficient 120

face-centered cubic lattice *38–9, 50*
Fermi energy 100, *103*
ferromagnetic materials *114–18*
ferromagnetic resonance *312–13*
Fick's laws 68–70
four point probe *305–6*
Fourier transform infrared spectroscopy *292–5*
Frank-van der Merwe growth *see* layer by layer
  growth
free electron gas 99–101
friction *337–8*

Garnett model *126–7*
Gibbs free energy 6, 81–3, 85, 90, 155–6, 161
Gibbs-Duhem equation 82
grain boundary *46–7*, 141
 two dimensional 141
grain size *172–3*

Hall effect *303–4*
hard sphere model 51, 91, 142
harmonic oscillator 5, 59, 133
helicon 25
Helmholtz free energy 83
heterogeneous nucleation 161–7
hexagonal close packed structure *49–50*
homogeneous nucleation *152–7*
hot wire CVD 220–1
hysteresis loop *116*

ideal gas law 4, 9–10, 13, 139, 351
incommensurate structures *142–3*
index of refraction 120
inductively coupled plasma 24
infrared spectroscopy *see* Fourier transform
  infrared spectroscopy
intensive thermodynamic variables 81–2
interferometry *322–4*
ion assisted deposition 196
ion scattering *265–6*

ionization coefficient 18
ionization energy **17–20**
island coalescence *169–70*
island growth *167–8*

jellium *101–3*

kinetic theory of gases 4, *9–17*, 138–9, 157, 213
Knudsen cell *182–3*

laminar flow *213–14, 216*
Langmuir isotherm 138
Larmor radius 22
laser ablation *see* pulsed laser deposition
laser enhanced CVD 220
lattice *35–6*
layer by layer growth *167–8*
Lennard-Jones potential 17, *44–6*, 128
light scattering *321–2*
Lindemann criterion 60
line-of-sight deposition 180, 195
longitudinal wave *61–2*, 65
Lorentz-Lorenz model 126
low energy electron diffraction *256–62*
low energy electron microscope 239–40
low energy ion scattering 265, 292
low pressure CVD 218–19

magnetic force microscopy *309–10*
magneto-optical Kerr effect *307–8*
magnetron sputter deposition *195–6*
Maxwell velocity distribution *11–12*
mean free path
 electrons 2, 19, 106, *233–4*, **257**
 atoms **13–14**, 18–19, 180
 phonons 128
melting *60–1*
 measurement 262, 331–2
metalorganic CVD 221
micro-thermal microscopy 333
microindentation *338–9*
Miller indices 38, *40–1*
molecular adsorption *144–5*
molecular beam epitaxy 197–9
molecular flow 14, 213, 218, 221
monolayer formation time 15

nanoindentation *338–9*
near-edge X-ray absorption fine structure *287–8*
nearly free electron model *101–3*
Nernst-Einstein relation 72
non-linear optics 123
nucleation *152–67*

optical microscope *229–32*
Ostwald ripening *169*

paramagnetic materials 114–15
partial pressure 10, 83–4
phase diagram *85–90*
    adsorbates *139–40, 144*
phonon **65–7**, 297
photothermal analysis *334*
physical adsorption (physisorption) *135–6, 144,*
    159
physical vapor deposition *see* evaporation;
        sputter deposition; molecular beam
        epitaxy; cathodic arc deposition;
        pulsed laser deposition
plasma 17–26
plasma chemistry 25–6
plasma enhanced CVD 219–20
plasma frequency 20–1, 113, 121
plasmon *113, 297*
point defects *46–8*
pole figure *254–5*
primitive lattice cell *35–7*
primitive translation vectors *35–6*
pull-off test *340–1*
pulsed laser deposition *201–3*

quartz crystal microbalance 268–9

Raman spectroscopy *295–6*
Rayleigh limit *229–30*
reactive deposition 196–7
reciprocal lattice *43*, 254–6
reciprocal space 43, 100, 254–5, 259–60
reflectance *315–17*
reflection high energy electron diffraction
        262–3
residence time 136, 159
resistivity 106–7, *304–7*
RF sputter deposition *194–5*
Richardson-Dushman equation 107
rocking curve *252–3*
Rutherford backscattering spectroscopy 265,
        290–2

scanning Auger microscopy *279–80*
scanning electron microscope **229**, *232–7*
scanning electron microscopy with polarization
        analysis *309*
scanning probe microscopes *240–6*
scanning tunneling microscope *241–3*
scratch test *341–2*
secondary electrons *234–6*

secondary ion mass spectrometry *288–90*
segregation: chemical *2–3*, 93–5, 235
semiconductor *105–6, 110–11*
sensitivity factor **277–8**
sheet resistance 306
sintering *169*
skin depth 121
Snell's Law 314–5
specific heat 128
spin wave *118–19*
spin-polarized low energy electron microscope
        308–9
sputter deposition 188–96
sputter yield *28–9*, 189–90, 193–5
stagnant layer 214, *216–17*
states of matter 7
sticking coefficient *137–8*, 157
strain 76, 198
Stranski-Krastanov growth *167–8*
stress measurement 335–7
stress-strain curve *339*
stylus profilometer *266–8*
sub-monolayer optical properties 123–4
superconducting quantum interference device
        magnetometer *311–12*
surface definition *49*
surface energy 52, 89, 91, 155, 162
surface roughness 53–4, 74–6
    measurement 246, 264, 266–7, 322
surface state *110–12*
surface stress 89–91
surface structure
    ideal *49–50*
    reconstruction *50–2*
surface tension 89–93, *327–8*
surface wave *66–7*

temperature-programmed desorption
        329–31
thermal accommodation *158–9*
thermal conductivity 128, 333–4
thermal expansion 128–9
thermogravimetric analysis 331–2
thin film definition 1, 151
tight binding model *103–5*, 112
Townsend equation 18–19
transmission electron microscope **229**, *237–9*
transverse wave *61–2, 64–5*
    acoustical mode *64–5*
    optical mode *64–5*
two-dimensional crystal structures *139–44*

ultraviolet photoelectron spectroscopy 286–7

vacancies *47–8, 51–2*
vacuum 9–10
valence band *105–6*
Van der Pauw method *306–7*
Van der Waals bond *see* bonds
vapor pressure *16*, 155, 179, 349
vibrating sample magnetometer *311–12*
vibration
    atomic 5–6, 60, 128, 145–7, 159, 165, 261
    electron 120–1
    molecular *293–4, 297–8*
viscous flow 13–14, 213, 221
Volmer-Weber growth *see* island growth

Walton-Rhodin model 166–7
wavelength dispersive X-ray analysis 281
Wigner-Seitz primitive cell *36–8*, 43
work function 109
Wulff plot *92–3*

X-ray diffraction *249–56*
X-ray magnetic circular dichroism *313–14*
X-ray photoelectron spectroscopy *281–6*
X-ray reflectivity *263–4*

Young's equation *168*

zone structure models *170–2*

For Product Safety Concerns and Information please contact our
EU representative GPSR@taylorandfrancis.com Taylor & Francis
Verlag GmbH, Kaufingerstraße 24, 80331 München, Germany